The Ecology and Natural History of Chilean Saltmarshes

José Miguel Fariña • Andrés Camaño
Editors

The Ecology and Natural History of Chilean Saltmarshes

Springer

EDICIONES UC

Editors
José Miguel Fariña
Center of Applied Ecology and Sustainability-CAPES
Pontificia Universidad Católica de Chile
Santiago, Chile

Department of Ecology and Evolutionary Biology
Brown University
Providence, RI, USA

Andrés Camaño
Camaño, Andres (editor) Golder Associate
ERM, Santiago, Chile

Translation from the Spanish language edition: Humedales Costeros: Aportes científicos a su gestión sustentable by José Miguel Fariña (eds), © Ediciones Universidad Católica de Chile, 2012. Original Publication ISBN 978956141259-0. All rights reserved.

ISBN 978-3-319-63876-8 ISBN 978-3-319-63877-5 (eBook)
DOI 10.1007/978-3-319-63877-5

Library of Congress Control Number: 2017952193

Printed on acid-free paper

This Springer imprint is published by Springer Nature
The registered company is Springer International Publishing AG
The registered company address is: Gewerbestrasse 11, 6330 Cham, Switzerland

We dedicate this book to our families,
collaborators, and students.

Prologue

The book *Coastal Wetlands Chile: Scientific Contributions for the Sustainable Management* is the first of its kind published in Chile, aiming to bring together ecosystem specialists to stand for all of its biophysical, ecological, and biogeographical diversity, with all of the knowledge gained, and above all for the scientific challenges of the future. The Government of Chile, as a signatory of the Ramsar Convention (1971) on wetlands, designed a specific policy to study and protect national coastal wetlands called "National Strategy on Coastal Wetlands" (CONAMA 2006). This at least delivers a referential framework and stresses the importance of these ecosystems for the country. However, experts say that in recent years, the generational progress of new knowledge has been slow, so incentives and catalysts are needed to achieve stronger boosts.

But boosts in what direction? Answers may be varied and multiple, and I only intend to explore three of them.

The first one has to do with recognizing the scientific potential and the low-cost accessibility to science offered by these ecosystems (interacting and functional dynamic biophysical units), which unlike many other ecosystems have very or at least relatively marked environmental outlines, albeit dynamic. When working in them, the ecosystem objective is "relatively delimited," and even more so, in general it can be successfully replicated in space. This is not easy to find in nature and carries comparisons when designing experiments and testing hypotheses. Therefore, in Chile, an important leap in science and understanding of spatial-temporal dynamics of the coastal wetlands (and others) is needed to apply experimental ecological science, beyond just the purely descriptive. It is the field experiment science (scientific method), so accessible in wetland ecosystems, which will allow taking new important steps in the future leading to understanding of its complexities and dynamics. In Chile there is a certain level of research, following this line (and there are examples of it in the book), even though it is still in its infancy. Experimental wetlands ecology, in its many aspects, has boosted significant advances in understanding the dynamics and better conservation approach (a vague concept that everybody defines differently) in other Latin American countries. Something cannot be adequately preserved if we do not know how it works. The mere description of the structures of an ecosystem does not allow the implementation of policies for a long-term conservation. In Chile, we are in this process, and the incentives are needed to make the leap into the myriad of the different types of wetlands that grace the country.

Secondly, wetland systems are extensive and immediately recognized by the population as ecosystems that harbor high biodiversity (a vague concept that everybody defines differently), in particular birds, and there are examples in this book. This is a constant in all of the world's wetlands, and in Chile, the biggest challenge is the maintenance of long-term monitoring systems and comparable methodologies. Here, the most important step is to have a network of continuous biodiversity observations – not only of birds but also throughout the national territory – that would include different types of wetland ecosystems. There are some admirable efforts concerning this, but these are still too confined to specific sites and focused primarily on birds. In other parts of the world, the long-term biotic monitoring has focused first on developing reliable observation methodologies, which eventually end up being transferred to and including the inhabitants of the areas, taking into account their local knowledge of the ecosystems. In Chile we are still far from reaching this point: there is still deep "scientific mistrust" when it comes to biological methodology, even though it is well designed and simple, and it can be transferred and used by the local population to integrate their own responsibility with monitoring and maintaining wetland systems. This is yet another issue and a great challenge in Chile.

Thirdly, wetlands, just as other ecosystems, do not exist in a "natural gap" and are actually social-ecosystems that should be conserved and focused on as such. It is a serious ecological, social, and ethical mistake to think that these systems exist in natural voids, in which the human being is not present. The human being is part of each and every one of the planet Earth's ecosystems, even the isolated and distant ones. The environment delivers varied benefits and services to the population, and ecosystems are directly related to human welfare. Human beings cannot exist without the ecosystems. These services, by and large, just as in wetlands can be procurement services (food, water, and resources), regulatory services (air quality, water quality, and erosion control), cultural services (aesthetic, spiritual, religious, and reflective), and support service systems (soil formation, oxygen production, and biodiversity).

The scientific approach to socio-ecological system of wetlands, which entails integrating wealth of biophysics systems, in harmony with the human presence, is a must in Chile. Wetlands, in general, continue to be seen primarily through the prism of biodiversity and not as socio-ecological systems. Therefore, in my opinion, the wetland conservation strategy (coastal or other), as it exists in Chile, should radically refocus and be based on modern conservation and sustainability guidelines, such as those promoted by the Millennium Ecosystem Assessment (2005), which offers a solution to the central equation of human welfare and ecosystem services. This is the biggest challenge for a modern scientific and social approach in the wetlands of Chile (and there are no examples in the book) – a giant step yet to be taken.

But great strides and progress require first steps and first moves. I have no doubt that this book is the first step toward the right direction. Furthermore, cooperation between the industry and academia shown during the first meeting on coastal wetlands of Chile in 2011, as well as publication of this book, is highly stimulating and unique and shows that the "child has stood up and is learning to walk." Hopefully, "the child" will soon walk straight without swaying.

Professor Juan Carlos Castilla, Professor at the Universidad Católica de Chile (Catholic University of Chile) and a winner of the National Award for Applied Science and Technology 2010.

Juan Carlos Castilla

Associate Professor at the Catholic University of Chile and winner of the National Award for Applied Sciences and Technology 2010

Preface

Definition and Components of Coastal Wetlands

According to the Ramsar Convention (Ramsar 1971), wetlands correspond to the "areas of marsh, fen, peatland or water-covered surfaces, whether these salted natural or artificial, permanent or temporary, static or flowing, fresh, brackish or salt, including areas of marine water the depth of which at low tide does not exceed six meters," and "Wetlands may incorporate riparian and adjacent coastal areas and islands or bodies of marine water deeper than six meters at low tide lying within the wetlands." In addition, the same Convention originally proposed a classification system for wetland type which considers three main categories: (1) marine and coastal wetlands, (2) inland wetlands, and (3) artificial wetlands.

For each of these categories, the Convention lists a number of subcategories, and in the case of marine and coastal wetlands, these include permanent shallow marine waters (in most cases less than 6 meters deep at low tide; bays and straits), subtidal seabed (fields of kelp, sea grass, and mixed tropical marine meadows), coral reefs, rocky marine shores (rocky islets and cliffs), sandy or pebble/rock beaches (barriers, banks, points, and sandy islets, as well as the systems of dunes), estuaries (permanent water of estuaries and delta systems), intertidal mud, sand or saline soils ("salt flats"), intertidal swamps and marshes (marshes and elevated areas flooded with salt water, halophyte meadows, salt marshes, fresh and brackish water areas flooded by the tide), intertidal forest wetlands (mangroves, "nipa" swamps, flooded forests, and floodable freshwater zones), brackish and saline coastal lagoons (lagoons with at least one relatively narrow connection to the sea), coastal freshwater lagoons (freshwater delta lagoons), karst, and other subterranean hydrological systems (marine and coastal).

The Ramsar categorization and definition of wetlands is extremely wide, and for it to be applied in Chile to define, for example, marine and coastal wetlands, it would include virtually all areas of the coastline. In fact, since the Ramsar definition is the official definition of wetlands in our country (Contreras 2006), some early efforts made by the government agencies, such as the Servicio Nacional de Pesca

(National Fisheries Service – SERNAPESCA 1999), applied the definition generating a list which has included virtually all coastal bays in the northern part of the country.

Apart from the definition of wetlands by the Ramsar Convention, there are a number of definitions (see Contreras 2006, Mitsh and Gosselink 2011), all of which differ in their operational degrees. For example, the National Academy of Sciences of the United States (NRC 1995) proposed a definition of "reference" for wetlands, which states that "wetlands are ecosystems that depend on the inundation or saturation of the substrate surface, be it constant or temporary, which determines the physical, chemical and biological characteristics of the ecosystem." In reference to this definition, the Academy established diagnostic characters of the state of the ecosystem soil (flooded) and the presence of hydrophytic vegetation. This definition contains the three common components that were recognized by Mitsh and Gosselink (2007) for almost all definitions of wetlands:

- The presence of water either at surface level or in the root zone.
- Unique conditions of different soil types and the rest of the emerged substratum.
- The presence of the characteristic biota adapted to wet conditions (e.g., hydrophytic plants) and therefore the absence of biotas intolerant to floods. The above three conditions arise from the interaction between (a) hydrology (i.e., water level, flow, frequency, etc.), (b) physicochemical environment (soil type, pH, potential redox, etc.), and (c) biota (vegetation, animals, bacteria, etc.). In the case of our country, La Comisión Nacional de Medio Ambiente (National Environmental Commission, CONAMA 2006), in the National Strategy on Coastal Wetlands, recognizes how small is the operational definition by RAMSAR and proposes, as an alternative, to consider as wetlands those "ecosystems associated with substrates of temporarily or permanently saturated water, which allows the existence and development of aquatic biota." In analyzing the requests for the incorporation of the Chilean wetlands to the RAMSAR Convention List (Valverde 1996a, b, c; Rojas and Tabilo-Valdivieso 2004; Vilina et al. 2004; Rodríguez and Contreras 2008a, b; Pool 2010a, b) and perhaps because one of the specific goals of the RAMSAR Convention is to protect the avifauna, it is clear that the most valued components of the aforementioned "aquatic biota" have been primarily birds, not making any mention of groups or important conservation efforts of fish, mammals, and, even less, the plant species.

On the other hand, the studies of wetland environment in general (Mitsh and Gosselink 2007) and of the coastal wetlands in particular (Silliman and Bertness 2009; Alberti et al. 2011; Botto et al. 2011) have focused on the analysis of ecosystems and communities. At the ecosystem level, it is known that plants in coastal wetlands have zonal spatial distributions that arise in response to marked gradients of physical factors (e.g., humidity, salinity, nutrient availability, temperature, etc.) that are especially seen from the lowlands (permanent flooding) to the upper parts of the coastal wetlands. At a community level, studies are focused on the interactions that determine the diversity and species composition and successional and production processes of the organisms living in coastal wetlands. At both ecosystem and community levels, the main objective is to determine whether there are simple general rules governing the functioning of wetlands, which can be used for the management and conservation of these environments (Bertness et al. 2002; Silliman and Bertness 2003; Bertness et al. 2004). It has become apparent in this line of study that the particular characteristics of the hydrological regime, the soil or the biotic components of wetlands, determine an idiosyncratic ecosystem functioning which cannot be ignored, if we are to manage or conserve the ecosystem (see Farina et al. 2009).

In the particular case of our country, a series of indicators make clear that either hydrology, soil characteristic, or biotic composition determines the degree of idiosyncrasy of coastal wetlands, which can be very high and which should be taken into account when creating policies and management and conservation strategies. For example, in terms of hydrology and soil characteristics, it is well known that on the Southern Pacific Coast, the absence of an extended continental platform restricts the presence of coastal wetlands in tidelands, at the river mouths (West 1977 Bujalesky et al. 2001), and all throughout the region. Due to the lack of rainfall in the north, the river mouths remain closed for a long time, creating hypersaline conditions in wetland soils (Arana and Salinas 2003; San Martin et al. 2001). Below 28° S, the rain increases so the salt load is reduced, which is reflected in the composition of plants (Ramirez and Añazco 1982; San Martin et al. 1992; Hauenstein et al. 2002). In addition, it has been observed that unlike in Europe or on the Atlantic coast of South America, the fluctuations in the water level of some coastal wetlands of Chile do not depend on the changes in the tide but rather on the wind direction and dynamics (Dussailliant et al. 2009). The set of conditions described above may be the reason that, from a biotic point, unlike what is observed in other latitudes (Costa et al. 2003), coastal wetlands of Chile practically lack or have very low percentage of marine invertebrates (Farina et al. 2009).

Moreover, some earthquakes associated with the subduction of the Nazca tectonic plate, under the South America plate, which have occurred along the entire north-central coast of Chile, have generated significant changes in geomorphology, erasing sandbars and coastal dunes and/or producing significant elevations and sinking of the coastline (Castilla 1988), which undoubtedly affects greatly the coastal wetlands of our country.

With reference to the above, as a way of contributing to the sustainable management of coastal wetlands of Chile, in this book's edition, we have brought together as many scientists as possible from different disciplines who are actively conducting research in coastal wetlands of the country. As a result, we deliver the first analysis of the geographical representation and the environmental sensitivity degree of the coastal wetlands of Chile (Chap. 1). Afterward, we continue with the analysis of the available information on the specific components of the coastal wetlands of our country (first part of the book), and in addition, we also deliver case studies on particular wetlands of the northern, central, and southern coasts (second part of the book) to have a very good idea about geomorphology and hydrology (Chaps. 1, 2, 7, and 9), vegetation and flora (Chaps. 4, 5, 6, and 10), and the birds (Chaps. 3, 5, and 6) of the coastal wetlands of our country. Being acquainted with the impact of the tsunami, associated with the earthquake on February 27, 2010, in a series of chapters, we implicitly or explicitly analyze the effect of this type of natural disturbance on biotic and abiotic components of the coastal wetlands of Chile (Chaps. 2, 6, 9, and 10). Thus, we leave registered the unique opportunity that was experienced by many of us working and conducting research on the coastal edge of our country, having been able to demonstrate, compare, and even contrast the magnitude of the impact, but even more importantly to show the dynamics of recovery of our natural systems to drastic disturbances associated with it.

This book is a result of a series of seminars on the effects of the February 2010 earthquake and tsunami on the coastal edge of our country, the first meeting of Management of Wetlands (2010), and the first seminar on Coastal Wetlands of Chile (2011), in which both editors had actively participated as organizers. In those instances, we have converged academics and researchers from different the universities in the country with representatives of the private companies and public environmental authorities. Then and there, it has become clear that for coastal wetlands, one of the most unique environments, there is only a lot of scattered information, but no consolidated, modern research. The second issue that became apparent was that the policies and management of wetland conservation in Chile have not developed sufficiently. For example, although in the past the CONAMA has issued a "National Strategy for Conservation and Rational Use of Wetlands" (CONAMA 2005), currently, to recognize any significant change in the conservation of these ecosystems has not been possible beyond the publication of inventories, land cadasters, and environmental assessment guides (e.g., CONAF-CONAMA 1999, CEA 2006a, b), which may have a political and/or media value, but questionable worth and technical grasp.

The general objective of the "National Strategy for Conservation and Rational Use of Wetlands" (CONAMA 2005) is "to promote the conservation of the primary wetlands of Chile, its features and benefits in a framework of sustainable development." To achieve this goal, six specific objectives are proposed and are due to be implemented on the basis of 41 lines of action. Out of the six goals, perhaps the most challenging, in terms of management, is the fifth: "To develop and implement participatory planning tools and management for the conservation and sustainable use of priority wetlands." The 11 lines of this target action point to the management and conservation of priority wetlands, and/or representative wetlands of our country, specifying that a priority wetland must be defined by its "biodiversity and ecological system services" or that it shows "specific environment or condition." Both statements are very good in their intention but are absolutely irrelevant. To recognize any given ecosystem for its biodiversity or even only in environmental terms, we need a much deeper understanding rather than just delivering a land cadaster or a list of wetlands. This line of action could be applied in countries with a long history of research and naturalism, based on which it is possible to have an estimate of the conditions and representation of different physical and biological components of wetlands. Our country is far from that. In the Action Line 5 of this objective, the promotion of the new Ramsar sites is incorporated. In our country, beyond supporting the nomination of the new Ramsar sites, after the nomination and even the addition of sites to the National System of Protected Areas (SNASPE), the government management has not gone beyond generating a large information vacuum on these wetlands. Therefore, the aim of adding more Ramsar sites in our country is not clear, and on top of that, what is even more worrying is the nonexistence of the necessary information to effectively determine whether wetlands on the Ramsar List actually correspond to the areas and have all the characteristics to be considered priority wetlands of the National Strategy. Based on this, the other lines of action, such as the assessment of the management of river basins containing wetlands (Action Line 2) and the implementation of mechanisms to monitor the evolution of wetlands (Action Line 7), both of which require a great economic effort and a high degree of political will to be implemented and maintained over time, clearly cannot be implemented in our country.

When comparing management tools and the scope of the "National Strategy for Conservation and Rational Use of Wetlands" (CONAMA 2005) with other government initiatives such as those proposed by the US Environmental Protection Agency (US EPA), some historical facts could be used to improve our environmental management in relation to wetlands. First of all, the strategy of the US agency is not directed toward wetlands as a complex natural system, but toward particular components of these ecosystems. For example, one of the mainstays of management and environmental management is maintaining the balance of natural systems and nutrients (EPA 2008). The overall goal of this program is to provide scientific guidance to assist states, tribes, territories, and other jurisdictions in assessing the nutrient status of wetlands and provide technical assistance to develop quantitative criteria for wetland systems. To do this, the agency has hired scientists and technicians who participated in the preparations and discussions of 20 modules, including the implementation, and who are serving as technical guides for the various aspects involved in the action (from design sampling, chemical analysis of the samples, to the prioritization and decision making).

The differences in outlook, when it comes to wetland management, by various environmental agencies, have been discussed at meetings and seminars that gave incentive for this book to be materialized. Parallel with the dialogue and recognition of the researchers, as well as the interaction between public and private actors involved in the research and management of wetlands in our country, the second objective of this book is to appeal to not only a more active approach in terms of discussion but also contribution toward improving environmental management in Chile. In those terms, this book marks the beginning of a "Research Network, Management and Conservation of Coastal Wetlands," consisting of representatives from the country's several universities and the company Arauco.

We hope that this book, beyond delivering a summary of the available information, will be a motivating factor for the academia, business, government, and public world representatives, all of which are invited to encourage study and research and to improve management and advance conservation of the coastal wetlands of our country.

Santiago, Chile

José Miguel Fariña
Andrés Camaño

References

Alberti, J., J. Cebrian, A. Mendez-Casariego, A. Canepuccia, M. Escapa, and O. Iribarne. 2011. Effectos for nutrient enrichment and crab herbivory on a SW Atlantic salt marsh productivity. *Journal of Experimental Marine Biology and Ecology* 405: 99–104.

Arana, C., and L. Salinas. 2003. Flora vascular de los Humedales de Chimbote. *Perú.Revista Peruana de Biología* 10: 221–224.

Bertness, M.D., P. Ewanchuk, and B.R. Silliman. 2002. Anthropogenic modification of New England salt marsh landscapes. *Proceedings of the National Academy of Science* 99: 1395–1398.

Bertness, M.D., G.C. Trussel, P.J. Ewanchuk, B.R. Silliman, and C. Mullan. 2004. Consumer controlled alternate community states on Gulf of Maine Rocky Shores. *Ecology* 85: 1321–1331.

Botto, F., E. Gaitan, H. Mianzan, M. Acha, D. Giberto, A. Schiarti, and O. Iribarne. 2011. Origin of resources and trophic pathways in a large SW Atlantic estuary: an evaluation using stable isotopes. *Estuarine, Coastal and Shelf Science* 92: 70–77.

Bujalesky, G.G., A.M. Coronato, and F.I. Isla. 2001. Ambientes glacifluviales y litorales cuaternarios de la regióndel río Chico, Tierra del Fuego, Argentina. *Revista de la Asociación Geológica Argentina* 56 (1): 73–90.

Castilla, J.C. 1988. Earthquake-caused coastal uplift and its effects on rocky intertidal kelp cimmunities. *Science* 242: 440–443.

Cea. 2006a. *Protección y manejo sustentable de humedales integrados a la cuenca hidrográfica,* 114. Centro de Ecología Aplicada Ltda, Gobierno de Chile: Comisión Nacional del Medio Ambiente.

Cea. 2006b. *Conceptos y criterios para la evaluación ambiental de humedales*, 81. Centro de Ecología Aplicada Ltda.

Conaf-conama. 1999. Catastro evaluación de recursos vegetacionales nativos de Chile. Proyecto Conaf-Conama-Birf, 89 pp.

Conama. 2005. Estrategia Nacional para la Conservacion y Uso Racional de los Humedales en Chile, 30 pp.

Costa, S.B., J.C. Juliano, and A.M.G. Azevedo. 2003. Plant zonation in irregularly flooded salt marshes: relative importance of stress tolerance and biological interactions. *Journal of Ecology* 91 (6): 951–965.

Dussaillant, A., P. Galdames, and C.l. Sun. 2009. Water level fluctuations in a coastal lagoon: El Yali Ramsar wetland, Chile. *Desalination* 246: 202–214.

EPA, enviromental protection agency, USA. 2008. *Nutrient criteria technical guidance manual.* Wetlands.

Fariña, J., B.R. Silliman, and M.D. Bertness. 2009. Can Conservation Biologists Rely on Established Community Structure Rules to Manage Novel Systems? ... Not in Salt Marshes. 2009. *Ecological Applications* 19: 413–422.

Hauenstein, E., M. González, F. Peña-Cortés, and A. Muñoz Pedreros. 2002. Clasificación y caracterización de la flora y vegetación de los humedales de la costa de Tolten, (IX Region, Chile). *Gayana Botánica* 59: 87–100.

Mitsh, W.J., and J.G. Gosselink. 2007. *Wetlands*, 295 pp. John Wiley & Sons.

NRC, national research council. 1995. *Wetlands, characteristics and boundaries*, 306. Washington DC: National Academy Press.

Pool, N.A. 2010a. Ficha Informativa de los Humedales RAMSAR: Sistema hidrológico de Soncor del Salar de Atacama, CONAF. Ficha Informativa de los Humedales de RAMSAR (FIR).

Pool, N.A. 2010b. Ficha Informativa de los Humedales RAMSAR: Salar de Tara, CONAF. Ficha Informativa de los Humedales de RAMSAR (FIR).

Ramírez, C., and N. Añazco. 1982. Variaciones estacioneal en el desarrollo de *Scirpus californicus*, *Typha angustifolia* y *Phragmites communis* en pantanos Valdivianos, Chile. *Agro Sur* 10: 111–123.

Ramsar. 1971. Convención relativa a los humedales de importancia internacional especialmente como hábitat de aves acuáticas. Ramsar, Iran 2.2.1971. Modificada según el protocolo de París,

2.12.182 y Enmiendas de Regina, 28.5. 1987. Copia Certificada París, 3.7.94 Director, Oficina de Normas Internacionales y Asuntos Legales, Organización de las Naciones Unidas para la Educación, la Ciencia y la Cultura (UNESCO).

Rodriguez, R.E., and J.P. Contreras. 2008a. Ficha para la designación de nuevo sitio RAMSAR: Salar de Aguas Calientes IV, CONAF. Ficha Informativa de los Humedales de RAMSAR (FIR).

Rodriguez, R.E., and J.P. Contreras. 2008b. Ficha para la designación de nuevo sitio RAMSAR: Salar de Pujsa, CONAF. Ficha Informativa de los Humedales de RAMSAR (FIR).

Rojas, M.E., and E. Tabilo-valdivieso. 2004.fichaparaladesignacióndenuevositioramsar: Santuario de la Naturaleza Laguna Conchali, Union de Ornitólogos de Chile/ Corporación Ambientes Acuáticos de Chile, Centro Neotropical de Entrenamiento en Humedales. Ficha Informativa de los Humedales de RAMSAR (FIR).

San Martin, D. Contreras, J. San Martin, and C. Ramirez. 1992. Vegetación de las marismas del centro-sur de Chile. *Revista Chilena de Historia Natural* 65: 327–342.

San Martin, C. Ramírez, J. San Martín, and R. Villaseñor. 2001. Flora y vegetación del estero Reñaca (V Región, Chile). *Gayana Botánica* 58: 31–46.

Schlatter, R.P. and Y. Mancilla. 1998. Ficha Informativa de los Humedales RAMSAR: Santuario de la Naturaleza e Investigación Científica “Carlos Andwandter” del río Cruces, Valdivia. Universidad Austral de Chile. Ficha Informativa de los Humedales de RAMSAR (FIR).

Serna pesca. 1999. Los Humedales en Chile. Base de Datos 64, 22 pp.

Silliman, B.R. & M.D. Bertness. 2009. Human impacts on Salt Marsh Ecosystems: causes, consequences and solutions. University of California Press.

Valverde, V. 1996a. Ficha Informativa de los Humedales RAMSAR: Laguna del Negro Francisco y Laguna Santa Rosa, CONAF. Ficha Informativa de los Humedales de RAMSAR (FIR).

Valverde, V. 1996b. Ficha Informativa de los Humedales RAMSAR: Humedal Salar de Surire, CONAF. Ficha Informativa de los Humedales de RAMSAR (FIR).

Valverde, V. 1996c. Ficha Informativa de los Humedales RAMSAR: Humedal El Yali, CONAF. Ficha Informativa de los Humedales de RAMSAR (FIR).

Vilina, Y., J. Gibbons, and N. Nuñez. 2004. Ficha informativa de los Humedales RAMSAR: Propuesta para la inclusión de Bahía Lomas, Isla Tierra del Fuego, Chile, como sitio RAMSAR. Universidad Santo Tomas, Instituto de la Patagonia y Comisión Nacional del Medio Ambiente. Ficha Informativa de los Humedales de RAMSAR (FIR).

West, C. 1981. *The nature and future of the coastal salt marshes of South America*, 21–28. Sydney: XII International Botanical Congress.

Contents

Contributors

Sebastián Abades Centro de Genómica y Bioinformática, Facultad de Ciencias, Universidad Mayor, Huechuraba, Chile

Juan Aguirre Universidad de Chile, Facultad de Ciencias Forestales y Conservación de la naturaleza, Laboratorio de Ecología de Vida Silvestre, Santiago, Chile

Rafael Almar UMR LEGOS, University of Toulouse/CNRS/IRD/CNES, Toulouse, France

Miguel Álvarez F. InRES, Ecología Vegetal, Universidad de Bonn, Bonn, Germany

Nuria Aragoneses Center for Advanced Studies in Ecology and Biodiversity (CASEB), Pontificia Universidad Católica de Chile, Santiago, Chile

Dagoberto Arcos Regional Center for Environmental Studies (CREA), Universidad Católica de la Santísima Concepción (Catholic University of the Sacred Conception), Talcahuano, Chile

Iván Barría Departamento de Ecología, Facultad de Ciencias Biológicas, Pontificia Universidad Católica de Chile, Santiago, Chile

Mark D. Bertness Department of Ecology and Evolutionary Biology, Brown University, Providence, RI, USA

José Rafael Campino Department of Hydraulic and Environmental Engineering, School of Engineering (Departamento de Ingeniería Hidráulica y Ambiental, Escuela de Ingeniería), Pontificia Universidad Católica de Chile, Santiago, Chile

Centro de Investigación para la Gestión Integrada del Riesgo de Desastres (CIGIDEN - Fondap 15110017)

Rodrigo Cienfuegos Department of Hydraulic and Environmental Engineering, School of Engineering (Departamento de Ingeniería Hidráulica y Ambiental, Escuela de Ingeniería), Pontificia Universidad Católica de Chile, Santiago, Chile

Centro de Investigación para la Gestión Integrada del Riesgo de Desastres (CIGIDEN - Fondap 15110017)

Alejandro Dussaillant Water Resource Associates LP, King's College of London, London, UK

Patagonia Research Center, Patagonia, Chile

Cristián F. Estades Universidad de Chile, Facultad de Ciencias Forestales y Conservación de la naturaleza, Laboratorio de Ecología de Vida Silvestre, Santiago, Chile

José Miguel Fariña Center of Applied Ecology and Sustainability-CAPES, Pontificia Universidad Católica de Chile, Santiago, Chile

Department of Ecology and Evolutionary Biology, Brown University, Providence, RI, USA

Sergio Figueroa Regional Center for Environmental Studies (CREA), Universidad Católica de la Santísima Concepción (Catholic University of the Sacred Conception), Talcahuano, Chile

Rosa Fuentes Department of Marine Sciences, Universidad Arturo Prat, Iquique, Chile

Departamento de Ciencias del Mar, Universidad Arturo Prat, Iquique, Chile

Eugenia Gayo Instituto de Ecología y Biodiversidad (IEB), Santiago, Chile

Center for Advanced Studies in Ecology and Biodiversity (CASEB), Pontificia Universidad Católica de Chile, Santiago, Chile

Jorge Gironás Department of Hydraulic and Environmental Engineering, School of Engineering (Departamento de Ingeniería Hidráulica y Ambiental, Escuela de Ingeniería), Pontificia Universidad Católica de Chile, Santiago, Chile

Centro de Investigación para la Gestión Integrada del Riesgo de Desastres (CIGIDEN - Fondap 15110017)

Pablo González Regional Center for Environmental Studies (CREA), Universidad Católica de la Santísima Concepción, Talcahuano, Chile

Rodolfo Jerez Regional Center for Environmental Studies (CREA), Universidad Católica de la Santísima Concepción (Catholic University of the Sacred Conception), Talcahuano, Chile

Vinko Malinarich Agricultural and Livestock Service, Ministry of Agriculture, Tarapacá Region, Iquique, Chile

Servicio Agrícola y Ganadero I, Ministerio de Agricultura, Región de Tarapacá, Iquique, Chile

Pablo A. Marquet Departmento de Ecología, Facultad de Ciencias Biológicas, Pontificia Universidad Católica de Chile, Santiago, Chile

Instituto de Ecología y Biodiversidad (IEB), Santiago, Chile

Laboratorio Internacional en Cambio Global (LINCGlobal) y Centro de Cambio Global UC. Pontificia, Universidad Católica de Chile, Santiago, Chile

Flavio Olivares Tierra del Sol – Research and Development consultores, Region of Atacama, Vallenar, Chile

Tierra del Sol – Investigación y Desarrollo consultores, Región de Atacama, Vallenar, Chile

Viviana Olmos Laboratory of Biodiversity and Conservation of Aquatic Resources. Center for Environmental Sciences EULA, University of Concepción, Casilla 160 -C, Concepción, Chile

Jose Ortiz Regional Center for Environmental Studies (CREA), Universidad Católica de la Santísima Concepción (Catholic University of the Sacred Conception), Talcahuano, Chile

Marcelo Pavez Regional Center for Environmental Studies (CREA), Universidad Católica de la Santísima Concepción, Talcahuano, Chile

Ronny Peredo Arica, XV Región de Arica y, Parinacota, Chile

Carlos Ramírez G. Departamento de Ecología, Pontificia Universidad Católica de Chile, Santiago, Chile

Natalia Sandoval Laboratory of Biodiversity and Conservation of Aquatic Resources. Center for Environmental Sciences EULA, University of Concepción, Casilla 160 -C, Concepción, Chile

Walter Sielfeld Department of Marine Sciences, Universidad Arturo Prat, Iquique, Chile

Departamento Ciencias del Mar, Universidad Arturo Prat, Iquique, Chile

Brian Silliman Department of Zoology, University of Florida, Gainesville, FL, USA

Patricio Torres-Ramírez Regional Center for Environmental Studies (CREA), Universidad Católica de la Santísima Concepción (Catholic University of the Sacred Conception), Talcahuano, Chile

Claudio Valdovinos Laboratory of Biodiversity and Conservation of Aquatic Resources. Center for Environmental Sciences EULA, University of Concepción, Casilla 160 -C, Concepción, Chile

Daniela Vásquez Laboratory of Biodiversity and Conservation of Aquatic Resources. Center for Environmental Sciences EULA, University of Concepción, Casilla 160 -C, Concepción, Chile

Mauricio Villagrán Department of Hydraulic and Environmental Engineering, School of Engineering (Departamento de Ingeniería Hidráulica y Ambiental, Escuela de Ingeniería), Pontificia Universidad Católica de Chile, Santiago, Chile

Departamento de Ingeniería Civil, Universidad Católica de la Santísima Concepción, Chile

M. Angélica Vukasovic Universidad de Chile, Facultad de Ciencias Forestales y Conservación de la naturaleza, Laboratorio de Ecología de Vida Silvestre, Santiago, Chile

Chapter 1
Distribution and Conservation of Coastal Wetlands: A Geographic Perspective

Pablo A. Marquet, Sebastián Abades, and Iván Barría

Abstract This chapter provides a synoptic view of the distribution, conservation and importance of the connectivity of coastal wetlands of central Chile. For this, a methodology was developed to identify coastal wetlands present in central Chile (30 °S and 41 °40'S), through the analysis of 13 multispectral images Landsat ETM + for the spring-summer seasons of 2000 and 2001. Furthermore, we analyzed the degree in which the identified wetlands are contained within the Chilean protected area network, and carried out a connectivity analysis using graph theory. This analysis, despite being preliminary, allow us to draw a number of simple conclusions, of great importance for the management and protection of these ecosystems. On one hand, the latitudinal tendencies in distribution and area of these ecosystems suggest that their average number and extension increase in the north-south direction, and that the degree of protection these sites have is relatively acceptable, but that this would greatly increase if the Priority Sites were included in the analysis. On the other hand, the connectivity analysis suggests that those organisms whose movement distances are below 10–20 km, perceive the landscape as highly fragmented and slightly heterogeneous. Finally, the analysis of the contribution of different coastal wetlands to the connectivity of the system allows us to highlight that, in general, most of the wetlands are important for maintaining connectivity of species with movement capacity below the critical limit of 10–20 km.

P.A. Marquet (✉)
Departmento de Ecología, Facultad de Ciencias Biológicas, Pontificia Universidad Católica de Chile, Alameda 340, Santiago, Chile

Instituto de Ecología y Biodiversidad (IEB), Casilla 653, Santiago, Chile

Laboratorio Internacional en Cambio Global (LINCGlobal) y Centro de Cambio Global UC. Pontificia, Universidad Católica de Chile, Alameda 340, Santiago, Chile
e-mail: pmarquet@bio.puc.cl

I. Barría
Departmento de Ecología, Facultad de Ciencias Biológicas, Pontificia Universidad Católica de Chile, Alameda 340, Santiago, Chile

S. Abades
Centro de Genómica y Bioinformática, Facultad de Ciencias, Universidad Mayor, Camino La Pirámide 5750, Huechuraba, Chile

J.M. Fariña, A. Camaño (eds.), *The Ecology and Natural History of Chilean Saltmarshes*, DOI 10.1007/978-3-319-63877-5_1

Keywords Coastal wetlands • Distribution • Conservation • Connectivity

Introduction

Coastal wetlands are the environment interface between land and marine ecosystems highly sensitive to the characteristics of the water masses moving in and out of them (Niering 1985). Not only do they represent unique and important habitat for many species of vertebrates and invertebrates (eg, Vilina and Cofré 2000, 2006; Valdovinos 2004; Gonzalez and Victoriano 2005; Vila et al. 2006; Estades et al. 2009), but they also provide a number of beneficial ecosystem services, among which stand out (see Zedler and Kercher 2005; Engle 2011) the retention and removal of nutrients, stabilizing the shoreline, carbon sequestration, sediment containment, provision and quality improvement of water and lessening the effect of storms, floods and other natural disasters; the latter for the role they play in mitigating the intensity and wave height (e.g., Kirwan et al. 2011). It is to be noted as well that its distribution along the coastline represents a linear array of habitats that serve as a corridor for the migratory movement of large numbers of species (e.g., Aparicio 2006), so keeping its connectivity is essential.

The unique ecological character of coastal wetlands, combined with their high degree of vulnerability and threat from human activities - more than 10% of the human population lives within 10 m of sea level (McGranahan et al. 2007)-, have made them the focus of conservation efforts to ensure the sustainability of the services they provide (e.g., Engle 2011), especially in a context where a number of aspects that make them particularly sensitive are combined, such as: (1) the outlook for these ecosystems in a climate change scenario is not encouraging (e.g., Nicholls et al. 1999; Craft et al. 2009); (2) they have a high degree of deterioration, and (3) these are relatively rare ecosystems, as they represent less than 9% of the overall global area.

Zedler and Kercher (2005) proposed that one of the challenges for the conservation of coastal wetlands is the lack of inventories that are regularly updated on their distribution and status. This is undoubtedly a major challenge, however, it is necessary to undertake actions that would correct this important information gap. In this chapter, a geographical analysis of the distribution and conservation of coastal wetlands in central Chile is presented, with the aim of providing a synoptic view of their distribution, connectivity and importance in an geographic area where these ecosystems are little known and highly vulnerable to the human actions (e.g., Stuardo and Valdovinos 1989), because most of the country's population is concentrated in this area, and where threats arising from the increasing use of the coastal area foresee an uncertain future for its persistence.

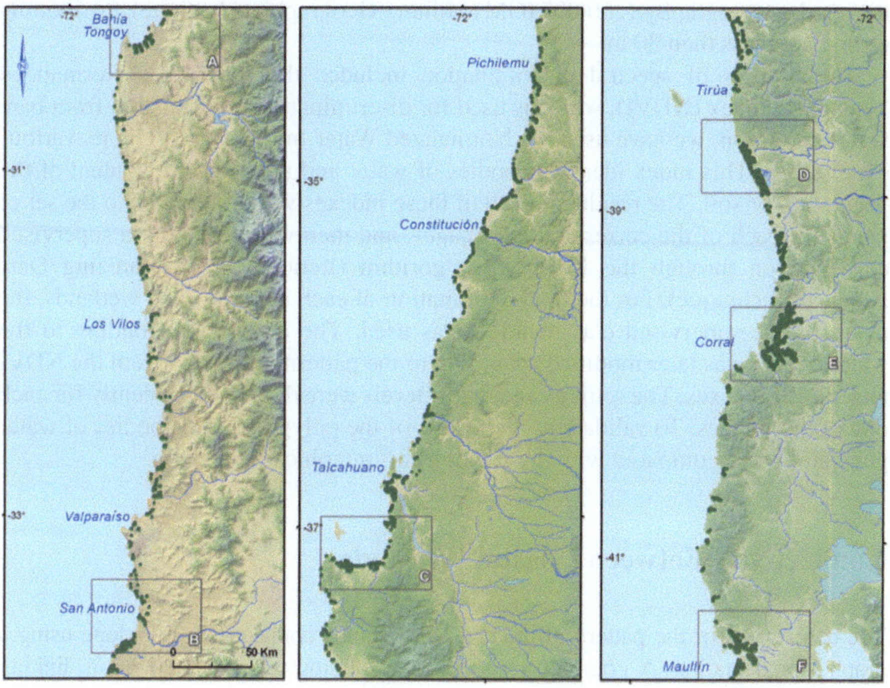

Fig. 1.1 Distribution of coastal wetlands identified in this study. The *boxes* identify wetlands presented in Fig. 1.2

Materials and Methods

The geographic area studied includes a strip of 5 km from the coastline, which stretches between parallels 30 °S and 41 °40'S (Fig. 1.1). Within this area the analysis of the distribution of coastal wetlands and their identification was based on the interpretation of a set of satellite images in order to evaluate the spectral behavior of elements characteristic to this landscape.

Identification of Coastal Wetlands

The identification of wetlands was carried out by digital processing of a total of 13 multispectral images Landsat ETM + in 2000 and 2001, during the spring-summer season. Previous to their analysis images were radiometry as well as atmospherically corrected. The identification of bodies of water by remote sensing, requires that the calibration significantly reduces atmospheric effects on the image data. The data used in this occasion were corrected by the atmospheric correction method used in the model COST (Chavez 1996). Georeferencing of images was based on the regular cartography 1: 50.000 IGM (Military Geographical Institute). Positioning errors were less than 30 m.

The process of spectral discrimination included the use of the Normalized Vegetation Index (NDVI), which is used for discrimination of vegetation from bare soil. In addition, we have used the Normalized Water Index (NDWI), in its various expressions. This index identifies bodies of water and the moisture content of the vegetation or soil. The results of each of these indexes were integrated to the set of bands for each of the corresponding images, and then we carried out a supervised classification through the ISODATA algorithm (Iterative Self-Organizing Data Analysis Technique). For the final delimitation of each of the coastal wetlands, the result of the supervised classification was used. The class corresponding to the water bodies was later modified according to the patterns extracted from the NDVI and NDWI indexes. The settings and cutoff levels were done independently for each one of the images. To validate it, the results of the polygons of the bodies of water generated were compared with the existing cartographic information.

Connectivity Between Coastal Wetlands

The description of the pattern of the coastal wetlands connectivity was done using a graph representation. A graph corresponds to a topological characterization, linking landscape elements (e.g., habitat patches) through connections (i.e. corridors), in order to provide a parsimonious description of landscape connectivity (Cantwell and Forman 1993; Urban and Keitt 2001). Formally, a graph G consists of a set of nodes or vertices V (G) connected by edges or arcs G (E), such an edge $e_{ij} = v_i v_j$ joins the nodes v_i and v_j (Harary 1969; Chartrand 1977). The existence of a connection between a pair of nodes implies a certain degree of potential ecological flow between them (e.g., dispersion of individuals, flow of materials and energy, migration). Depending on the research context, the estimated connections between the nodes are based on a distance matrices of different nature. In this analysis Euclidean geographical distance matrices were used, in order to explore simple spatial relationships and its effects on potential connectivity patterns for the complete system of habitat patches. However, more sophisticated approaches may include aspects of the biology of species of interest, degree of hostility matrix surrounding habitat patches, etc. (Schumaker 1996; Hanski 1999; Bunn et al. 2000; Urban and Keitt 2001; Vos et al. 2001; D'Eon et al. 2002; Jordan 2003).

A path within a graph corresponds to a sequence of interconnected nodes, in such way so that each node is unique along its route, that is, each node and edge are visited only once. A closed path including three nodes forms a cycle. A graph with paths that do not include cycles forms a tree. A tree that includes all available nodes forms a connected tree (Bunn et al. 2000; Urban and Keitt 2001). Obviously, if we define d_{max} as the maximum distance allowed to connect two patches of habitat, the topology of the resulting graph will be affected as d_{max} changes and so will be the landscape perceived by a species whose potential dispersion is approaching d_{max}.

Therefore, the simulation of graphs with different values of d_{max} can be used to assess the potential fragmentation that a species could perceive given the landscape provided. This approach is especially useful when you want to know the critical distance at which the landscape losses it connectivity for a wide range of organisms with different capacities and dispersal strategies.

Results and Discussion

Distribution

In total we have identified 412 coastal wetlands, with a total area of 38,167 ha (Fig. 1.1). The same as the distribution of wealth in our country, the distribution of the coastal wetlands areas is highly biased to small sizes, so much so that 63% (n = 260) of wetlands have an estimated area lower than 10 ha, and the 10 wetlands covering larger areas (among which are: Río Cruces, the Bío Bío river mouth, Puerto Saavedra; see Fig. 1.2) represent 66% of the total area of wetlands in the study area.

Latitudinally, the area of wetlands per degree of latitude does not exceed 2000 ha, except from 37 °S (Fig. 1.3). It is also noted in this figure that the average area of wetlands follows a similar trend, with a maximum around 700 ha at 40 °S.

Unlike what happens with other conservation targets, such as species or vegetation formations, coastal wetlands better represented within Protected Areas reaching a percentage close to 20%. This result includes considering the wetland area that overlaps with the SNASPE (National System of Protected Areas), PPP (Private Protected Areas) and Ramsar sites (Fig. 1.4) sites. Although this percentage appears high and above the threshold recommended by international bodies, it is necessary to know if we are protecting what really is the most important, in terms of the connectivity they provide, and ecosystem processes and services they are supporting, and considering the degree of threat to which they are being subjected to.

Regarding the latter, the particularly important ones are the coastal wetlands located in areas with high population density, such as Viña del Mar, Valparaíso, Talcahuano, Coquimbo and Valdivia, each of which has more than 140,000 inhabitants (near a million in total) and where about 25% of the total area of wetlands, present in the study, is concentrated (see section on the relative importance of coastal wetlands below). These, presumably, would be highly threatened as a product of human activities, especially if one takes into account the non-resident population using the coastal zone in these areas. Finally, it should be noted that, considering the priority sites for conservation proposed by the Ministry of Environment, the percentage of protection would increase by an additional 43.9%. This indicates a greatly important increase of protection for this types of ecosystems, to the extent that these sites would become part of the National Network of Protected Areas.

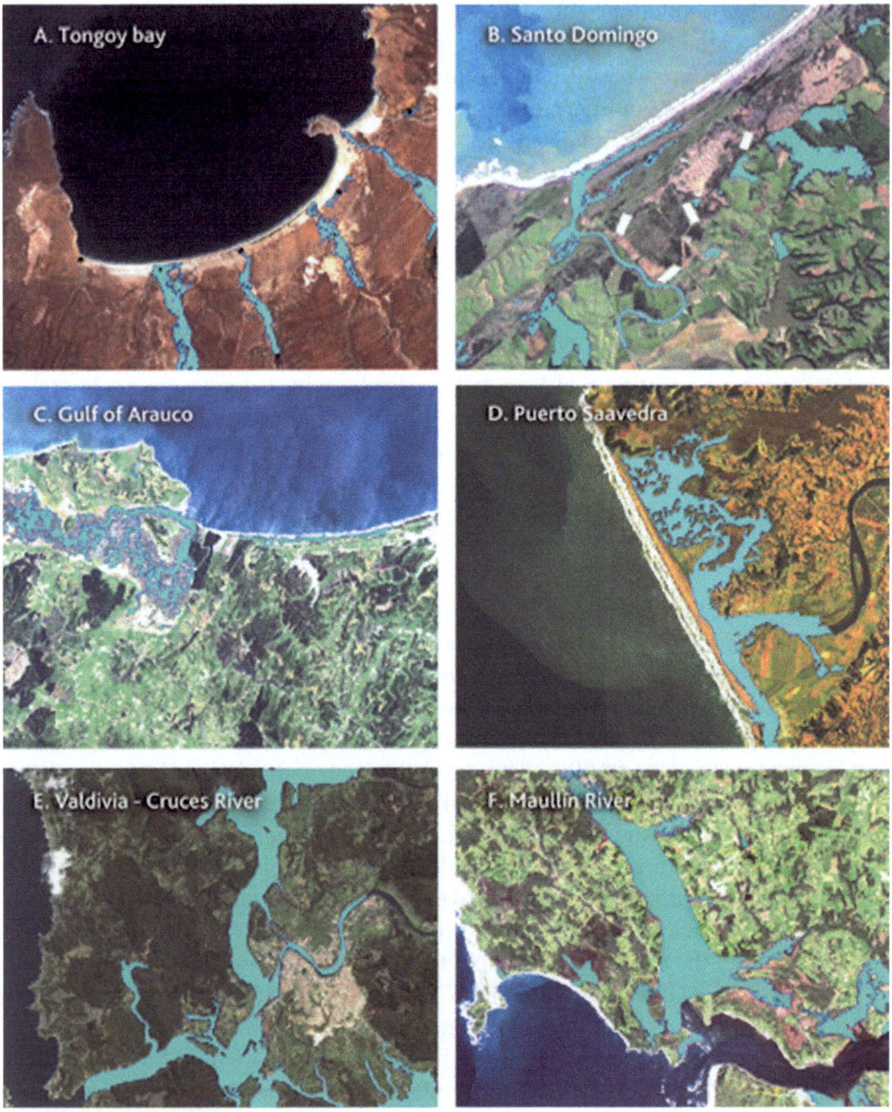

Fig. 1.2 Coastal wetlands identified (in *boxes*) in Fig. 1.1

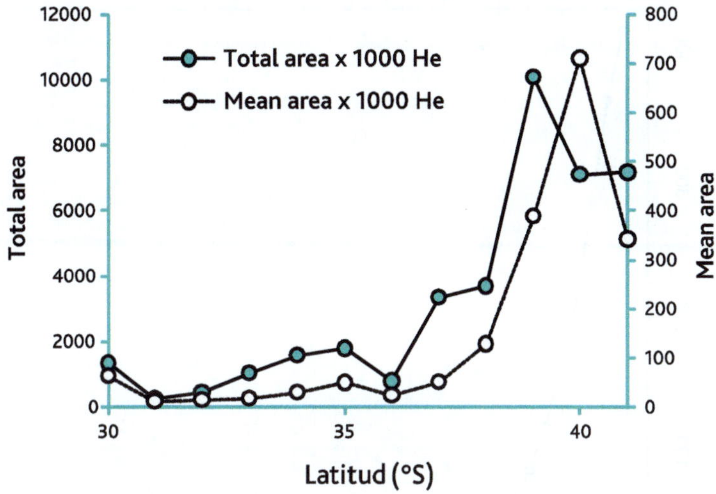

Fig. 1.3 Latitudinal tendency in the total accumulated area (*solid symbol*) and average area (*empty symbol*) of coastal wetlands present in each latitudinal band identified in this study

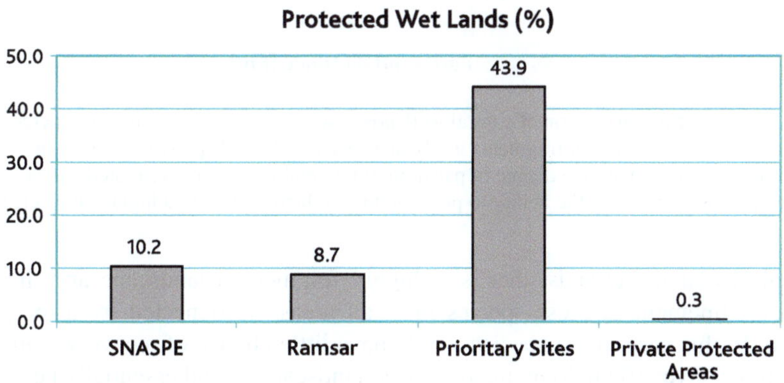

Fig. 1.4 Level of protection (as percentage) of coastal wetlands in the different types of protected areas established in the country

Connectivity

Regarding the connectivity of the system composed of the 412 identified coastal wetlands, Fig. 1.5 shows the connectivity variation obtained simulating different values of d_{max}, defined as the perceived distance by a theoretical species whose maximum dispersion oscillates around this value.

It is recognized that given the current topology of wetlands, there would be a rapid loss of connectivity for species whose dispersion does not exceed 10 km. Specifically, species with low dispersal capacity perceive the landscape as a set of

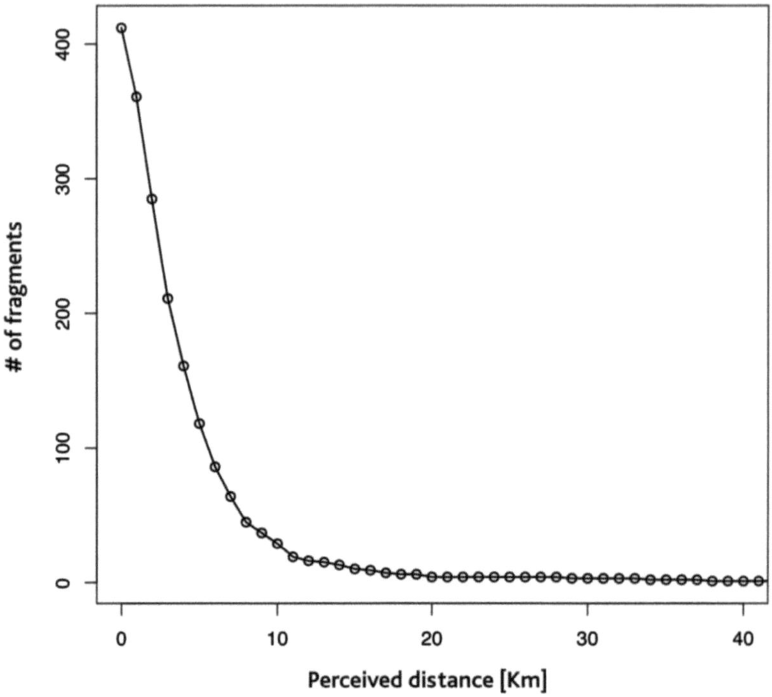

Fig. 1.5 Level of fragmentation of coastal wetlands based on the maximum distance perceived by a species. A higher number of fragments imply that species whose dispersion is below the distance perceived, experience the high degree of partition of the total landscape as isolated habitats. A low number of fragments entail the ability to perceive the wetland system as a highly connected set

multiple isolated wetlands, that is, a highly fragmented landscape and thus, the likely confinement of these species in small areas disconnected from the global system. In this case, these species would have little chance of recovery from local extinctions. Over 10 km long, the perceived landscape would essentially be unitary, because the high local connectivity between neighboring wetlands provide potential access to the entire network of habitats available in the existing system of coastal wetlands. Note that the decay curve is steep, which is expected, since the spatial arrangement of the system of coastal wetlands is essentially linear.

Figure 1.6 shows the effects of d_{max} in an average area of perceived clusters at this scale of connection, that is the amount of available habitat for the species. Leading up to a distance of approximately 20 km, the area of connected wetland clusters shows a continuous gain. Above this distance, a step-wise pattern in mean fragment area or increase in connectivity is observed. This pattern is attributed to the specific characteristics of the geographical distribution of coastal wetlands, and implies the existence of a significant degree of risk against potential losses of patches of habitat that may cause spatial gap by about 35 km, given that the species that exploit the landscape at this scale would perceive a loss of potentially available area by approximately 50%.

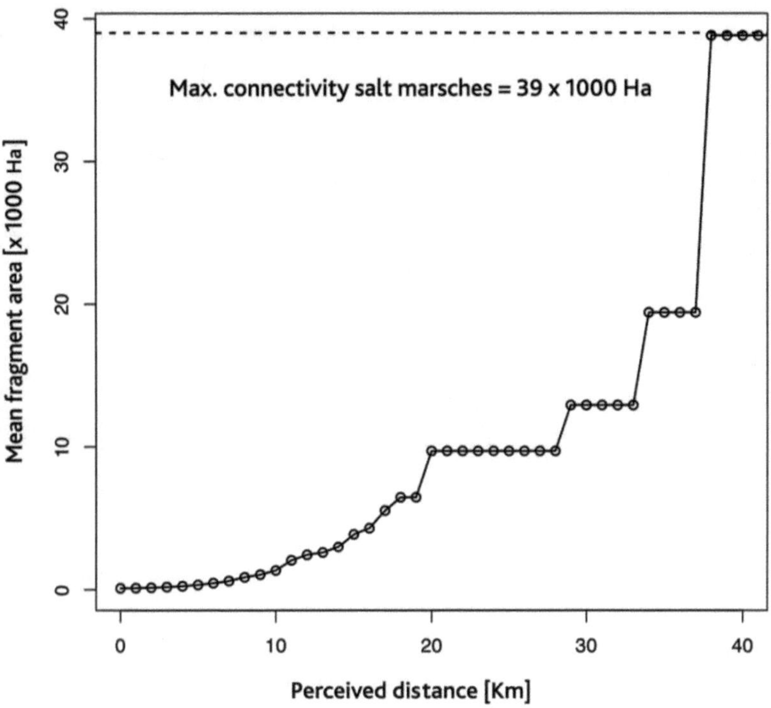

Fig. 1.6 Loss of area for species that perceive the landscape at a scale lower than the distance indicated in abscissa

In the case of the analysis presented in Fig. 1.6, it is important to consider that the available area could be made up of fragments whose areas are very heterogeneous, which could be a relevant factor to take into account, because the heterogeneity in the area is associated with diversity of environmental resources and situations which, usually in a positive way, impact the persistence of the metapopulations that inhabit them (reducing the risk of extinction, Hanski 1991). Figure 1.7 shows the heterogeneity in area of the connected coastal wetlands that are generated at distances less than d_{max}, simply defined as the standard deviation of the average values obtained for the graph, and for different scales of perception. Interestingly enough, it is noted that at 20 km is verified the highest degree of heterogeneity, which implies a connected landscape composed of coastal wetlands of widely different sizes. Above this distance, a rapid loss of variability is observed as a product of the coalescence of differents clusters of connected wetlands in a large geographical extent.

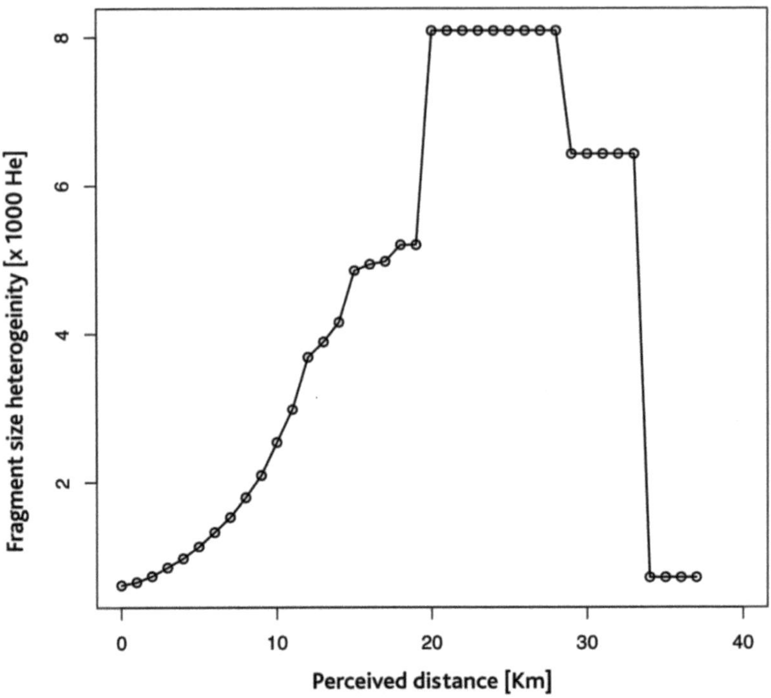

Fig. 1.7 Heterogeneity in the sizes of areas available for species that perceive the landscape at a scale lower than the distance indicated in the abscissa

Assessment of the Relative Importance of Coastal Wetlands

The relative importance of coastal wetlands was assessed by a simple procedure based on the recursive removal of neighboring wetlands: evaluating the size of the gap generated by eliminating a focal wetland and a progressive number of neighboring wetlands. Figure 1.8 shows the effect that would produce the loss of a focal wetland and its closest N neighbors, generating a distance gap. The diagram shows on the ordinate axis, each one of the coastal wetlands analyzed, sorted from north to south according to latitudinal position. The abscissa shows the number of nearest neighbors removed to the north and to the south of the disturbed wetland. The response shows on a color scale the distance gap created by removing these wetlands from the coastal network.

The impact of the removal varies depending on the latitudinal location of the wetland, which is accentuated as the number of neighbors of the focal wetland that are lost increases. The inferior diagram on the binary color scale shows the resulting pattern by examining gaps of size d_{max} <20 km. It is generally shown that for most of the wetlands, the loss of one or two close neighbors would create a large enough gap to reach a critical distance global loss of connectivity (according to the general pattern observed in Fig. 1.6).

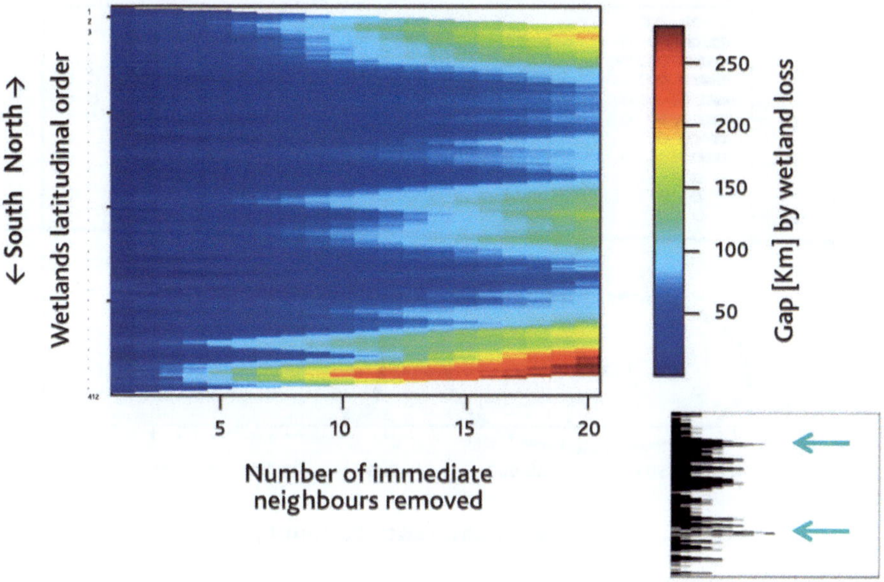

Fig. 1.8 Sensitivity diagram showing coastal wetland loss. *Y* axis shows the latitudinal ranking of the 412 coastal wetlands analyzed (without associated geographical coordinates). *X* axis shows the number of immediate neighbors removed, both to the north and to the south, of a focal wetland evaluated. *Color scale* indicates the size of the gap generated by the loss of the neighboring wetland collection. Lower diagram shows, in binary scale, the removals that imply the creation of a critical gap of 20 km. *Arrows* signal robust wetlands that require a significant loss of close neighbors to cause a notable change in the global connectivity of the wetland system. Most of the wetlands show a high sensitivity, since removing a few neighbors seems to be enough to trigger effects of global disconnection

This implies that the loss of a relatively low set of neighbors could be sufficient to cause an effect of global loss of connectivity in the current arrangement of coastal wetlands. This result is expected because of the linear arrangement of this habitat type that makes global connectivity highly sensible to the loss of local elements, which break the linearly concatenated arrangement. However, it should be noted that in some latitudes (arrows on the bottom panel) there is a significant degree of robustness, and therefore a large number of neighboring wetlands need to be lost in order to generate a critical gap in connectivity. This is possibly due to the high density of small wetlands concentrated at these latitudes. A simple and useful way to assess the priorities for conservation of coastal wetlands is to jointly consider their importance for connectivity (Fig. 1.8) and the degree of threat they are subjected to (this is assessed through evaluation of the number of inhabitants in municipality where the wetland is located). This is presented in Fig. 1.9, where you can see that the Tongoy and Maullin wetlands are of great importance, and at the same time, are under high potential threat, which render them as of high priority for conservation.

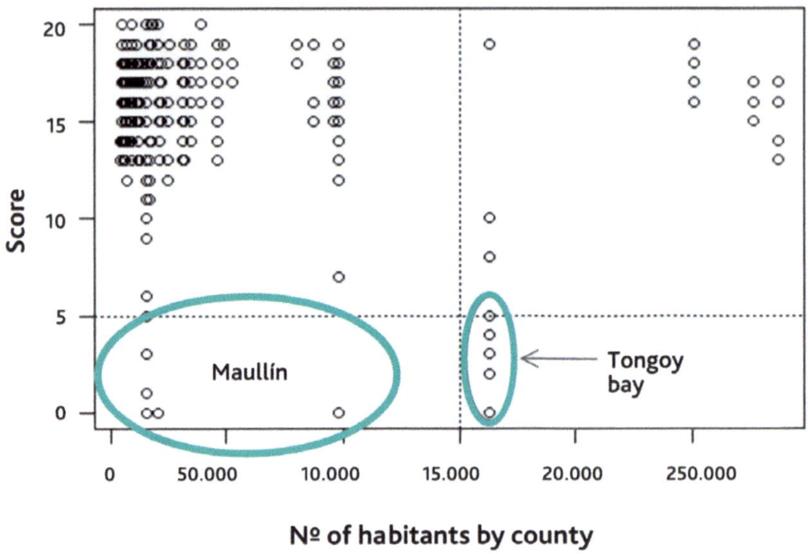

Fig. 1.9 Ordination diagram between wetland importance and the potential threat of human effects. The wetland's importance corresponds to the score value obtained in Fig. 1.8, where values close to zero would entail a high vulnerability associated to the loss of the focal wetland and its closest neighbors

Conclusions

This chapter has attempted to provide a synoptic view of the distribution, conservation and importance for the connectivity associated to the coastal wetlands of central Chile. This view, despite being preliminary, allows us to draw a number of conclusions that are of great importance for the management and protection of these ecosystems. On one hand, the latitudinal distribution trends and areas of these ecosystems, point out that their number and average size are increasing in the north south direction, that the degree of protection they have is relatively acceptable, and that this would markedly increase if Priority Sites where become part of the National Network of Protected Areas (SNASPE). On the other hand, the connectivity analysis suggests that those organisms, whose movement distances are below 10-20 km, perceive the landscape as highly fragmented and with little heterogeneity. Further, the analysis of the contribution of various coastal wetlands to the connectivity of the regional system analyzed, points out that in general, most wetlands are important to maintain connectivity for species with movement capabilities under the critical threshold of 10-20 km.

Acknowledgements This project was funded by FONDECYT-FONDAP 1501-0001 for the ICM P05-002 project and the PFB-23 contract, CONICYT, Chile. We appreciate any feedback, suggestions and the ongoing concern of the editors of this volume, without whose tenacity this chapter would not have been completed.

References

Aparicio, A. 2006. Abundance, distribution and migration chronology of shorebirds on exposed sandy beaches of south central Chile. In *Waterbirds around the world*, ed. G.C. en Boere, C.A. Galbraith, and D.A. Stroud, 188. Edinburgh: The Stationery Office.

Bunn, A.G., D.L.Y. Urban, and T.H. Keitt. 2000. Landscape connectivity: a conservation application of graph theory. *Journal of Environmental Management* 59: 265–278.

Cantwell, M.Y., and R.T.T. Forman. 1993. Landscape graphs: ecological modeling with graph theory to detect configurations common to diverse landscape. *Landscape Ecology* 8: 239–255.

Chartrand, G. 1977. *Introductory graph theory*. New York: Dover Publications.

Chavez, P.S., Jr. 1996. Image-based atmospheric corrections- revisited and improved. *Photogrammetric Engineering and Remote Sensing* 62 (9): 1025–1036.

Craft, C., J. Clough, J. Ehman, S. Joye, R. Park, S. Pennings, H. Guo, and M. Machmuller. 2009. Forecasting the effects of accelerated sea-level rise on tidal marsh ecosystem services. *Frontiers in Ecology and the Environment* 7: 73–78.

D'eon, R.G., S.M. Glenn, I.Y. Parfitt, and M.-J. Fortin. 2002. Landscape connectivity as a function of scale and organism vagility on a real forested landscape. *Conservation Ecology* 6: 10–23.

Engle, V.D. 2011. Estimating the provision of ecosystem services by Gulf of Mexico coastal wetlands. *Wetlands* 31: 179–193.

Estades, C.F., M.A. Vukasovic, and V. López. 2009. Las aves acuáticas del río Itata. In *La cuenca hidrográfica del río Itata. Aportes científicos para su gestión sustentable*, ed. O. Parra, J.C. Castilla, H. Romero, R. Quiñones, and A. Camaño, 213–228. Concepción: Editorial Universidad de Concepción.

González, A., and P. Victoriano. 2005. Aves de los humedales costeros del Centro-Sur de Chile. Capítulo 35 En. In *Biodiversidad y Ecología de la Cordillera de la Costa*, ed. C. Smith, J. Armesto, and C. Valdovinos. Santiago: Editorial Universitaria.

Hanski, I. (1991). Single-species metapopulation dynamics: concepts, models and observations. Biological Journal of the Linnean Society 42: 17–38.

Hanski, I. 1999. *Metapopulation ecology*. Oxford: Oxford University Press.

Harary, F. 1969. *Graph theory*. Reading: Addison-Wesley.

Jordán, F. 2003. Quantifying landscape connectivity: key patches and key corridors. En. In *Ecosystemand sustainable development IV*, ed. E. Tiezzi, C.A.Y. Brebbia, and J.-L. Usó, 883–892. Southampton: WIT Press.

Kirwan, M.L., E. Wolanski, E.B. Barbier, and B.R. Silliman. 2011. The present and future role of coastal wetland vegetation in protecting shorelines: answering recent challenges to the paradigm. *Climatic Change* 106: 7–29.

Mcgranahan, G., D. Balky, and B. Anderson. 2007. The rising tide: assessing the risks of climate change and human settlements in low elevation coastal zones. *Environment and Urbanization* 19: 17–37.

Nicholls, R.J., F.M.J. Hoozemans, and Y.M. Marchand. 1999. Increasing flood risk and wetland losses due to global sea-level rise: regional and global analyses. *Global Environment Change* 9: S69–S87.

Niering, R. 1985. Wetlands. The Audubon society nature guides, 638 p. Alfred A. Knopf Inc. New York, NY.

Schumaker, N.H. 1996. Using landscape indices to predict habitat connectivity. *Ecology* 77: 1210–1225.

Stuardo, J., and C. Valdovinos. 1989. Estuarios y lagunas costeras: Ecosistemas importantes de Chile Central. *Ambiente y Desarrollo* 5: 107–115.

Urban, D.L.Y., and T.H. Keitt. 2001. Landscape connectivity: a graph-theoretic perspective. *Ecology* 82: 1205–1218.

Valdovinos, C. 2004. Ecosistemas estuarinos. En. In *Biología marina y oceanografía: Conceptos y procesos*, ed. C. Werlinger, vol. 2, 395–414. Chile: Trama Impresores S. A.

Vila, I., A. Veloso, R. Schlatter, and C. Ramirez (Eds). 2006. Macrófitas y vertebrados de los sistemas límnicos de Chile. Editoral Universitaria, Santiago, Chile.
Vilina, Y.A., and H. Cofré. 2000. El Niñoeffectsontheabundanceandhabitatassociation patterns of four grebe species in Chilean wetlands. *Water birds* 32: 95–101.
Vilina, Y., and H. Cofré. 2006. Aves acuáticas continentales. P 270-279 en Biodiversidad de Chile: Patrimonio y Desafíos. In *Comisión Nacional del Medio Ambiente*, ed. P. Saball, M.K. Arroyo, J.C. Castilla, C.F. Estades, J.M. Ladrón de Guevara, S. Larraín, C. Moreno, F. Rivas, J. Rovira, A. Sánchez, and L. Sierralta. Santiago: Editorial Ocho Libros.
Vos, C.C., J. Verboom, P.F.M. Opdam, and C.J.F. Yter Braak. 2001. Toward ecologically scaled landscape indices. *The American Naturalist* 183: 24–41.
Zedler, J.B., and S. Kercher. 2005. Wetland resources: status, trends, ecosystem services and restorability. *Annual Review of Environment and Resources* 30: 39–74.

Chapter 2
River Mouths and Coastal Lagoons in Central Chile

Rodrigo Cienfuegos, José Rafael Campino, Jorge Gironás, Rafael Almar, and Mauricio Villagrán

Abstract River mouths and coastal lagoons in central Chile are some of the most dynamic ecosystems in the country. They are controlled by a great variety of natural and anthropic factors, which have confer these ecosystems a high environmental and human value. The diversity of these ecosystems is mainly due to a complex set of parameters that define them: from the hydrological and climatic variability, the heterogeneity of the river mouths, the high energy of the coasts, to different annual sediment availability, etc. These parameters, as well as meteorological and hydro-morphodynamics, all create a very unique combination, even at a global scale, being similar only to some sectors of South Africa, Australia and New Zealand.

Additionally, the seismic characteristics of Chile commonly cause morphological alterations in coastal areas, and these changes range from subsidence or uplifts to dramatic sedimentary changes caused by tsunamis that hit the central coast every 50–100 years. This uniqueness of the river systems of central Chile, enhances its scientific interest within the international community. Moreover, the predominant dynamics of central Chile are quite different than those traditionally studied in the northern hemisphere.

R. Cienfuegos (✉) • J.R. Campino • J. Gironás
Department of Hydraulic and Environmental Engineering, School of Engineering (Departamento de Ingeniería Hidráulica y Ambiental, Escuela de Ingeniería), Pontificia Universidad Católica de Chile, Santiago, Chile

Centro de Investigación para la Gestión Integrada del Riesgo de Desastres (CIGIDEN - Fondap 15110017)
e-mail: racienfu@ing.puc.cl

R. Almar
UMR LEGOS, University of Toulouse/CNRS/IRD/CNES, Toulouse, France

M. Villagrán
Department of Hydraulic and Environmental Engineering, School of Engineering (Departamento de Ingeniería Hidráulica y Ambiental, Escuela de Ingeniería), Pontificia Universidad Católica de Chile, Santiago, Chile

Departamento de Ingeniería Civil, Universidad Católica de la Santísima Concepción, Chile

J.M. Fariña, A. Camaño (eds.), *The Ecology and Natural History of Chilean Saltmarshes*, DOI 10.1007/978-3-319-63877-5_2

This chapter presents the general characteristics that govern the geomorphological processes of the central coast of Chile, such as hydrology, climate, prevailing rise, geographic characteristics and impacts caused by the tsunamis. Additionally, a description and classification of some of these phenomena is presented for different basins in Chile, paying attention to some particular cases used as examples for the description of the coastal processes that govern coastal morphodynamics. Finally, it is concluded that the coastal formations in central Chile are strongly controlled by hydrology and typology of their river mouths, with emphasis on the size of the basin, as well as on the seasonal precipitations. These last two variables are fundamental in explaining and characterizing the coastal geomorphology of central Chile.

Keywords Hydrodynamics • Basin • Geomorphology

Introduction

The morphodynamic evolution of the Chilean coast is determined by complex interactions between waves, tides and wind, on one hand, and the geomorphological and hydro-sedimentary basins that provide overflow and solid material being transported to the Pacific Ocean, on the other. (Araya-Vergara 1971; Benítez 1994). In addition, the highly active tectonic movements of the Nazca and South American plates, can cause significant changes in coastal systems due to overflows and turbulences (Weischet 1963; Farías et al. 2010).

River mouths are complex and fragile systems where a dynamic interplay between physical, chemical and biological variables (Kennish and Paerl 2010) occur. They also represent very active areas from the hydro-sedimentary point of view, where the sediment pattern depends on the relative influence of the ebbs and flows of the tidal currents, the flow and the runoff, the waves and the wind, and the physical characteristics of sediments. Generally, these systems can differentiate between estuaries and coastal lagoons. According to Bird (2007), estuaries are influenced by the dynamic fluvial and marine actions, while coastal lagoons are seawater bodies generally separated by sedimentary barriers that can be connected to the sea intermittently.

Deltas can be classified according to the degree of tidal influences in their dynamics (Davies 1964).

A microtidal estuary (tidal range less than 2 m) is generally dominated by the waves and river currents.

A mesotidal estuary (tidal range of 2 and 4 m) has more intense tidal currents, with a wider intertidal zone and multiple winding canals. A macrotidal or megatidal estuary (tidal range greater than 4 m and 6 m respectively) has dominant tidal currents, shaped like a funnel and a wide mouth. In wide mouth deltas, the effect of the tides favors marine sediment deposits forming banks and dunes, sometimes shaping deltas which are submerged at high tide with divergent channels, upstream and downstream (Davies 1964; French 1997).

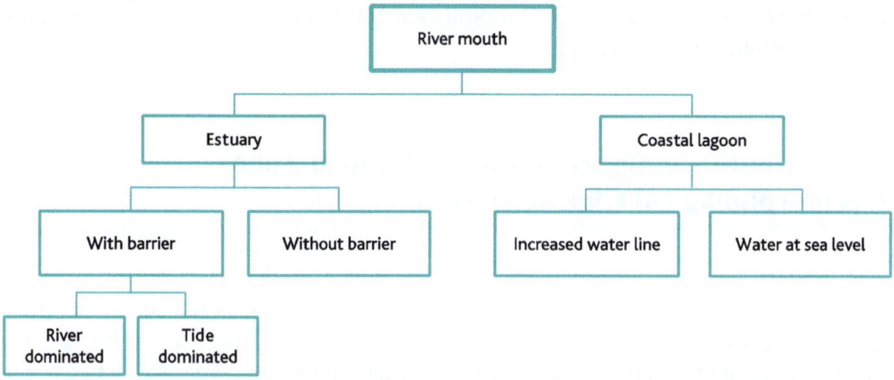

Fig. 2.1 Classification of estuaries in microtidal coasts (Adapted from Cooper 2001)

Cooper (2001) studies the microtidal estuaries of the South African coast, very similar to the systems present in central Chile. Like the South African coast, the Chilean coast is mostly microtidal with the exception of estuaries located south of Puerto Montt, Chiloe and the Chacao channel, which highlights the Reloncavi estuary that is dominated by the tides. Given these similarities, the classification of Cooper (2001) will be used to describe the river mouths located in central Chile. In the Fig. 2.1 this classification that distinguishes between estuaries (whose mouth is open more than 70% of the time) and coastal lagoons (with open mouths less than 30% of the time), is presented.

In turn, river mouths can be divided into those with a bar or sedimentary barrier mouth and those without. Estuaries with a barrier may be dominated by runoffs from either the rivers or the ebb and flow of the tidal currents. On the other hand, coastal lagoons, usually separated from the sea, may present a stilted/heavily banked water levels (above sea level) or be at sea level.

In Chile, studies on configuration of the river mouths in central area are relatively few and recent. The highlighted contributions are those of Araya-Vergara (1971, 1981, 1982, 1996), Stuardo and Valdovinos (1989) and Andrade and Grau (2005). Araya-Vergara (1981, 1982) proposes a zoning of the delta systems according to morphoclimatic characteristics, distinguishing the area of the distal deltas of the desert, the area of the Norte Chico prograded rivers (from the Copiapo River to Maipo), the area of the estuarine deltas in the transition zone between semi-arid and humid climate (from the Rapel river to the Biobio river), and the area of the estuaries of humid and very humid climate in the Lakes Region. Meanwhile, Stuardo and Valdovinos (1989) and Andrade and Grau (2005) distinguish between estuaries and coastal lagoons, incorporating in their classification the seasonal nature of the hydrological regimen forming the fresh water courses. All of these works provide valuable information, which, once consolidated, can serve as a starting point and reference for future studies of river mouths and coastal lagoons of central Chile.

In this chapter, the basic background for understanding of the processes that govern hydro-sedimentary dynamics of the river mouths in the central zone of Chile is exhibited, in order to establish common elements to classify the different morphological configurations observed.

River Mouth Configuration in the Central Zone – Geomorphological Characteristics

The western edge of the South American continent is characterized by a narrow steep continental shelf (Araya-Vergara 1971). This feature is the result of active tectonic processes taking place in the vicinity of the submarine pit that borders the meeting point between the Nazca and South American plates. From Arica to Chacao channel, there are four main distinguished geomorphological units: the Coastal Plains (las Planicies Litorales), the Coastal Range (Cordillera de la Costa), Intermediate Depression (Depresion Intermedia) and the Andes (Benítez 1994). Hydrographic features described below are largely determined by this geomorphological configuration.

The sediments that form the coastal systems of Chile have several features that can largely be explained by their geological origin and form (Paskoff and Manríquez 2004). They are distinguished by sediments example of continental origin transported by surface runoff and winds, where sediments dunes or beaches can be moved by the wind, and those that originate in situ, under the erosive action of waves on cliffs or rocky edges (Bird 2007).

In central Chile, Paskoff and Manríquez (2004) highlight the high sediment of fluvial origin that occurs at the river basins due to erosion caused by rain. Apparently, these contributions have grown over the last century by the gradual decline of vegetation, increasing deforestation and intensification of long, dry summers that favor the removal of solid material. To the south of the Imperial River (Araucanía Region), the dominance of large rivers linked to the disappearance of the Coastal Range (Cordillera de la Costa) and increasing tidal scope inhibits the development of dunes and promotes the formation of estuaries (Araya-Vergara 1981).

Furthermore, the sediments of continental origin carried by water are subjected to the action of coastal drift currents when they reach the Pacific Ocean. In the south-central coast of Chile, this coastal current transports sediments from river mouths or estuaries towards north as a product of reigning South-West waves (Araya-Vergara 1971). This is the case of the El Yali wetland (Valparaiso Region), whose sands are not coming from the neighboring beaches and boulders, but from the deflation of ancient marine deposits of the Punta Torro terraces (Paskoff and Manríquez 2004). Similarly, the Ritoque dune field (Valparaiso Region) is supplied by sands of the Aconcagua River, which flows into the sea a few kilometers further south. The sediments are transported by a coastal drift, and unloaded in the northern sector of this dune field, as the Quintero elevation puts obstacle to this action. The Concon sand dunes (Valparaíso Region) have their origin in the marshlands of Vina del Mar (Marga-marga) and Renaca, transported north along the littoral drift (Paskoff and Manríquez 2004).

The Santo Domingo dune ridges (Valparaiso Region) have a clear relationship with sediments carried by the Maipo River, which are also transported over 35 km to the north by the coastal current. Moreover, the Putu dune ridges located north of Constitucion (Maule) converge towards the mouth of the Maule River, demonstrating that its sediments have been supplied by the river and aligned by the littoral drift on the proximal convex part (Araya-Vergara 1981); the large volume of sediment shows considerable force of the river. Further south, the Itata River (Biobio Region) is an important source of sediment and sand for the dunes, marshes and beaches. According to Paskoff and Manríquez (2004), the Chanco dunes have their origin in the solid material carried by the river.

The above mentioned examples demonstrate the role of littoral drift and river sediments in the dynamic, and source of sand for littoral coastal systems of the central zone. However, although the material carried out by the rivers is one of the most important sources of beach and dune sediment on the central coast of Chile, there are also structures of different origins. For example, the Gulf of Arauco dunes in the Bio-Bio Region – the largest dune complex in the country (30,000 hectares approximately) - is not located north of any major delta that could explain its origin. In this case, it is thought that the continental shelf, due to its low slope and large width, could be an important source of sediments (Araya-Vergara 1985). These sediments have accumulated during glacial periods of low sea level. Tavares-Correa (1996a, b) has noted that the average diameter of the sand grows despite the dominant coastal current coming from the south. According to him, this increase in size is due to a process of selective removal of the finer grains through the wave action. However, the detour to the south of most of the stream and river mouths that cross this dune field, remains to be explained (Paskoff and Manríquez 2004). Araya-Vergara (1985) also considers that the tectonic rising trend in this area can promote the formation of the dunes. The phenomenon was confirmed during strong earthquakes in 1835, 1960 and 2010 (Darwin 1846; Weischet 1963; Farías et al. 2010). The positive co-seismic movements have the effect of reducing coastal depths, and thus facilitating the transfer of submerged sediments to the coast by the sheer power of waves. Those were probably accumulated by the estuaries on the emerged continental shelf during the last glacial period. Proof of this is in the remarkable presence of characteristic metamorphic rock minerals around regional neighborhood of the Coastal Range Mountains and in the dune sands of the Arauco coast (Tavares-Correa 1996a, b). According to Paskoff and Manríquez (2004), the sand forming dunes, beaches and marshes near the Gulf of Arauco range from fine grain with good selection and light color in the south, to medium grain, well selected and with yellow to brown tones in the north. The origin of the sediments ranges from metamorphic rocks basseting of the Coastal Range Mountain range (quartz, feldspar, and magnetite), to volcanic rocks of the Andes (olivine, pyroxene, and amphibole).

Even though there is not much information on the temporal variability of sediments in estuarine systems in Chile, this seems not to be very significant. Bertrán et al. (2001) have shown that the variability of sediment at the Biobio River (Biobio Region) between winter and summer is not significant, having the particle size a termporal variability greater at the mouth. The greater temporal variability was detected in the area nearest to the mouth, probably because of the greater relative influence of waves and tides with respect to the runoff of the river. Sediments observed in all monitoring stations (up to 5 kilometers upstream of the mouth) correspond to coarse sands.

In the estuaries throughout the country, the average size of sediments typically ranges from 0.1 mm (fine sand) to 2 mm (very coarse sand). An exception are the sediments of the Elqui River mouth, which can be classified as rocks, with a representative diameter of 13.5 mm. Table 2.1 summarizes the size of sand located in the mouths of different rivers along the area of central Chile. In the case of Longotoma and Ligua Rivers, Vina del Mar estuary and Rapel River (Valparaiso Region), these sands have been found in dunes north of their mouths. For coastal rivers of Arauco (Lebu, Paicavi, Lleulleu and Tirua, located in the Biobio Region), it is assumed that the sands have been created by the continental shelf, and are equivalent to those that form the dunes of Arauco. For coastal rivers Lenga, Tubul and Raqui (Biobio Region) it can be assumed that the sands located at their mouths must be of the same origin as those of other coastal rivers in the Gulf of Arauco.

Table 2.1 Granulometric characteristics of sands in estuaries

River mouth (Delta, Estuary)	Average diameter (mm)	Average river gradient (%)	Sediment type
Copiapo River	0.4	1.5	Medium Sand
Elqui River	13.6		Boulders
Ligua and Longotoma Rivers	0.1	1.1	Fine Sand
Viña del Mar Tideland	0.1	1.1	Fine Sand
Rapel River	0.2	1.2	Medium Fine Sand
Cahuil			Medium Fine Sand
Mataquito	0.1		Fine Sand
Maule	0.5	1.2	Medium Coarse Sand
Itata	1.8		Very Coarse Sand
Lenga			Fine Sand
Tubul-Raqui			Medium Fine Sand
Biobio	0.8		Medium Coarse Sand
Ríos costeros de Arauco			Arena media-fina

Fuente: Paskoff and Manríquez (2004), Stuardo et al. (1993), Bertrán et al. (2001), Niño (2002)

Hydrology

Weather and Rainfall

The mainland of Chile stretches over 4000 km with an average width between 150 and 200 km. Throughout the country there is a variety of climates, morphological and lithological features, which affect different behaviors regarding its hydrology. In particular, the range of climates observed in the country ranges from the desert climates with warm anticyclone influences in the north to the oceanic climates with cold cyclone influences in the south (Fig. 2.2). This climate variability translates into large differences in annual rainfall and average annual temperature (Fig. 2.3).

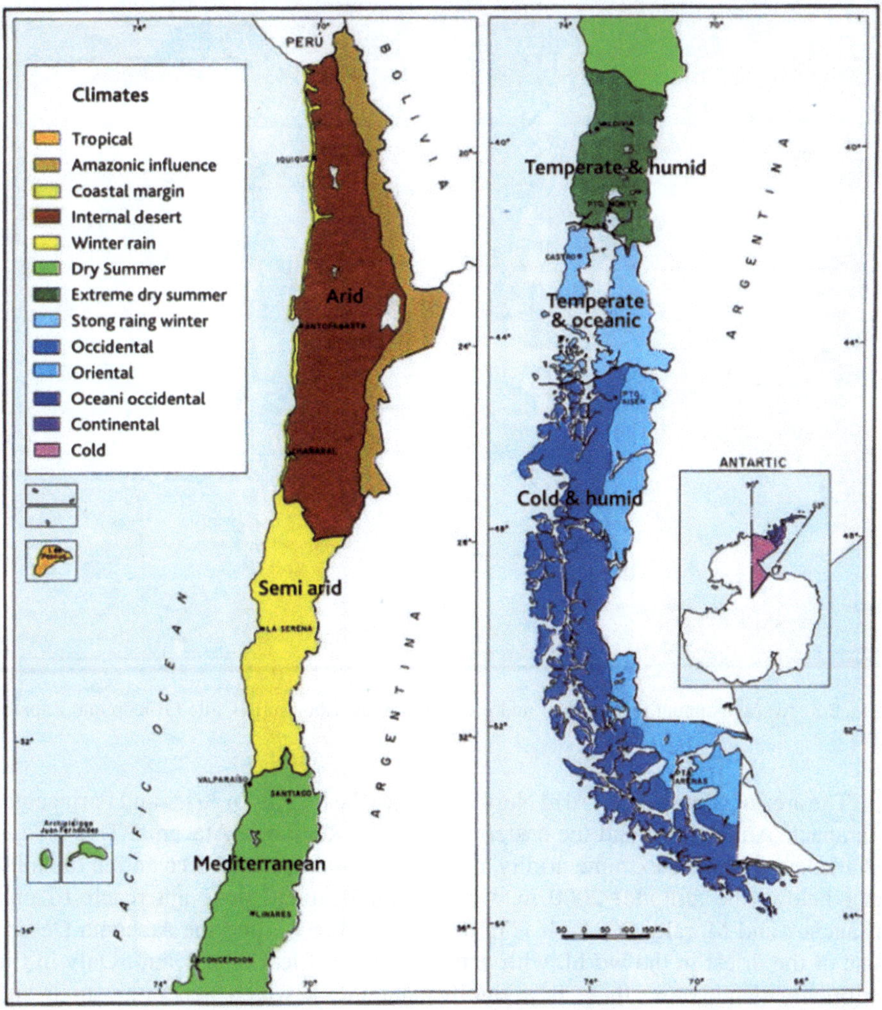

Fig. 2.2 Classification of climates in Chile (Taken from Toledo and Zapater 1989). To be compared with hydrologic characteristics presented in Tables 2.10 and 2.11

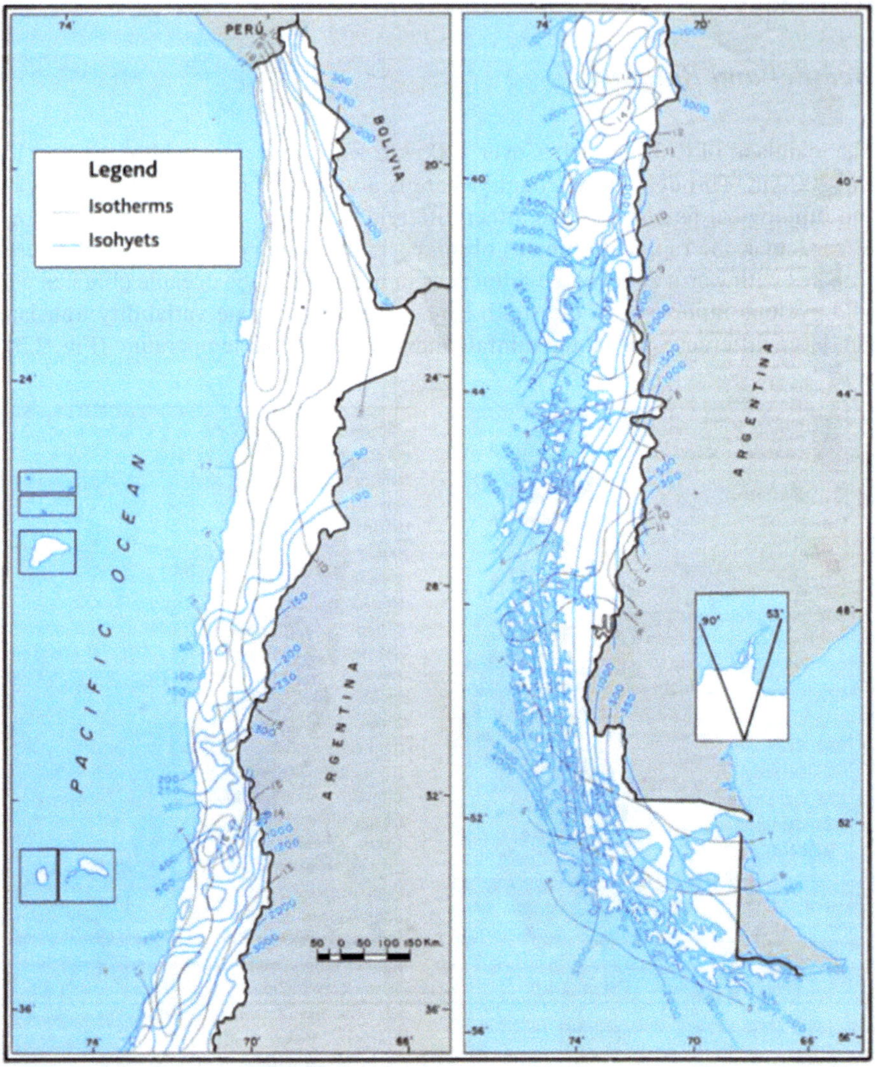

Fig. 2.3 Average annual temperature and precipitation distribution in Chile (Toledo and Zapater 1989)

The area of Norte Arido (Arid North) includes the regions of Arica and Parinacota, Tarapaca, Antofagasta and the eastern part of the Region of Atacama. This area is characterized by the extreme aridity, particularly in the coastal area and in the interior below (the altitude) 2000 m, where annual rainfall does not reach 10 mm (Sánchez and Morales 2004). It is in this area where we find the Atacama Desert, one of the driest in the world, with annual rainfall of less than 1 mm. Only in the Altiplano (Highlands) there is a significant rainfall, which generates small and sporadic runoff throughout the year, and significant flooding in summer. To the south of Norte Arido area is a semiarid climate, with rainfall ranging from 25 mm per year in the Atacama region, to 250 mm in the Valparaiso Region. These precipitations fall mostly in the winter months, when it rains up to 2500–3000 msnm. Above this elevation there is a significant nival precipitation significant for the annual hydrological regimen.

The central region is divided into a north-central area between the mountain range of Chacabuco to the Biobio River basin, and the south-central area from the Basin to the Chacao Channel and within Reloncavi. The first area has a Mediterranean climate, with rainfall concentrated in the winter months, while the south-central area is temperate, with annual rainfall of greater magnitude and distributed throughout the year.

To the south is the southern zone or Chilean Patagonia, which has an oceanic humid climate. This sector of archipelagos and fjord networks of the Pacific coast has one of the highest rainfalls in the world, and there are areas with annual rainfall greater than 6000 mm, while in the continental sector they decrease. An example of the magnitude of rainfall in this region is the annual rainfall of 4266 mm on the Island of San Pedro in the Gulf of Penas (Sánchez and Morales 2004).

The water balance in Chile (DGA 1987) shows the remarkable increase in rainfall with latitude, with the precipitation in southern Chile nearly 40 times higher than in the north of the country (Table 2.2[1]). In its entirety, Chile presents precipitations equivalent to South America, but it almost doubles the precipitations on a global level (Table 2.3).

Table 2.2 Annual water balance in Chile per administrative region

Region	Precipitation	
	m3/s	mm
I	157	93.6
II	182	44.5
III	205	82.4
IV	281	222
V	211	434
RM	335	650
VI	508	898
VII	1347	1377
VIII	2467	1766
IX	1451	2058
X	6319	2970
XI	11,763	3263
XII	11,748	2713

Source: DGA (1987)

[1]The 1st regiǿn, in 1987, was considering the actual I and XV Regions. On the other hand, the Region X was considering the actual regions X and XIV. This is also valid for the Figure 2.4.

Table 2.3 Annual water balance per macro-zone in Chile, and comparison with other global balances

	Precipitation	
Macro zone	m3/s	mm
I, II y XV Region	340	59
III a X, y XIV Region	13,120	1246
XI a XII Region	23,490	2363
Continental Chile	36,950	1522
South America	888•103	1564
Global	3522•103	746

Source: DGA (1987)

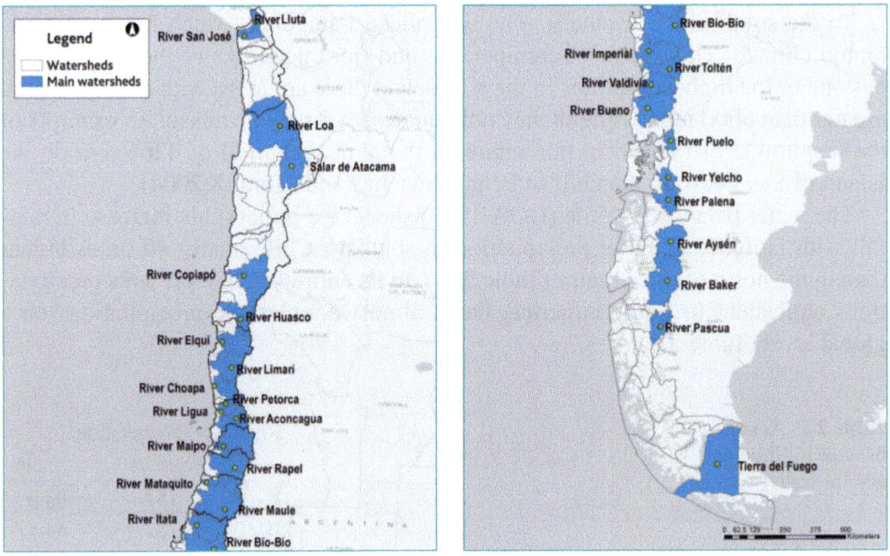

Fig. 2.4 Main rivers and watersheds in Chile (Taken from Niemeyer and Cereceda 1984)

Runoff (Overflow, Surplus)

Water basins and their respective hydrological regimens throughout the country exhibit significant variability due to spatial diversity of climate, morphological and lithological features, vegetation, topography and other relevant factors. Within the national territory there is a number of basins contributing to the major rivers in the country (Fig. 2.4). In Chile, it is possible to identify the following macro-areas with similar hydrographic and fluvial characteristics (Niemeyer and Cereceda 1984): northern arid, semi-arid, central Mediterranean, central temperate and southern zone.

Most of the basins of the northern arid zone (Regions I, XV and II) are the endorheic type, and rivers Lluta, San Jose and Loa with their major surface watercourses drain into the sea. However, as it is shown in Table 2.4, the average flow

Table 2.4 Flows in the main hydraulic systems in Chile (Modified from Peña and Salazar 1993)

Region	Basin	Medium entrance flow rate (m^3/s)	Medium flow rate exit (m^3/s)	Minimum flow rate exit (m^3/s)
I	LLuta River	2.3	1.4	0.09
I	San Jose River	1.5	0.0	0.0
I	Tamarugal Pampas	1.0	0.0	0.0
II	Loa River	2.8	0.6	0.04
III	Copiapo River	2.9	0.1	0.04
III	Huasco River	3.5	1.7	0.3
IV	Elqui River	8.1	1.0	0.15
IV	Limarí River	15.1	7.5	0.0
IV	Choapa River	13.2	12.8	0.0
IV	Aconcagua River	38.0	30.0	0.0
RM	Maipo River	116.0	100.0	1.0
VI	Rapel River	130.0	174.0	1.0
VII	Maule River	257.0	569.0	58.0
VIII	Bio Bio River	639.0	1000.0	120.0

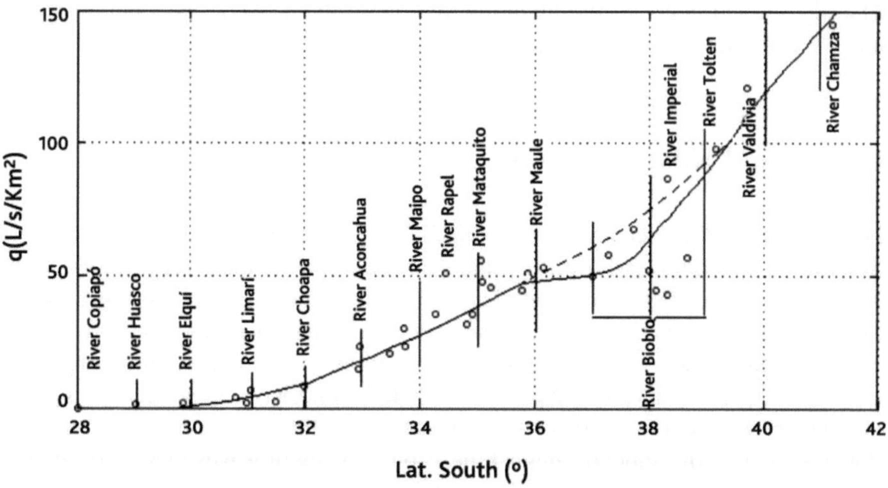

Fig. 2.5 Variation of the mean flow as a function of latitude (Peña and Salazar 1993)

rates in the areas close to the river mouth (river mouth exit flow) are very low or nonexistent.

Figure 2.5 show the variations of average annual flow rate for the semiarid, central mediterranean, temperate central and the southern areas, depending on the latitude. The semiarid zone located from the Copiapo River to the Chacabuco Ridge (regions III, IV and V) is characterized by torrential rivers, with steep slopes on the east-west conditioned by local tectonics. A direct relationship between the volumes of the runoff surface and annual rainfall, which tends to increase with latitude, is observed. The rivers in this area have a mixed system, where the importance of regimen component depends on the proportion of the basin located at higher elevations. They originate in the Highlands and manifest runoff throughout the year. The main watercourses are, from north to south: Copiapo, Huasco, Elqui, Petorca, La Ligua and Aconcagua.

Central mediterranean area (from Chacabuco Ridge to the Biobio River) presents river regimen and/or mixed, thanks to the seasonal variability of rainfall and temperatures, which favors overflows throughout the whole year. The main courses in this area are the rivers Maipo, Rapel, Mataquito, Maule, Itata and Biobio. Just as is in the case of water courses in the north, most of the rivers in the central area have hydraulic torrent regime relating to speed-altitude of the fairly high runoffs. This type of system is caused by the high slopes of its basins and whose origins are in the high mountains of the Andes.

On the other hand, the temperate central zone (from the Biobio to the Reloncavi River mouth) contains calm rivers with lake system, which presents a more stable runoff, with peak flows during the winter time. Majority of waterways of this area have a hydraulic river regime (i.e. speed vs. height reasons rather than low runoff), due to the smaller gap between the lakes that gives birth to these rivers and their balance base, as well as to the greater paths of these (Niemeyer and Cereceda 1984). The highlighted ones are the Imperial, Tolten, Valdivia, Well, and Maullin Rivers and their tributaries.

Coastal and Mountain Rivers

For the study of the central zone estuaries, it is useful to consider the work of Stuardo et al. (1993), who differentiate the behavior of the river mouths of the Mediterranean and temperate central area of Chile according to their origin and type of basin. On one side, they consider the rivers of Andean origin, which flow into the estuaries that remain open all year and, depending on the degree of silting at the mouth, can present an estuarine behavior. The Biobio River is a clear example of this type of course. On the other side, coastal rivers have smaller water basins and flow rate, and to disembogue against the strong energy of waves they favor the formation of barriers, particularly in summer. This typically generates a narrow channel and, in some cases, complete closure, with the consequent formation of coastal lagoons. Only in those channels where the mouth is partially protected from the littoral drift and wave energy, the mouths are kept open. Examples of the rivers like that are Carampangue, Lebu and Tirua.

Similar phenomena occur in estuaries linked to Coastal Plains with large areas of marshland, as in the estuaries of Tubul-Raqui and Lenga. However, the deltas of coastal rivers generally remain closed most of the year.

The above classification may be extended to certain rivers of Andean origin of the semiarid area to. In the case of the Copiapo River, for example, the average monthly flow ranges from 0.63 m^3/s in January, and 0.26 m^3/s in October, which would explain how this river mouth can be closed in spring, even more so if its waters are removed for productive activities. Moreover, the Elqui and Choapa Rivers have minimum average flow rate of 3.11 m^3/s in October, and 4.65 m^3/s in April respectively (DGA 1987), which would explain their deltas remaining open most of the year. The distinction between rivers of the coastal and Andean origin is therefore relevant for understanding the dynamics of the river mouths.

Seasonality

The evolution and temporal stability of an estuary are strongly conditioned by the annual variability of the hydrologic regime, which influences the coastal morphology and determines whether the mouth of an estuary is open or closed. Then we quantify this variability by seasonality of monthly flows around the annual average flow (Qanual), characterized by the coefficient of variation CV = /Qanual where corresponds to the standard deviation of monthly average flows with respect the Qanual. Table 2.5 shows these values calculated for a number of rivers in the central area with fluviometric records that have been updated at the river mouth.

Being that the coefficient variant is the measure of the uscaled invariant dispersion, the seasonality in various waterways can be directly compared. It is observed that there is a high variability for coastal rivers in the central area, such as Nilahue and Loanco, with variation coefficients greater than unity.

Moreover, the Itata River, of Andean origin, has a lower variability (CV = 0.84), despite the mainly rainfall regime. Without a doubt its mouth remains open even in the summer months due to its high discharge. The Lleulleu River, of coastal origin, does not present a very high seasonality, because it is located further south, where rainfall is more uniform throughout the year.

The hydrograph of monthly flows normalized by the average annual flow allows for comparing of the hydrological regimes of each river. Figures 2.6 and 2.7 show the comparison for Andean rivers and coastal rivers respectively. Andean rivers of

Table 2.5 Statistics of mean monthly flows in central Chile (by the authors, based on information from DGA 1987)

River	Station			*Qanual* (m3/s)	σ (m3/s)	σ/Qanual
	latitude S	longitude W	Altitude			
Nilahue	034° 34′ 08	071° 47′ 27	50	10.9	14.6	1.3
Mataquito	034° 59′ 04	072° 00′ 36	20	130.2	78.8	0.6
Maule	035° 24′ 25	072° 12′ 30	20	505.5	309.2	0.6
Loanco	035° 34′ 08	072° 35′ 00	20	4.5	4.5	1.0
Itata	036° 28′ 05	072° 45′ 05	10	352.7	295.4	0.8
Biobio	036° 50′ 14	073° 05′ 11	5	982.0	633.9	0.7
Lleu-Lleu	038° 07′ 00	073° 24′ 00	10	20.1	15.3	0.8

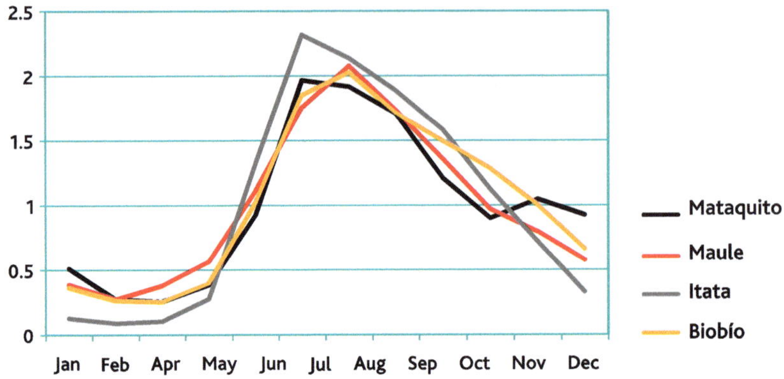

Fig. 2.6 Monthly normalized flows for Andean rivers (by the authors, based on information from DGA 2009)

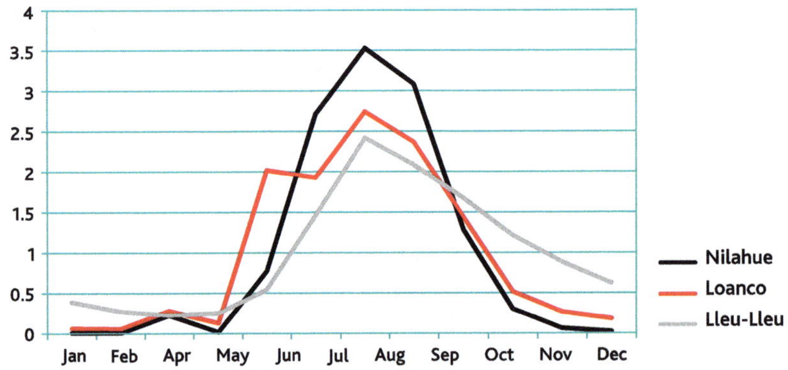

Fig. 2.7 Monthly normalized flows for coastal rivers (by the authors, based on information from DGA 2009)

central and south Maule Region have a very similar seasonal behavior reflected in the essentially pluvial character of these watercourses. However, coastal rivers Loanco and Nilahue have a marked seasonality, but with a superficial overflow only during the winter months. The Lleulleu River, located further south, has a similar seasonality, but with a permanent flow throughout the year, as rainfall it receives its basin is a more frequent contributor.

The coefficient of variation can adequately characterize the seasonality of the monthly flows of the rivers in the central area. Therefore, the behavior of the river mouth will depend on this parameter and the average annual flow. For rivers with high seasonality and relatively low average annual flow, it is very likely that their river mouths are closed during dry seasons. Similarly, river mouths with low seasonality and high average annual flow tend to have their mouths open all year, although there could be ravines forming in its mouth.

Tides

Tides play an important role in the morphology of river mouths since the sea level restricts the area where the energy of waves and coastal currents are present. Besides the ebb and flow of the tidal currents within the estuary, tides may also have an impact on erosion, sediment deposition and the degree of saline intrusion into the estuary (Prandle 2009).

The tidal wave that travels the coast of Chile spreads in the north-south direction. The tide observed is of mixed semidiurnal-type with two low tides and two high ones every day, and considerable differences of water levels reached in the tides every morning and afternoon.

Maximun amplitudes of the syzygy tides are variable along the coast, ranging between 1.5 and 1.9 m between Arica and Cape Quedal (Lake Region). Further south, in the bosom of Reloncavi and the gulfs of Ancud and Corcovado, the amplitudes have a significant increase, ranging between 5 and 8 m, due to the lower depths of the fjords areas that favor the establishment of resonant conditions. Table 2.6 presents a summary of typical tidal ranges in some ports.

Based on the information gathered it can be established that a large coastal area of Chile between Arica in the north, and Puerto Saavedra in the south, is subject to a microtidal regime according to the classification proposed by Davies (1964). In microtidal environments (ranges less than 2 m tide), it is often observed the formation of sand ravines at the river mouths under the effect of waves, as energy dissipation and the contribution of river sediments are concentrated in this sector. It is interesting to note that this type of tidal regime, coupled with a high seasonal variability in river hydrology, has helped explain the dynamics of closing and opening of some coastal lagoons in India (Bruun 1986), Australia (Ranasinghe and Pattiaratchi 1999), South Africa (Cooper 1990; Largier et al. 1992), United States (Elwany et al. 1998), and Brazil (Moller et al. 1996), among others.

From ecological and environmental point of view, the influence of the tides in upstream rivers coming from their mouths can be very relevant. This situation affects the development of a salt wedge modulated by the river runoff and fluctuations at sea level.

Table 2.6 Typical tidal ranges in some ports of chile (SHOA 2010)

Port	Latitude	Average sea level (m)
Coquimbo	29° 56′	0.9
Valparaiso	33° 02′	0.8
San Antonio	33° 35′	1.4
Talcahuano	36° 41′	1.1
Corral	39° 52′	1.1
Puerto Montt	41° 29′	4.7
Ancud	41° 52′	2.1
Chacabuco	45° 28′	2.8

The river mouths that have a lower slope generally have a greater influence on the tides penetration. Table 2.7 shows the penetration level for some rivers on the central coast of Chile, according to the studies of Araya-Vergara (1981).

Surf (Swell, Breaking Waves)

The Chilean coast is affected by the waves of local origin (wind sea) and remote origin (swell). Typical periods depend on the type of generation, distinguishing themselves in general to periods between 4 to 8 s to wind sea and between 8 and 20s to swell (Scott et al. 2002).

The surf waves reaching the coast of Chile originate in distant storms in the Pacific Ocean and therefore correspond to swell. Dominating surf is of south-west direction, with low seasonal variability, energy conditions and relatively long periods (Acuña and Monárdez 2008). During the winter, in front of the coast of Chile may occur intense storms of cyclonic origin. These conditions are often accompanied by strong winds and affect the wave occurrence of great intensity that reaches the central southern coast in the north-west direction.

The statistics on surf climates on the Chilean coast is generally scarce. There is little information by direct observations, so it is necessary to use global statistical data generated by numerical models as WaveWatch III (Tolman 1991) or WAM (Hasselmann et al. 1988).

Table 2.7 Average slope of inferior courses of rivers in central chile (Araya-Vergara 1981)

River	Latitude river mouth	Secant slopes (%)	Average penetration distance (km)	Average tide penetration level starting at low water level (m)
Maipo	33° 37′	4	20	1.2
Rapel	33° 54′	2	41	2.5
Mataquito	34° 58′	2	52	3.1
Maule	35° 19′	1 a 2	62	3.7
Itata	36° 23′	1 a 2	60	3.5

Table 2.8 Climate summary statistics for the annual medium wave climate in the south central coast of chile (CEPAL 2009)

	Height Hmo (meters)		Period Tm (seconds)		Direction
Coordinates	Average maximum	Maximum	Average 10% exc.	10% exc.	Average
31°S	2.5	5.9	8.9	11.1	226°
33°S	2.5	6.8	9.0	10.5	243°
35°S	2.6	8.0	9.0	11.2	235°
37°S	2.9	8.5	9.2	11.2	237°
39°S	2.8	9.2	9.2	11.1	242°

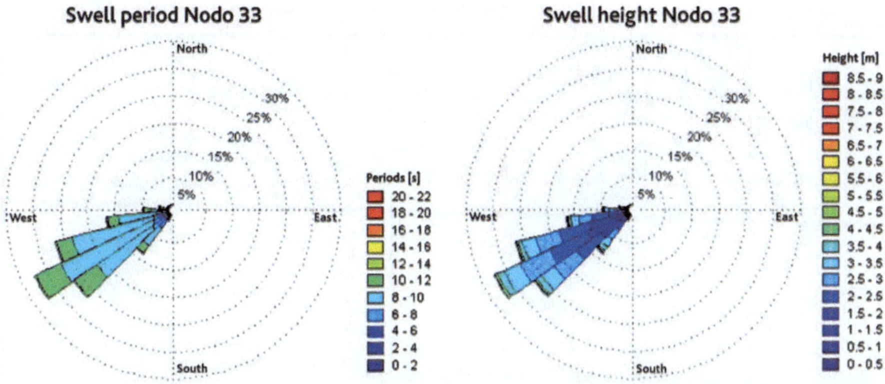

Fig. 2.8 Frequency-direction for swell periods and height in the coastline of parallel 33°S (CEPAL 2009)

The Oceanic Engineering Group of the University of Valparaiso has conducted an analysis of artificial wave during a period of twenty years, generated by these models (Cepal 2009). Table 2.8 shows the direction and two statistical waves parameters presented: the significant wave height (Hmo), corresponding to the average one third of wave heights in a sea (state), and the mean period (Tm), obtained in deep 5 nodes waters located off the south-central coast of Chile, between Los Vilos and Puerto Saavedra. It is noted that the annual significant average heights are between 2.5 and 3.0 m increasing gradually to the south. The maximum heights comfortably exceed 5 m. The average wave direction is south-west, while the average period is close to 9 s. These values confirm the prevailing south-west direction and highly energetic character of the waves reaching the coast of Chile. In Fig. 2.8 it is shown, for example, the wind rose of frequency-direction periods and wave heights at 33 ° S parallel that is in front of Valparaiso.

Tsunamigenic Earthquakes

The confluence of very active tectonic plates against the Chilean coast has historically produced large-scale seismic events that have generated destructive tidal waves. From colonial times until today, historical records (Lagos 2000; Monge 1993; Lomnitz 1970) indicate the occurrence of hundreds of tsunamigenic earthquakes near the area to the north and the central coasts of Chile, out of which at least one third has produced considerable damage. Figure 2.9 shows the main events which are on record. Besides the loss of human lives and enormous direct economic damages, these events can generate changes in ecosystems and in coastal morphology, and alter the dynamics of the coastal processes (Darwin 1846; Weischet 1963).

In the national historical literature, there is a wealth of information on seismic events and tsunamis that have occurred since the time of the Spanish Conquest. The

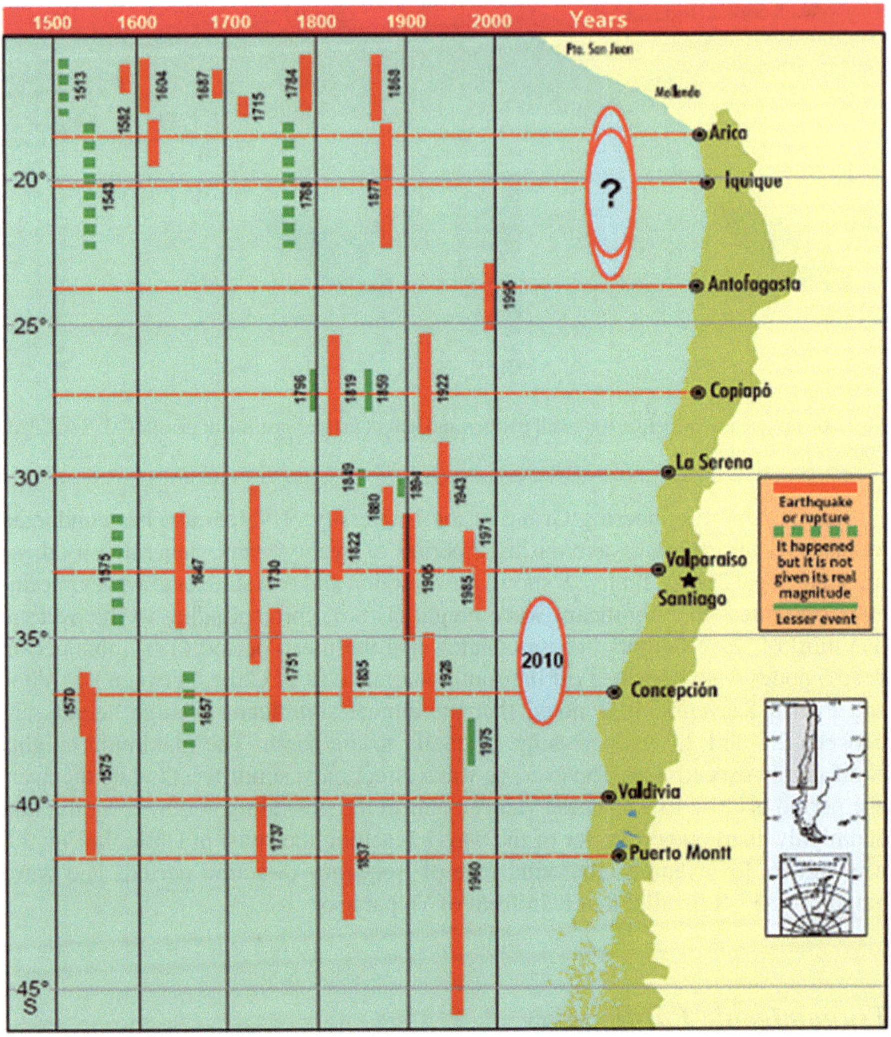

Fig. 2.9 Map with localization and dates of Tsunamigenic seismic events recorded in Chile (Source: Geophysics Department (Departamento de Geofísica), Universidad de Chile)

first historical event which was record corresponds to the earthquake and tsunami of February 8, 1575, that occurred near the city of Concepcion and which has affected greatly much of the city.

Although no deaths were recorded, the city of Concepcion had to be re-founded (Lomnitz 1970). Table 2.9 shows the main tsunamigenic events in the region over the past 500 years.

Similarly, Lagos (2000) classified historical information of tsunamis that have occurred over the past 450 years and produced a graph in which the degree of tsu-

Table 2.9 Summary of tsunamigenic events in Chilean coasts

Date	Hour	Epicenter	Richter scale	Comments
08-02-1570	09:00	Penco	8.5	It's hard to stand. Destruction of property, successive waves devastate the coast of the city of successive waves devastate the coast of the city of Concepcion (three waves). No fatalities were recorded.
16-12-1575	18:00–19:00	Corral	8.5	Great earthquake and tsunami in southern Chile that has affected the areas from the Araucania to Chiloe.
24-11-1604		Arica	8.7	Major disaster in northern Chile and southern Peru. 16 m flood in Arica.
15-03-1657	19:30–20:00	100 km. to the sea from Penco	8.0	Destruction of the city in few seconds. It is impossible to stand. Three successive waves arrive two hours after the devastating earthquake. Forty dead.
08-07-1730	04:45	Valparaiso	7.5	Tsunami destroyed the coast of Valparaiso. There were three consecutive waves which swept the city.
25-05-1751	01:00	Penco	8.0–8.5	Total destruction of the city. The tsunami arrived half an hour after the earthquake. Gap and a collapsed cavern on the coast.
11-05-1819	23:00	Caldera	8.5	4 m flooding of Caldera. Tsunami raged from Huasco to Constitution.
19-11-1822	22:15	Valparaíso	8.3	3.5 m flooding in Valparaiso.
20-02-1835	11:30	30 km. from Concepcion	8.2	Destruction of the city of Concepcion.
17-11-1849	06:00	Coquimbo, La Serena	7.5	5 m flooding in Coquimbo.
05-10-1859	08:00	Copiapo-Caldera	7.0–7.5	6 m flooding in Caldera.
13-08-1868	16:45	Arica	8.8	20 m flooding in Arica and the entire nothernorthernst northern coast of Chile and southern Peru.
09-05-1877	20:30	Mejillones	8.8	Grand Tsunami in the Pacific Rim Basin. 20 m flooding in Mejillones. Iquique y Antofagasta also affected.
10-11-1922	23:53	Copiapo-Coquimbo	8.4	9 m flooding at Chañaral.
22-05-1960	14:55	Valdivia	9.5	It lasted 10 min. Tsunami devastated Chilean coast and caused deaths in Japan.
30-07-1995	01:11	Antofagasta	8.0	It affected the Antofagasta area with 2.8 m flooding and was felt in all of the Pacific.
27-02-2010	03:34	Cobquecura	8.8	The tsunami caused great destruction along the coast of Vl and Vll Regions, Few hundreds dead.

Based on various sources: Lomnitz 1970; Monge 1993; Lagos 2000

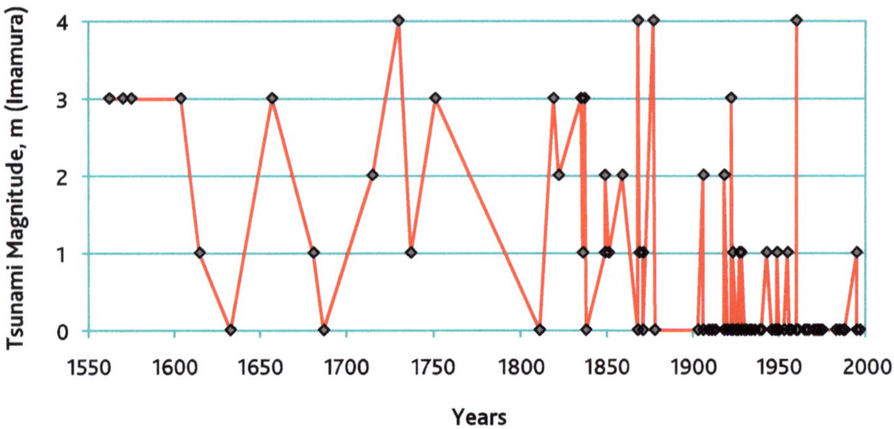

Fig. 2.10 Chronology of Tsunamis that have occurred along the coast of Chile with its associated magnitude (Adapted from Lagos 2000)

nami related to a scale from 0 to 4, where zero is inconsequential and 4 is catastrophic consequences. This classification is presented in Fig. 2.10. It is noted that the events of magnitude m = 0 represent approximately 65% of reported cases. The increase in frequency in the last century is mainly due to the increase and improvement of specialized instruments and a greater occupation of the coast. Additionally it shows that between 2 and 3 events or the tsunamis of great magnitude (about magnitude 3) have occurred in Chile in every century. This suggests that the coastline has been permanently shaped by these phenomena, being affected by at least one major event every 100 years. This fact can make us think that coastal configurations of the Chilean coast, especially sedimentary base, are relatively recent formation, since tidal waves tend to modify coastal morphological conditions.

Classification of the River Mouths of Central Chile

The morphology of the river mouths depends largely on the availability of sediment (geomorphology), wave energy and the direction of dominant attack, the effect of tidal flow and seasonal surface runoff from rivers (hydrology) and the slope (Bird 2007; Cooper 2001). These factors have a major influence on the morphology of coastal systems, both in plan and profile, and can be used to classify the different configurations found along the central Chile. In the previous sections the main features of these factors and a general classification have been summarized for some of the river mouths located in central Chile.

The central zone of Chile has a microtidal tidal regime type, ranging generally at less than 2 m. However, when it comes to surf/wave climate, it is in the south-west direction, with significant heights of waves on highly energetic annual average (over 2 m), which allows us to deduct that the estuaries of this zone are dominated by

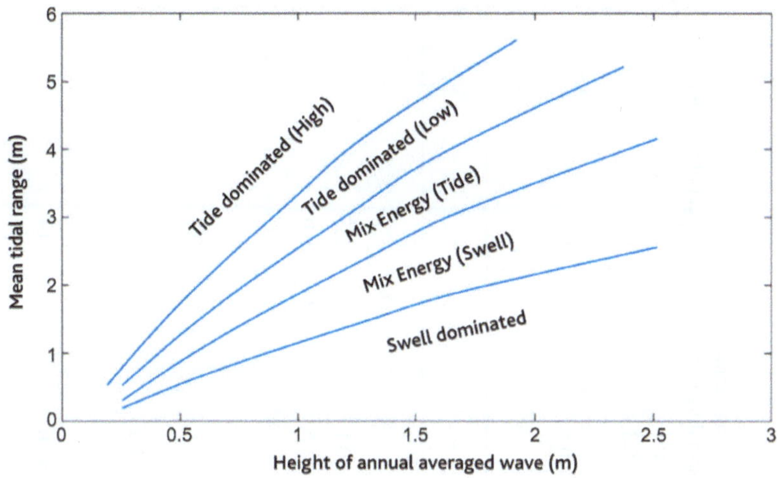

Fig. 2.11 Relationship between the average tidal level and the height of an annual averaged wave, outlining the degree of influence of the tide/swell on the morphology of a beach (Modified from Davis and Hayes 1984)

waves according to characterization proposed by Davis and Hayes (1984) and Bird (2007), presented in Fig. 2.11.

Table 2.10 shows the summary of the hydrological characteristics of the semiarid region of Chile. The rivers are differentiated by Andean and coastal origin, and their seasonality has been quantified through the coefficient of variation. It is important to note that in this case the average annual flow is the dominant element. Rivers with the low average annual flow and relatively high coefficient of variation tend to have closed mouths and form coastal lagoons, since sediment transport caused by the dominant waves favors the ravine formation that the river's superficial runoff fails to break. This is the case of the Copiapo River mouth, which despite being a mountain river, has an extremely low average annual flow, causing the landlocked river under normal conditions (Fig. 2.12). As latitude increases, the average flow of the Andean rivers also increase, and their mouths can remain open, despite having major barriers generated under the influence of waves. The Limari and Aconcagua rivers, with extensive basins, are two examples of this type of behavior (Fig. 2.12) and can be classified according to the Cooper conceptual model (Cooper 2001), as open barrier estuaries, dominated by the river.

On the other hand, as a general rule, coastal estuaries tend to form lagoons or wetlands at their mouths, because their surface runoff is low and highly variable. This is the case of Pachingo marshlands, in the vicinity of Tongoy, and Catapilco estuary near Maitencillo (Fig. 2.12). These river mouths are connected to the sea only in exceptional circumstances, under the effect of flash floods generated by heavy rain. They can be classified as coastal lagoons according to the Cooper model (Cooper 2001). It is important to note that the configuration of these systems favors the proliferation of great ecological biodiversity whose value has been widely recognized but poorly studied (Figueroa et al. 1999).

Table 2.10 Summary of hydrologic characteristics for estuaries in the semiarid zone

Zone	Region	Mountain Rivers	Tributaries	Coastal Rivers	Cavity (km^2)	Average annual spending (m3/s)	Coefficient of variation
Semi Arid	III	Copiapo	Varied		18,407	0.4	0.3
		Huasco	Transito y Carmen		9850	6.3	0.5
	IV			Punta Teatinos			
		Elqui	Turbio & Claro		9657	6.8	0.3
				Estuary Culebron			
				Adelaida			
				EstuaryTongoy	233		
				Salinas Grande			
				Salinas Chica			
				Pachingo	487		
		Limari	Hurtado & Grande		11,760	10.1	0.5
		Choapa	Cuncumen & Totoral		8124	13.8	0.6
				Laguna de Conchali			
	V	Petorca	Pedernal & Sobrante		2669	3.8	0.7
		La Ligua	Alicahue & Est. Cajon		1900	5.9	0.7
			Los Angeles				
		Aconcagua	Varied		7163	42.6	0.4

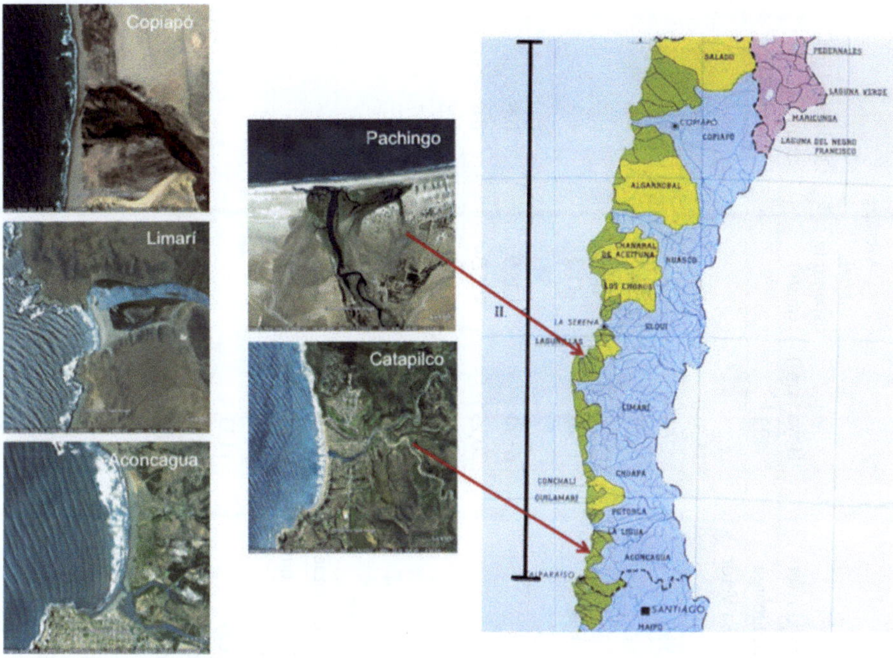

Fig. 2.12 Some examples of estuaries along the semiarid zone of Chile (Google Earth images and map by the Military Geographical Institute {Instituto Geográfico Militar})

Table 2.11 summarizes the main hydrological characteristics of rivers in the Mediterranean and wet temperate zones of Chile. In this area there are marked differences in behavior for the rivers of Andean and coastal origin, mainly associated with the value of average annual waste of its surface runoff.

When it comes to Andean rivers, a rising trend of flow with latitude is observed yet again, although the size of basins also affects this variable. In general, mountain rivers have coefficient variation lower than the coastal rivers, which together with major flows, explains why their mouths stay open permanently. The Rapel and Maule Rivers (Fig. 2.13) are clear examples of this type of morphological configuration. Furthermore, it is possible to see that in these estuaries, sediment transport by coastal currents is less intense because their mouths are protected by the rocky points located towards the south. This creates the less abundant sediment contributions of coastal origin. In these cases, the origin of the sediments is mainly fluvial thus generating a significant source of sediment for the beaches and dune systems located further north of the mouth. According to the Cooper classification (2001), these estuaries enter the way with no barriers, with permanently open mouths and at relatively large depths.

The case of the Imperial River is interesting, since being a mountain river with a relatively constant and high flow, it still presents a barrier at its mouth (Fig. 2.13). This situation is explained by the action of waves and coastal transport, given the

Table 2.11 Summary of hydrologic characteristics for estuaries in Mediterranean and wet temperate zones

Zone	Region	Mountain Rivers	Tributaries	Coastal rivers	Hoya (km^2)	Average annual spending (m3/s)	Coefficient of variation
Mediterranean	V			Margamarga	422		
				Laguna Verde			
		Maipo	Various		15,380	146.9	0.4
				El Yali	765		
	VI	Rapel	Cachapoal &Tinguiririca		14,177	195.5	0.6
				Nilahue	1770	11.1	1.3
				Llico			
	VII	Mataquito	Teno & Lontue		6190	125.9	0.6
		Maule	Various		20,295	495.3	0.6
	VIII	Itata	Various		11,090	351.6	0.8
		Biobio	Various		24,029	982	0.7
Wet template	VIII			Lenga			
				Andalien	780		
				Carampangue	1237	45	
				Tubul – Raqui			
				Lebu	813		
				Paicavi	1133		
				Lleulleu	670	19.4	0.8
				Tirua	400		
	IX	Imperial	Cautin& Cholchol		12,064	240	
				Lake Budi	51		
		Tolten	Various		7886	548.9	0.6
		Petrohue			2644	678	

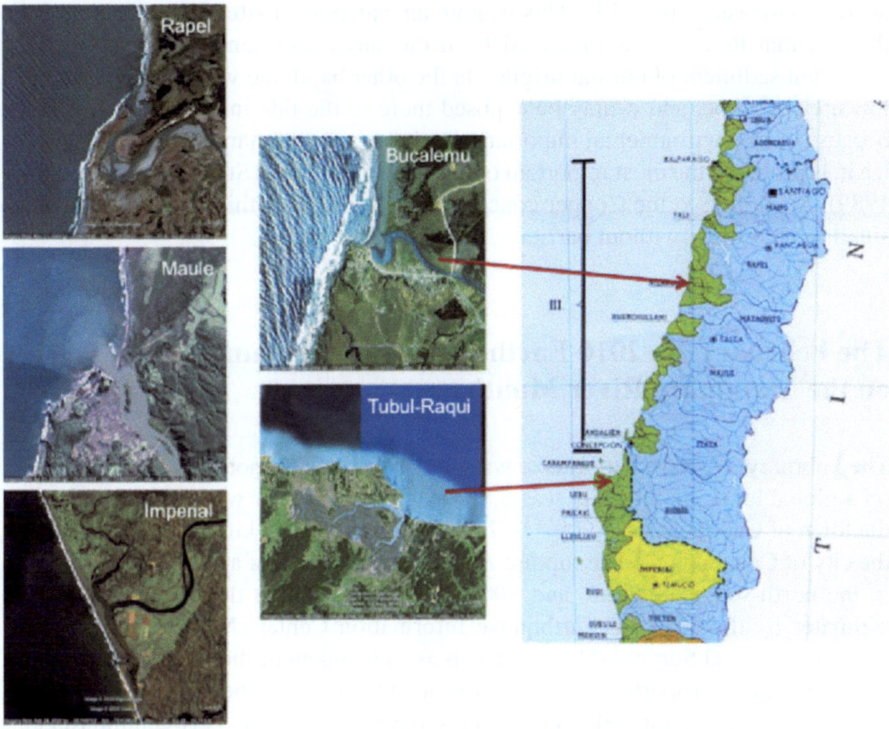

Fig. 2.13 Some examples of estuaries in Mediterranean and wet temperate zones in Chile (Google Earth images and map by the Instituto Geográfico Militar)

south-east orientation of the coastline, it has a predominant south movements. The surf energy together with availability of sediments of marine origin, produces a different morphological configuration of the Rapel and Maule's river mouths. Therefore, the importance of coastal processes, the orientation of the coast and geomorphology resulting in the type of configuration are revealed. According to the conceptual Cooper model (2001), this estuary can be classified as an open estuary with a barrier, mixed regimen, dominated by the river, but with a greater degree of influence over the tidal currents.

Regarding the rivers of coastal origin, with lower annual average flows and increased hydrological variability, it is usual to observe a configuration of coastal lagoons with closed mouths, when there is a high degree of wave exposure. This is the case of the El Yali wetland (Dussaillant et al. 2009) and the Marga-Marga, Nilahue tidelands (Andrade and Grau 2005), Bucalemu (Fig. 2.13) and Llico, among others. These examples can be classified as coastal lagoons following the Cooper model (2001), and being able to function as estuaries with barrier dominated by tides, when connected to the sea during the wet seasons.

The Tubul-Raqui wetlands show different characteristics from the aforementioned coastal rivers, because it has no barriers, and it can permanently stay connected to the sea (Fig. 2.13). This is quite an exceptional situation stemming from the fact that its mouth is protected from the surf action, and having no existing important sediment of coastal origin. On the other hand, the wetland is formed in a low area of slope, and it may be exposed more to the tide influence. This estuary, because of its environmental importance and vast areas of marshland that characterize it, is probably the most important one in the central zone (Stuardo and Valdovinos 1989). According to the Cooper conceptual model (2001), this system can be classified as an estuary without barrier.

The February 27, 2010 Earthquake and Tsunami impact on the Mataquito River Mouth

The February 27, 2010 earthquake was one of the most important in recent history, on a global level. Its intensity was 8.8Mw and its epicenter was located in front of the town of Cauquenes (35,909 ° S, 72.733°W), about 105 kilometers northwest of the city of Concepcion.[2] The rupture zone covered an area of approximately 600 km in the north-south direction, and 300 km in the east-west direction according to estimates by the National Earthquake Information Center (NEIC) of The United States Geological Survey[3] (USGS). The mass movement of the seafloor generated a tsunami of great proportions that then spread in the basin of the Pacific Ocean, causing major damage along 600 kilometers of the coast of Chile, between the Regions V, VI, VII and VIII, and in the archipelago of Juan Fernandez (Fritz et al. 2011).

One of the most dramatic consequences of this event, although little addressed so far, is related to the changes on the affected Chilean coast. The earthquake produced important horizontal and vertical displacements, highlighting the upheaval around the southern edge of the Gulf of Arauco, and subsidence in the Biobio and Mataquito mouths (Farías et al. 2010). This situation reinforces the need to incorporate tectonic variables in the analysis of morphological behavior of the coast of Chile. It is important to highlight that throughout Chilean history, there have been similar changes after each mega-earthquake. The effects on the coast after the 1835 Concepcion earthquake were described by Darwin (1846), while some of the morphological consequences of the earthquake and tsunami of 1960 have been reported by Weischet (1963). It is therefore reasonable to imagine that each of the major seismic events occurred off the coast of Chile has generated significant changes in the configurations of the river mouths located in the affected regions.

As an example of the aforementioned, it is interesting to describe the impact of the February 27, 2010 earthquake and tsunami on the mouth of the Mataquito River, located in the Maule Region (34 ° 58′S, 72 ° 2′W). Before the earthquake, this system was made up of a sedimentary bar several kilometers long, which accounts for

[2] http://earthquake.usgs.gov/earthquakes/eqinthenews/2010/us2010tfan/

[3] http://earthquake.usgs.gov/earthquakes/eqinthenews/2010/us2010tfan/finite_fault.php

the intense coastal traffic associated with dominant waves, in south-west direction. Coastal currents and winds are responsible for transporting sand brought by the Maule River and Putu dunes, located south of the Mataquito River mouth. The coastal traffic occurs in the south-north direction, with no points or rocky barriers that impede the progress of sediments to the mouth of river Mataquito.

According to hydrological analysis presented in Sect. 2.2, the Mataquito River is of the mountain origin, with an average annual flow rate of close to 126 m^3/s and a relatively low variation coefficient of 0.63, presenting permanent surface runoff throughout the year. Therefore, the surf that affects the river mouth is energetic,

Fig. 2.14 Estuary in Mataquito River and evolution of its bank before and after the earthquake and Tsunami on February 27, 2010 (satellite images obtained from Google Earth)

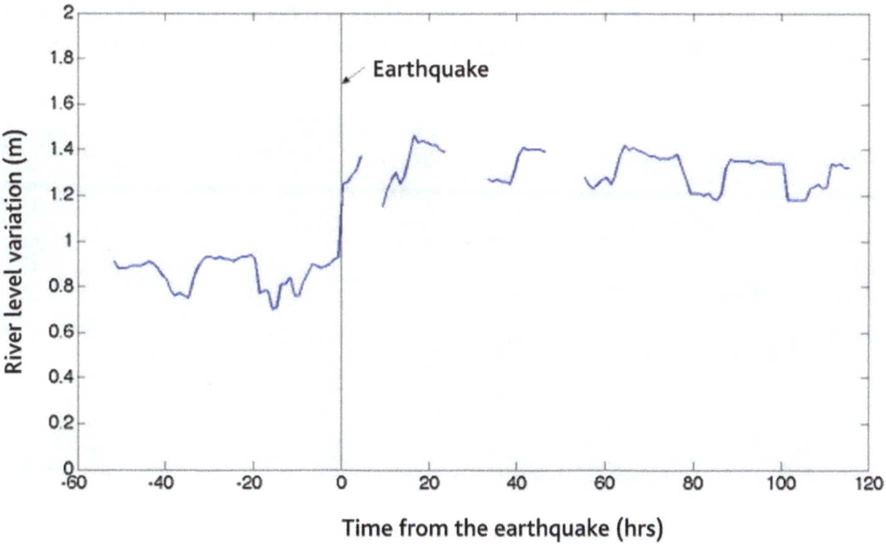

Fig. 2.15 Records from Licanten fluviometric station, part of the Dirección General de Aguas (DGA) de Chile

which together with the wide availability of sediment, generates intense hauling in the south-north direction. This combination of factors explains the formation of the bar at its mouth, which influences the river forcing it to turn north before reaching the ocean, at the proximity of the town of La Pesca. In Fig. 2.14 there are some satellite images that reflect the evolution of the bar before and after the February 27, 2010 earthquake and tsunami. The large size and stability of the bar between 2004 and 2008 is observed, and the impact of the mega-earthquake on it is from March 2010. According to the Cooper classification (2001), the Mataquito river mouth is an estuary with a barrier, dominated by the river.

After the earthquake, there was a subsidence of several tens of centimeters in the area of the river mouth (Farías et al. 2010), a situation that is confirmed by the recorded water level of Mataquito river at the fluviometric station Licanten DGA (Fig. 2.15), located several kilometers upstream from the mouth. So the first effect of the earthquake on the river mouth was to produce a relative increase of the sea level. A few minutes later, the first waves of the tsunami hit the bar, causing a major sediment transport. Several waves reached this sector in the next few hours, contributing to the disappearance of this sedimentary structure.

Despite the February 27 earthquake and tsunami great impact on the morphology of the Mataquito river mouth, during the remaining months of 2010 the river bank has begun to mend quickly. In Fig. 2.16 aerial photographs that were taken in April and September of 2010, it is possible to notice the changes that have occurred. Coastal currents have brought a significant amount of sediment in south-north direction, which explains the progress of the barrier to the north. It is also possible to observe the formation of a beach in the southern heading spike located near the area called La Pesca. Thus, it appears that the great influence of surf/waves, dominating in 2010, has not registered major floods of the river in this period. The prevailing direction of the surf generates currents that have been able to mobilize sediment in significant quantities, so it would be expected for the bar to reach the river mouth during 2011, a similar situation happening before the earthquake too. The seismic

Fig. 2.16 Evolution of the Mataquito river estuary after the earthquake and Tsunami of February 27, 2010 (Photographs courtesy of Arauco taken by natureFilms)

event of February 27, 2010 has therefore allowed to put in evidence the coastal processes responsible for the creation of the Mataquito River bar and to better understand the hydro-sedimentary dynamics of this system. Particularly striking is the great speed with which the morphology of the river mouth has evolved during the months following the earthquake and tsunami.

Conclusions

In this chapter, we have observed and reviewed the basic background information on the processes that govern hydro-sedimentary dynamics of the river mouths of the central zone of Chile, in order to establish common elements to classify the different morphological configurations.

The central coast of Chile is subject to one of the most energetic surf climates in the world, with significant annual average surf/wave heights of over 2 m. The predominant direction of wave attack is south-west, generally resulting in coastal currents carrying sediments from south to north. Additionally, the tidal range in the central area is microtidal, with less than 2 m of amplitudes. The combination of these oceanographic forces allows to conclude that the morphological configuration of the beaches and river mouths will generally be and is dominated by surf (Davis and Hayes 1984).

From the geomorphological point of view, the central-southern Chile has abundant sources of sediments, which are mainly provided by hydrographic basins and waterways that flow into the Pacific Ocean. However, there are some exceptions, such as in the Gulf of Arauco, where the origin of the sediments is rather marine. This situation can be explained by the pronounced tendency of upheaval under the effects of tectonic movements, which combined with a wider and less steep continental shelf, helps produce sediments that can then be carried by coastal currents. In short, the south-central coast of Chile is a subject to a sediment aggradation (rising), but records no events or tendencies towards coastal erosion.

The various morphological configurations observed in river mouths are mainly determined by the hydrological behavior of rivers that form them. Therefore it is possible to distinguish behaviors depending on the size of the basins (of coastal or Andean origin) and the seasonality of the hydrological regime (variability of surface runoff during the year). The average annual flow and the variation coefficient are the most important variables to explain the type of river mouth observed. In rivers and estuaries of coastal origin, or in the Andean rivers with low flow in the semiarid zone, there are often found coastal lagoons or wetlands with their water mouths remaining closed most of the time. The littoral drift produced by the surf dominates, creating sedimentary barriers that are transferred only during winter risings. The case of Tubul-Raqui wetland, located in the southern end of the Gulf of Arauco, is an exception, because even though it is a coastal river, is stays permanently open and connected to the sea. This configuration helps keep its (river) mouth protected from direct surf action.

Additionally, the Andean rivers of the Mediterranean and wet temperate zones have river mouths that are open throughout the year. The largest flow rate recorded and its minor seasonal variability, deliver the energy needed to overcome the surf action and open free to the sea. The existence of sandbars in their mouths depends on the degree of surf exposure, and the same goes for the slopes and the depth of the channel.

Finally, it is important to note that the tectonic events that take place off the coast of Chile, are responsible for generating large seismic occurrences that may even be accompanied by tsunamis. The historical records indicate that, on average, in Chile, there have occurred between 2 and 3 tsunamigenic earthquakes every century. These events produce major changes on the ground, as upheavals and subsidence, and can also change the morphology of the coast dramatically. An example of the above is the disappearance of the great sedimentary bar at the mouth of the Mataquito River after the earthquake and tsunami of February 27 2010. This structure was swept away almost completely, but has been rehabilitated quickly during the following months, highlighting the hydro-sedimentary dynamics responsible for shaping the river mouths of the central-southern Chile.

References

Acuña, H., and P. Monárdez. 2008. *Evaluación del potencial de la energía del oleaje en Chile. XVIII Congreso Chileno de Hidráulica*. Santiago de Chile: SOCHID.

Andrade, B., and S. Grau. 2005. La laguna de Cáhuil, un ejemplo de estuario estacional en Chile central. *Revista Geografía Norte Grande* 33: 59–72.

Araya-Vergara, J.F. 1971. Bases geomorfológicas para una división de las costas de Chile. *Revista de Geografía de Chile* 21-22: 5–3.

———. 1981. El concepto de delta en ría y su significado en la evolución litoral (ejemplo de Chile Central). *Información Geográfica de Chile* 28: 19–42.

———. 1982. Análisis de la localización y de los procesos y formas predominantes de la línea litoral de Chile. *Información Geográfica de Chile* 29: 35–55.

———. 1985. Trendanalysisofshorelinechangesandcoastalmanagement in Central Chile (33°-34.5°S). En: Union Géographique Internationale. Actes escursion symposium N°9: La cote Atlantique: Comission sur l'environnement cotier, 99–110.

———. 1996. Sistema de interacción oleaje-playa frente a los ergs de Chanco y Arauco, Chile. *Gayana Oceanol* 4: 159–167.

Benítez, A. 1994. Características climáticas y geológicas de Chile. En: Espinoza G., Pisani, P., Contreras, L. Y Camus. P. (eds.) Perfil ambiental de Chile: 19–37. Comisión Nacional del Medio Ambiente, Santiago, Chile.

Bertrán, C., J. Arenas, and O. Parra. 2001. Macrofauna del curso inferior y estuario del río Biobío (Chile): cambios asociados a variabilidad estacional del caudal hídrico. *Revta. Chilena de Hist. Natural* 74: 331–340.

Bird, E. 2007. Coastal geomorphology, an introduction. 2.da edición. Ed. Wiley.

Bruun, P. 1986. Morphological and navigational aspects of tidal inlets on littoral drift shores. *Journal of Coastal Research* 2: 123–145.

CEPAL. 2009. La economía del cambio climático. Capítulo sobre impactos por alza del nivel del mar. Winckler, P. Y Cienfuegos, R.

Cooper, J.A.G. 1990. Ephemeral stream-mouth bars at flood-breach river mouths on a wave-dominated coast: Comparison with ebb-tidal deltas at barrier inlets. *Marine Geology* 95: 57–70.

———. 2001. Geomorphological variability among microtidal estuaries from the wave-dominated South Africa coast. *Geomorphology* 40: 99–122.

Darwin, C. 1846. Geological observations on South America. Smith, Elder and Co, London, UK. Available at http://darwin-online.org.uk/

Davies, J.l. 1964. A morphogenic approach to world shorelines. *Zeitschrift für Geomorphologie* 8: 127–142.

Davis, R.A., and M.O. Hayes. 1984. What is a wave-dominated coast? *Marine Geology* 60: 313–329.

Dirección general de aguas (DGA). 1987. *Balance hídrico de Chile*. Santiago, Chile: Dirección general de aguas.

Dirección Nacional de Aguas (DGA). 2009. *Serie Caudales mensuales*. Santiago, Chile: Dirección general de aguas.

Dirección general de aguas (DGA). 2010. *Serie Caudales mensuales*. Santiago, Chile: Direccion general de aguas.

Dussaillant, A. P. Galdames, and C.l. Sun. 2009. Water level fluctuations in a coastal lagoon: El Yali Ramsar wetland, Chile. *Desalination* 246: 202–214.

Elwany, M.H.S., R.E. Flick, and S. Aijaz. 1998. Opening and Closure of a Marginal Southern California Lagoon Inlet. *Estuaries* 21: 246–254.

Farías, M., G. Vargas, A. Tassara, S. Carretier, S. Baize, D. Melnick, and K. Bataille. 2010. Land-Level changes produced by the Mw 8.8 2010 Chilean earthquake. *Science*. doi:10.1126/science.1192094.

Figueroa, R., M.L. Suárez, A. Andreu, V.H. Ruiz, and M.R. Vidal-Abarca. 1999. Caracterización ecológica de humedales en la zona semiárida de Chile Central. *Gayana* 73: 76–94.

French P.W. 1997. Coastal and estuarine management. 1.era edición, Routledge Environmental Management Series.

Fritz, H.M., C.M. Petroff, P.A. Catalán, R. Cienfuegos, P. Winckler, N. Kalligerisk, R. Weiss, S.E. Barrientos, G. Meneses, C. Valderas-Vermejo, C. Ebeling, A. Papadopouluys, M. Contreras, R. Almar, J. C. Domínguez, and C.E. Synolakis. 2011. Field Survey of the 27 February 2010 Chile Tsunami. Pure and Applied Geophysics*: en prensa.*

Hasselmann, S., K. Hasselman, E. Bauer, L. Bertotti, C.V. Cardone, J.A. Ewing, J.A. Greenwood, A. Guillaume, P.A.E.M. Janssen, G.J. Komen, P. Lionello, M. Reistad, and L. Zambresky. 1988. The WAM model – a third generation ocean wave prediction model. *Journal of Physical Oceanography* 18: 1775–1810.

Kennish, M.J., and H.W. Paerl. 2010. Coastal Lagoons: Critical Habitats of Environmental Change. CRC Press, Marine Science, 1.era edición.

Lagos, M. 2000. Tsunamis de origen cercano a las costas de Chile. *Revista de Geografía Norte Grande* 27: 93–102.

Largier, J.L., J.H. Springer, and S. Taljaard. 1992. The stratified hydrodynamics of the Palmiet - a prototype bar-built estuary. In *Dynamics and Exchanges in Estuaries and the Coastal Zone*, ed. D. Prandle, 135–154. Washington DC: American Geophysical Union.

Lomnitz, C. 1970. Casual ties and behavior of populations during earthquakes. *Bulletin of the Seismological Society of America* 60: 1309–1313.

Moller, O.O., J.A. Lorenzzentti, J. Stech, and M.M. Math. 1996. The Patos Lagoon summertime circulation and dynamics. *Continental Shelf Research* 16: 335–351.

Monge, J. 1993. Estudios de riesgo de Tsunami en costas chilenas. *Jornadas chilenas de Sismología en ingeniería antisísmica, Santiago de Chile* 2: 3–22.

Niemeyer, H., and P. Cereceda. 1984. Hidrografía. Geografía de Chile, Instituto Geográfico Militar.

Niño, Y. 2002. Simple Model for Downstream Varation of Median Sediment Size in Chilen Rivers. *Journal of Hydraulic Engineering*: 934–941.

Paskoff, R., and H. Manríquez. 2004. Las dunas de las costas de Chile. Instituto Geográfico Militar.

Peña, H., and C. Salazar. 1993. Calidad de aguas. Serie de documentos técnicos. SIT N°2. DGA.

Prandle, D. 2009. Estuaries: Dynamics, mixing, sedimentation and morphology. 1. Era edición, Cambridge University Press.

Ranasinghe, R., and C. Pattiaratchi. 1999. The seasonal closure of tidal inlets: Wilson Inlet- a case study. *Coastal Engineering* 37: 37–56.

Sánchez, A., and R. Morales. 2004. *Las Regiones de Chile*. Santiago: Editorial Universitaria.

Scott, D., D. Resio, and C. Pantoja. 2002. Swell propagation and near shore wave climate. 7th International Workshop on Wave Hindcasting and Forecasting, Meteorological Service of Canada, 13–24.

Servicio hidrográfico y oceanográfico de Chile (shoa). 2010. Nivel medio del mar, información disponible en www.shoa.cl.

Stuardo, J., and C. Valdovinos. 1989. Estuarios y lagunas costeras; ecosistemas importantes de Chile central. *Amb. Y Des.* 5 (1): 107–115.

Stuardo, J., C. Valdovinos, R. Figueroa, and A. Occhipinti. 1993. Los ambientes costeros del Golfo de Arauco y áreas adyacentes. EULA. Gestión de los recursos hídricos de la cuenca del río Biobío y el área marina costera adyacente.

Tavares-Correa, C.H. 1996a. Clasificación de la duna bordera artificial en el litoral de la provincia de Arauco, VIII Región, Chile. Anales XVII Congreso de Geografía, Universidad de La Serena, 71–78.

———. 1996b. Propuesta de uso del suelo en las dunas litorales de la provincia de Arauco, VIII Región, Chile. Tesis de Doctorado en Ciencias Ambientales, Universidad de Concepción, Chile.

Toledo, X., and E. Zapater. 1989. *Geografía general y regional de Chile*. Santiago, Chile: Editorial Universitaria.

Tolman, H.L. 1991. A third-generation model for wind waves on slowly varying, unsteady, and inhomogeneous depth and currents. *Journal of Physical Oceanography* 21: 678–6797.

Weischet, W.W. 1963. Further observations of geologic and geomorphic changes resulting from the catastrophic earthquake of May 1960. *Bulletin of the Seismological Society of America* 53: 1237–1257.

Chapter 3
Birds in Coastal Wetlands of Chile

Cristián F. Estades, M. Angélica Vukasovic, and Juan Aguirre

Abstract Between the extreme north of Chile and the island of Chiloe, there are approximately 300 coastal wetlands. Among them, there are estuaries where the largest waterfowl populations are concentrated, due to their high productivity and diversity of habitats. Around 125 species of birds, approximately 30% of the species that live regularly in Chile, use this type of environment, which makes them high-value sites for conservation of biodiversity. Because the Chilean coast is part of a major migratory route for birds in the Americas, the composition of the avifauna of these ecosystems is also highly dynamic. In addition, the uneven distribution of wetlands on the Chilean coast is reflected in the presence of a small number of sites with concentration of these birds, which are characterized by a good supply of food and the availability of safe resting places (e.g. sand bars). Population data of different bird species in four estuaries of the central zone (Itata, Reloca, Mataquito and Topocalma) indicate that, although the communities of aquatic birds (migratory and resident) are highly dynamic, they also have a very regular and predictable structure. The high degree of overlap in population trends of most species, even among non-migratory ones, suggests that these dynamics are largely due to regional trends rather than individual behavior of each estuary. Among the migratory species it is possible to detect some that settle throughout the summer in the estuaries of central Chile, while others use them only as a place of refueling during trips to and from more southern sites. Additionally, these estuaries have an important role as post-reproductive aggregation sites for many resident species, which apparently move from the country's central valley wetlands. Information on the waterfowl use of estuaries of central Chile clearly indicates that the main role of these environments is to serve as a resting place for most birds. In addition, many marine species gather in these places for bathing, and that way clean the salt of their plumage. A large proportion of the birds that are concentrated in the estuaries, forages at sea, except for some freshwater birds (less abundant) that make it into these wetlands. Reproductive activity of water birds in these estuaries is marginal, due to unfavorable factors for nesting and the high degree of human intervention.

Keywords Waterfowl • Water Birds • Estuary • Nesting

C.F. Estades (✉) • M.A. Vukasovic • J. Aguirre
Universidad de Chile, Facultad de Ciencias Forestales y Conservación de la naturaleza, Laboratorio de Ecología de Vida Silvestre, Santiago, Chile
e-mail: cestades@uchile.cl

J.M. Fariña, A. Camaño (eds.), *The Ecology and Natural History of Chilean Saltmarshes*, DOI 10.1007/978-3-319-63877-5_3

Introduction

Due to the particular geography of Chile, coastal environments have a high representation in the country (Gonzalez and Victoriano 2005). This means that a very high percentage of Chile's avifauna is composed of species that prefer using these ecosystems (Aparicio 2006). In addition, the Chilean coast is part of a major migration route in the Americas (Morrison and Myers 1987), which gives them a highly dynamic and seasonal communities of shorebirds character.

Among the different coastal environments, the wetlands that are located in this area are one of the most important ecosystems for birds, particularly migratory species (Page et al. 1997). Among the coastal wetlands, estuaries tend to concentrate the largest waterfowl populations due to its high productivity. In addition, the fact that these systems constitute a system interface between freshwater and sea, generating a salinity gradient that is positively influencing the diversity of birds that use them (González and Victoriano 2005).

Relative gridlock that the river water faces in the estuary makes its flow rate decrease significantly, so a new environment, more like a lagoon is generated. These areas of calm waters with high concentrations of nutrients attract many fish to spawn (Haedrich 1983), so there is a great deal of food for fish-eating birds.

However, estuaries are also highly dynamic systems due to the tide action as well as atmospheric factors such as precipitation (Colwell 1993; Chaparro et al. 2008), which significantly affects the supply of food for some water birds (Colwell 1993).

In Chile coastal lagoons are relatively few and not all rivers are clearly recognizable estuaries (Stuardo and Valdovinos 1989). Still, estuaries are valuable for the conservation of many species of waterfowl in the country (Victoriano et al. 2006).

Table 3.1 shows the list of species of water birds (i.e. fully or partially dependent on wet environments) that can be found in coastal wetlands of Chile, not including several species that have only one or two records in the country. The 125 species listed are approximately 30% of the species of birds that live regularly in Chile, which speaks of the importance of such environments as habitat for birds in the country. The background outlined above warrants a consolidated analysis of the available information on the birds of the coastal wetlands of Chile.

In this chapter the role of coastal wetlands in the composition and dynamics of Chilean avifauna is studied. The analysis presented is based on different sources of information about the use of these environments by the migratory and resident birds.

Role of Coastal Wetlands Migratory Birds Geographical Distribution

Long distance migration, on which many water birds embark, requires the existence of numerous rest and refueling sites along the way (Myers et al. 1987). Unlike birds migrating to inland areas, commonly using temporary or questionable quality

Table 3.1 Bird species present in chile's coastal wetlands, together with their protection status, the principal type of habitat associated with each species and a qualitative description of their abundance/ frequency ratio in three latitudinal zones along the country

Family/ Specie		Status	Main	Abundance		
		in Chile	environment	north	center	south
Podicipedidae						
Rollandia rolland	White-tufted grebe	Resident	Humedal	•	o	o
Podilymbus podiceps	Pied-billed grebe	Resident	Humedal		o	o
Podicephorus major	Great grebe	Resident	Humedal	•	o	o
Podiceps occipitalis	Silvery grebe	Resident	Humedal	o	o	o
Sulidae						
Sula variegata	Peruvian booby	Resident	Mar	•	o	•
Pelecanidae						
Pelecanus thagus	Peruvian pelican	Resident	Mar	O	O	O
Phalacrocoracidae						
Phalacrocorax brasilianus	Neotropic cormorant	Resident	Mar	O	O	O
Phalacrocorax bougainvilli	Guanay cormorant	Resident	Mar	O	•	•
Ardeidae						
Ixobrychus involucris	Stripe-backed bittern	Resident			•	•
Ardea cocoi	Cocoi heron	Resident	Humedal		•	•
Ardea alba	Great egret	Resident	Humedal	o	o	o
Egretta thula	Snowy egret	Resident	Humedal	o	O	O
Egretta tricolor	Tricolored heron	Errant	Humedal	•		
Egretta caerulea	Little blue heron	Resident	Humedal	o		
Bubulcus ibis	Cattle egret	Resident	Humedal	•	•	•
Butorides striata	Striated heron	Errant	Humedal/ Praderas		•	
Nycticorax nycticorax	Black-crowned night heron	Resident	Humedal	o	o	o
Threskiornithidae						
Platalea ajaja	Roseate spoonbill	Errant	Humedal	•	•	
Theristicus melanopis	Black-faced ibis	Resident	Humedal/ Praderas	•	•	o
Plegadis chihi	White-faced ibis	Resident	Humedal		•	•
Plegadis rigdwayi	Puna ibis	Resident	Humedal	•		

(continued)

Table 3.1 (continued)

Family/ Specie		Status	Main	Abundance		
		in Chile	environment	north	center	south
Phoenicopteridae						
Phoenicopterus chilensis	Chilean flamingo	Resident	Humedal	•	o	o
Phoenicoparrus andinus	Andean flamingo	Resident	Humedal	•	•	
Anatidae						
Coscoroba coscoroba	Coscoroba swan	Resident	Humedal		o	o
Cygnus melanocoryphus	Black-necked swan	Resident	Humedal	•	o	O
Chloephaga hybrida	Kelp goose	Resident	Coast			•
Tachyeres pteneres	Fuegian steamer duck	Resident	Coast			•
Tachyeres patachonicus	Flying steamer duck	Resident	Humedal			•
Speculanas specularis	Bronze-winged duck	Resident	Humedal			•
Anas platalea	Red shoveler	Resident	Humedal		o	o
Anas cyanoptera	Cinnamon teal	Resident	Humedal	o	o	o
Anas versicolor	Silver teal	Resident	Humedal		•	•
Anas puna	Puna teal	Resident	Humedal	•		
Anas sibilatrix	Chiloe wigeon	Resident	Humedal	•	O	O
Anas flavirostris	Yellow-billed teal	Resident	Humedal	•	O	O
Anas bahamensis	White-cheeked pintail	Resident	Humedal	o	o	•
Anas georgica	Yellow-billed pintail	Resident	Humedal	o	O	O
Anas platyrhynchos	Mallard	Exotic	Humedal		•	
Netta peposaca	Rosy-billed pochard	Resident	Humedal		•	•
Heteronetta atricapilla	Black-headed duck	Resident	Humedal		•	•
Oxyura vittata	Lake duck	Resident	Humedal		O	o
Oxyura ferruginea	Ruddy duck	Resident	Humedal		•	o
Pandionidae						
Pandion haliaetus	Osprey	Migratory	Humedal	•	•	•
Accipitridae						
Circus cinereus	Cinereous harrier	Resident	Humedal	•	•	•
Circus buffoni	Long-winged harrier	Errant	Humedal		•	

(continued)

Table 3.1 (continued)

Family/ Specie		Status in Chile	Main environment	Abundance north	center	south
Rallidae						
Laterallus jamaicensis	Black rail	Resident	Humedal	•	•	•
Pardirallus sanguinolentus	Plumbeous rail	Resident	Humedal	•	o	o
Gallinula chloropus	Common moorhen	Resident	Humedal	o		
Gallinula melanops	Spot-flanked gallinule	Resident	Humedal		o	o
Fulica armillata	Red-gartered coot	Resident	Humedal		O	O
Fulica leucoptera	White-winged coot	Resident	Humedal	o	O	O
Fulica rufifrons	Red-fronted coot	Resident	Humedal		O	o
Haematopodidae						
Haematopus palliatus	Pilpilén	Resident	Coast	O	O	o
Haematopus ater	Pilpilén negro	Resident	Coast	•	•	•
Haematopus leucopodus	Pilpilén austral	Resident	Coast			o
Recurvirostridae						
Himantopus melanurus	Perrito	Resident	Humedal	•	O	o
Recurvirostra andina	Caití	Resident	Humedal	•		
Charadriidae						
Vanellus chilensis	Queltehue	Resident	Humedal/ Praires		O	O
Pluvialis dominica	Chorlo dorado	Migratory	Coast	•	•	•
Pluvialis squatarola	Chorlo ártico	Migratory	Coast	o	o	o
Charadrius semipalmatus	Chorlo semipalmado	Migratory	Coast	•		
Charadrius alexandrinus	Chorlo nevado	Resident	Coast	•	•	•
Charadrius falklandicus	Chorlo de doble collar	Resident	Costa/ Humedal	•	•	o
Charadrius collaris	Chorlo de collar	Resident	Costa/ Humedal		o	o
Charadrius vociferus	Chorlo gritón	Resident	Coast	o		
Charadrius modestus	Chorlo chileno	Resident	Coast	•	o	O
Charadrius wilsonia	Chorlode picogrueso	Migratory	Coast	•		
Oreopholus ruficolis	Chorlo de campo	Resident	Dunes/ praires	•	•	

(continued)

Table 3.1 (continued)

Family/ Specie		Status in Chile	Main environment	Abundance north	center	south
Scolopacidae						
Tringa melanoleuca	Pitotoy grande	Migratory	Humedal	o	o	O
Tringa flavipes	Pitotoy chico	Migratory	Humedal	o	O	O
Catoptrophorus semipalmatus	Playero grande	Migratory	Coast	o	•	•
Actitis macularia	Playero manchado	Migratory		•		
Numenius phaeopus	Zarapito	Migratory	Coast	O	O	O
Limosa haemastica	Zarapito de pico recto	Migratory	Coast	•	o	O
Arenaria interpres	Playero vuelve piedras	Migratory	Coast	•	•	•
Aphriza virgata	Playero de las rompientes	Migratory	Coast	•	•	•
Caldera's canutus	Playero ártico	Migratory	Coast	•	•	o
Calidris alba	Playero blanco	Migratory	Coast	o	o	o
Calidris pusilla	Playero semipalmeado	Migratoria	Coast	•		
Calidris minutilla	Playero enano	Migratoria	Coast	•		
Calidris fuscicollis	Playero de lomo blanco	Migratoria	Coast	•		o
Calidris bairdii	Playero de Baird	Migratoria	Coast/ Humedal	o	o	o
Calidris melanotos	Playero pectoral	Migratory	Coast	•	•	•
Calidris himantopus	Playero de patas largas	Migratory	Coast	•		
Gallinago paraguaiae	Becasina	Resident	Humedal		o	o
Phalaropus lobatus	Pollito de mar boreal	Migratory	Mar	•	•	
Phalaropus fulicarius	Pollito de mar rojizo	Migratory	Mar	•	•	•
Phalaropus tricolor	Pollito de mar tricolor	Migratory	Coast/ Humedal ta/	o	•	•
Pluvianellidae						
Pluvianellus sociales	Chorlo de Magallanes	Resident	Coast			o
Thinocoridae						
Thinocorus rumicivorus	Perdicita	Resident	Dunes/ praderas	•	•	•
Laridae						

(continued)

Table 3.1 (continued)

Family/ Specie		Status in Chile	Main environment	Abundance north	center	south
Stercorarius chilensis	Salteador chileno	Resident	Mar	•	•	•
Larus serranus	Gaviota andina	Resident	Humedal	•		
Larus maculipennis	Gaviota cáhuil	Resident	Humedal		O	O
Larus pipixcan	Gaviota de Franklin	Migratory	Coast/ Humedal	O	O	O
Larus scoresbii	Gaviota austral	Resident	Coast			o
Larus dominicanus	Gaviota dominicana	Resident	Coast	O	O	O
Larus belcheri	Gaviota peruana	Resident	Coast	o	•	
Larus modestus	Gaviota garuma	Resident	Coast	O	O	
Larus cirrocephalus	Gaviota de capucha gris	Migratory	Coast	•		
Sterna hirundinacea	Gaviotín sudamericano	Resident	Mar/ Humedal	o	o	O
Sterna hirundo	Gaviotín boreal	Migratory	Mar	•	•	
Sterna paradisaea	Gaviotín ártico	Migratory	Mar	•	•	•
Sterna trudeaui	Gaviotín piquerito	Resident	Mar/ Humedal		o	o
Sterna lorata	Gaviotín chico	Resident	Mar	•		
Sterna elegans	Gaviotín elegante	Migratory	Mar/ Humedal	O	O	O
Sterna sandvicensis	Gaviotín de Sandwich	Migratory	Mar	•	•	•
Larosterna inca	Gaviotín monja	Resident	Mar	o	o	•
Rynchops niger	Rayador	Migratory	Wetland	O	O	O
Strigidae						
Asio flammeus	nuco	Resident	Dunes/ praderas	•	•	•
Alcedinidae						
Ceryle torquata	Martín pescador	Migratory	Humedal			•
Chloroceryle americana	Martín pescador chico	Errant	Humedal	•		
Furnariidae						
Cinclodes nigrofumosus	Churrete costero	Resident	Coast	o	o	o
Cinclodes patagonicus	Churrete común	Resident	Humedal		o	o
Cinclodes oustaleti	Churrete chico	Resident	Humedal	•	o	o

(continued)

Table 3.1 (continued)

Family/ Specie		Status in Chile	Main environment	Abundance north	Abundance center	Abundance south
Cinclodes fuscus	Churrete acanelado	Resident	Humedal		o	o
Phleocryptes melanops	Trabajador	Resident	Humedal	•	o	o
Tyrannidae						
Tachuris rubrigastra	Siete colores	Resident	Humedal	•	o	o
Pseudocolopteryx flaviventris	Pájaro amarillo	Migratory	Humedal		•	•
Lessonia rufa	Colegial	Resident	Humedal	•	o	o
Hymenops perspicillata	Run run	Migratory	Humedal		•	o
Trogloydytidae						
Cistothorus platensis	Chercán de las vegas	Resident			•	o
Motacillidae						
Anthus correndera	Bailarín chico	Resident	Dunes/ praderas		o	•
Anthus lutescens	Bailarín chico peruano	Errant	Dunes/ praderas	•		
Anthus hellmayri	Bailarín chico argentino	Errant	Dunes/ praderas			•
Icteridae						
Agelaius thilius	Trile	Resident	Humedal		O	O

*Asilvestrado
•Infrecuente/raro
°Escaso/normal
OComún/abundante

wetlands, birds that use the coast as a migration path found in coastal wetlands a highly stable and predictable resource, allowing them to inhabit it by relatively long periods of time (Skagen 1997).

Thus, along a coastal migration route, estuaries are a chain of sites that include not only centers of temporary re-supply, but also places to spend a winter in (Skagen 1997).

Among the factors that determine the selection of wetlands for migratory birds, abundance and quality of the food is probably the most important, but there is also the effect of physical variables of wetlands and the type of human intervention (Cole et al. 2002; Placyk and Harrington 2004). The coincidence of the ideal conditions for each of the above factors, in one place, is usually rare. This evidence makes up the fact that very few sites often concentrate a high proportion (e.g. more than 30-40%) of the world population of some species (Morrison and Harrington 1979).

Along the Chilean coast, between Arica and Chiloe, there are about 300 coastal wetlands, some of which receive substantial amounts of migratory birds, while others are marginally used by these species. Faced with this phenomenon we ask: What factors influence this pattern? Following, we try to partially answer this question by analyzing the case of two long-distance migratory species, the Elegant Tern (Sterna Elegans) and Black Skimmer (Rynchops Niger). Both species are concentrated in estuaries, although the Elegant Tern forages more on the coast (Schaffner 1986), while the Black Skimmer does it directly on the banks of wide rivers and estuaries (Gochfeld and Burger 1994).

Using different sources of information (censuses AvesChile, public data of the eBird system, and observations of the authors), we have determined that the average and maximum abundance recorded is between the months of October and April for both species in Chile, in different coastal wetlands in the country. We have obtained information (at least through one census) for 85 sites between the mouth of Lluta River and Chiloe, covering virtually the entire range of distribution of these species in Chile (Jaramillo 2003).

Figures 3.1 and 3.2 show the distribution of the Black Skimmer and Elegant Terns, respectively. Along with a description of the average abundance at each site (circle diameter) is also presented a characterization of the quality of information (number of surveys in which the data is based). Both figures show a very similar pattern of distribution observed ($r = 0.85$, $p < 0.0001$), which highlights a point of concentration at the mouth of the Lluta and San Jose (extreme north) rivers, followed by about 1600 km of absence of important records. Further south, the areas of greatest concentration of these species are found in the regions of Maule and Bio-Bio. The southernmost records of these species correspond to the northern part of the island of Chiloe.

To evaluate the existence of physical parameters of wetlands that influence the degree of aggregation of these species, we have estimated from Google Earth's total water surface and the total area of the isolated sand bars (i.e. surrounded by water) for each site. We have also determined that if the river mouth is adjacent to a sandy beach or not (1.0). Finally, based on the concentration of houses and other infrastructures around wetlands, we have defined the quality index of human presence in estuaries (0, 1 and 2). Through a step by step linear model (function stepAIC, R. Ihaka and Gentleman 1996), we have correlated these variables with the maximum abundance logarithm of each species.

Table 3.2 shows that neither the total surface of water nor the degree of human presence, have significantly influenced the abundance of species in different wetlands. By contrast, both the average and maximum abundance of Black Skimmer were explained significantly by the surface of sandbars, while the abundance of Elegant Terns was positively associated with the presence of beaches. However, in all of these cases the percentage of explained variance was very low (3–9%).

A visual inspection of the data identified the Biobio River as a special case, with a high amount of sand bars silting, posing in the final portion of its course, but with very low records of the species studied. This situation, coupled with the fact that the site has reported only two censuses, caused the data to be removed. The lower part of Table 3.2 contains an analysis as above, but without the Biobio River. It is noted that the significance of the sandbars as predictors of the abundance of species

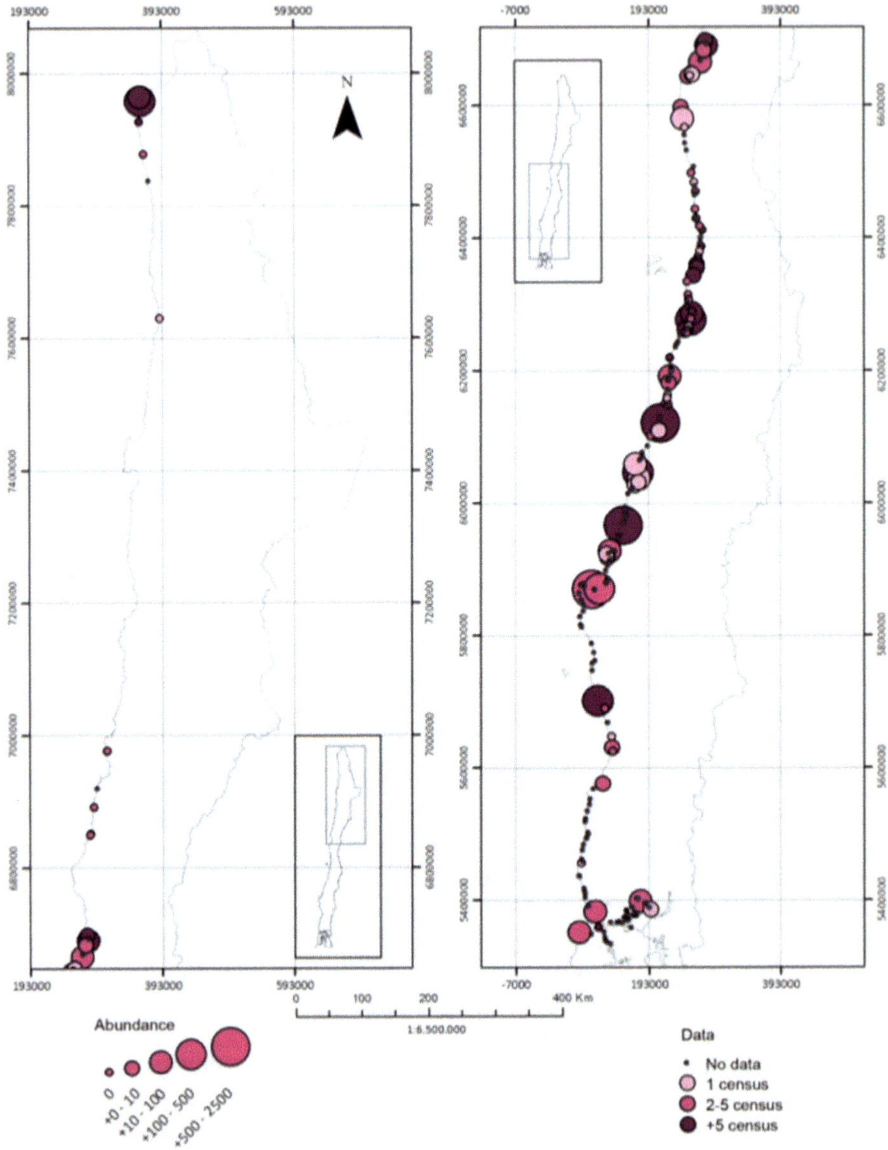

Fig. 3.1 Distribution of Black Skimmer (*Rynchops niger*) in Chile's coastal wetlands. The average size of those populations where a census was carried out in each site is graphically represented by the circle's diameter. The quality of the information (number of census) for each site is represented by the color of the figure

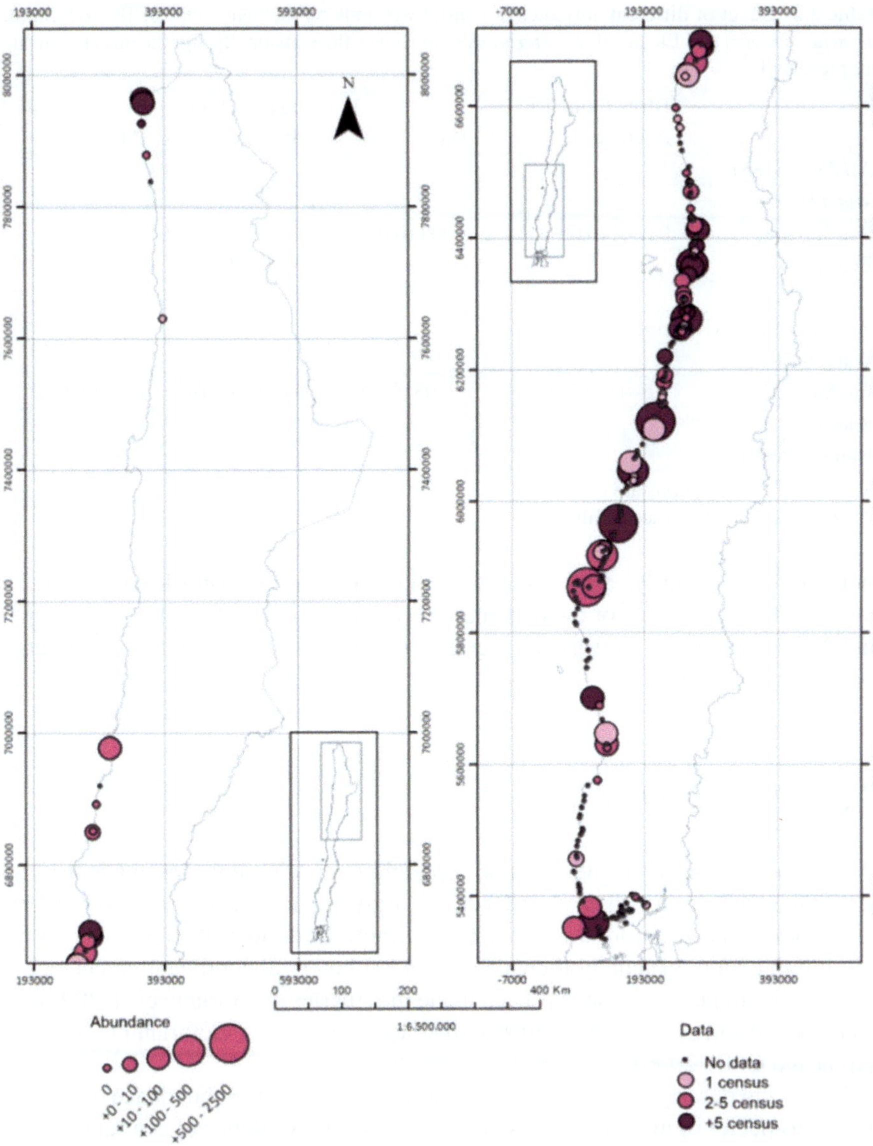

Fig. 3.2 Distribution of Elegant Tern (*Sterna elegans*) in Chile's coastal wetlands. The average size of populations where a census was carried out in each site is graphically represented by the circle's diameter. The quality of the information (i.e. number of census) for each site is represented by the color of the figure

Table 3.2 Effect of different attributes of coastal wetlands on the abundance of Black Skimmer (*Rynchops niger*) and Elegant Tern (*Sterna elegans*). Only the statistically significant relationships are presented

	Black Skimmer		Elegant Tern	
	Average	Max	Average	Max
All data (n=85)				
Water Surface				
Bar Surface	0.004* (0.09**)	0.008 (0.07)		
Beach			0.046 (0.03)	
Human Presence				
Without Biobio River(n=84)				
Water Surface				
Bar Surface	7 e–07 (0.25)	1 e-05 (0.20)	9 e-05 (0.16)	0.001 (0.15)
Beach				
Human Presence				

*value P. All the relations are positive
**R2 adjusted by the grade of liberty

increased significantly, also explaining the concentration of Elegant Terns. In addition, the percentage of the explained variance increased the values between 15 and 25%. In contrast, the variable presence of beaches had no significant effect on the species studied.

Population Dynamics

One of the obvious characteristics of populations of migratory species is the high seasonal variability of their abundances at a given site. Thus, apart from a large geographic concentration, the interaction of these birds with the coastal wetlands has a high degree of temporal concentration. Although a significant percentage of water birds studies, in Chile, has been made in estuaries (Victoriano et al. 2006), the available information on population dynamics is very low (Victoriano et al. 2006; Vilina and Cofre 2006).

Following description is of the population dynamics of the abundance of some of the main species of migratory birds visiting the coastal wetlands of central Chile. To this end, we rely on information collected by us as part of a program of long-term research developed in four estuaries of central Chile: Itata (36o23 'S, 72o51' W), Reloca (35o40 'S, 72o35' W), Mataquito (34o59 'S, 72o10' W) and Topocalma (34o07 'S, 71o59' W). The first three sites have been monitored since 2006, while the fourth was incorporated in 2008.

In all of the estuaries, the four seasonal censuses per year have been organized, focused on the most biologically significant periods for birds. Thus, these campaigns correspond to summer (boreal species wintering), autumn migration (return of boreal species), winter (southern wintering species) and spring migration (arrival of boreal species).

Since the study has focused on Itata, with the other three sites acting as population control trends, the sampling intensity was higher in the first. Thus, while for the control of each campaign there were considered two census visits (eight visits for eight annual surveys) in Itata migration campaigns there were three visits involved, each of which contains censuses for 3 days in a row, with a repeat in the morning and one in the evening (10 visits and 60 censuses per year). However, for purposes of this study, in the case of Itata, the average data per census visit census is presented.

Each census represents the total count of individuals present in each study site (each river estuary) from fixed observation stations. To make the count of individuals, binoculars and telescope are used.

Here on after, in this document, the estuaries of Itata, Reloca, Mataquito, and Topocalma rivers will be referred to as IRMT estuaries. Figure 3.3 shows the population trajectory of Black Skimmer and Elegant Tern for each of the individual sites and for all estuaries. In the case of Black Skimmer there can be seen a high degree of regularity in the patterns of abundance, particularly in the aggregate numbers. The data for this species indicate that there is a population tendency to concentrate on Itata (the southernmost site) at the start of the season and then to slowly move northward through Reloca and Mataquito.

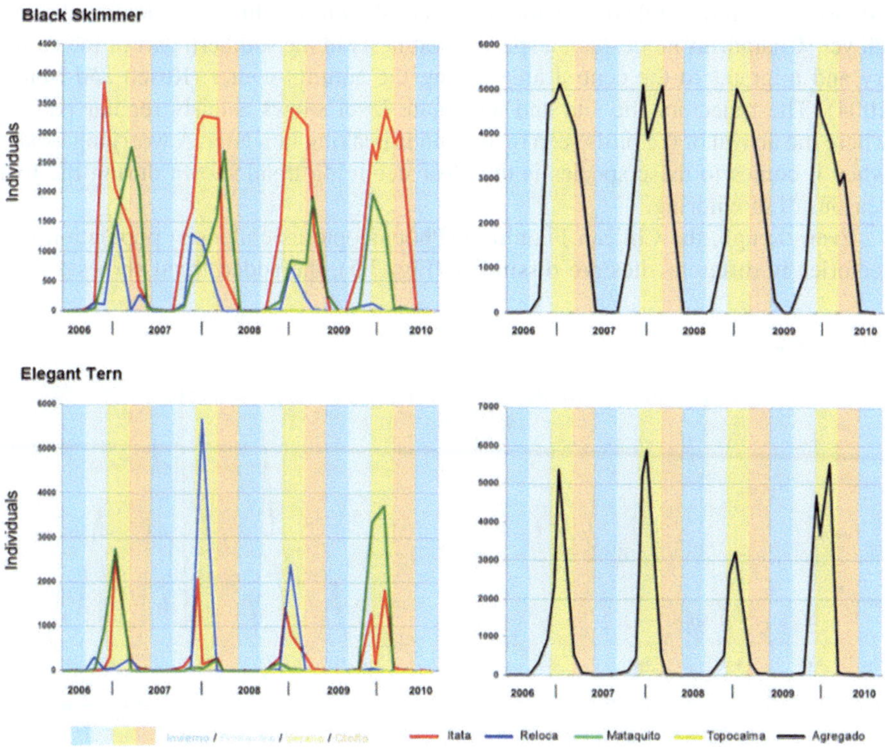

Fig. 3.3 Population trajectories for `Black Skimmer (*Rynchops niger*) and Elegant Tern (*Sterna elegans*) in four estuaries in central-south Chile (*color lines*) and for all these sites as a whole (*black lines*). Data shown correspond to the period of winter 2006 – winter 2010

The shape of the curve of the abundances accumulated suggest that Black Skimmer populations visiting the IRMT estuaries are a relatively closed group, where birds remain throughout the summer. This implies that these sites have an important role as a station stop for birds traveling further south (see the case of the Baird's Sandpiper), although there are indeed places further southern concentration (Fig. 3.1), but not of the same importance as the IRMT estuaries.

The case of Elegant Terns (Fig. 3.3) is relatively different. While it is true that the accumulated abundance has a high degree of regularity, the situation at the site has varied significantly during the study period, and a clear negative relationship between the abundance of the species in Mataquito (high in 2006–2007 and 2009–2010) and Reloca (high in 2007–2008 and 2008–2009) was observed. Both results suggests a redistribution of birds between sites in different seasons. Figure 3.4 shows the population trajectory of the Baird's Sandpiper (Calidris bairdii), another boreal migratory species that, unlike previous ones, uses the IRMT estuaries as a transit station, rather than wintering site. This follows from the clear and regular trend where two annual peaks are observed: one in spring, associated with the migration of individuals to the south, and one in summer, due to the migration of individuals to the north.

Lastly, in Fig. 3.5 two examples of short-distance migratory species that visit the estuaries of central Chile during the winter are shown. The first is the Rufus-chested Plover (Charadrius modestus), a species that nests in the southern part of the country and migrates to the central area during the autumn-winter (Kusch and Marin 2004). The trajectory the site and aggregate level shows a fairly regular pattern where the arrival of the birds seen in autumn is peaking in winter. A worrying aspect when it comes to these species is the clear declining trend shown during the five seasons of monitoring.

Even though, the Chilean Flamingo (Phoenicopterus chilensis) population trajectories in different sites are dissimilar (Fig. 3.5), the added trend shows a high

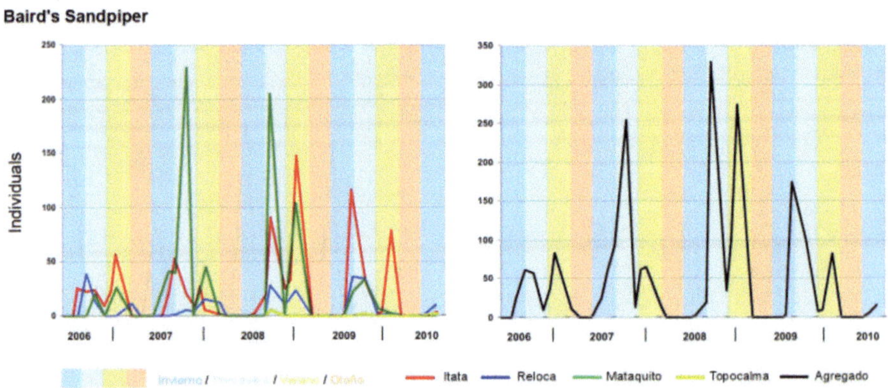

Fig. 3.4 Population trajectories of Baird's Sandpiper (*Calidris bairdii*) in four estuaries in central-south Chile (*color lines*) and for all these sites as a whole (*black lines*). Data shown correspond to the period of winter 2006 through winter 2010

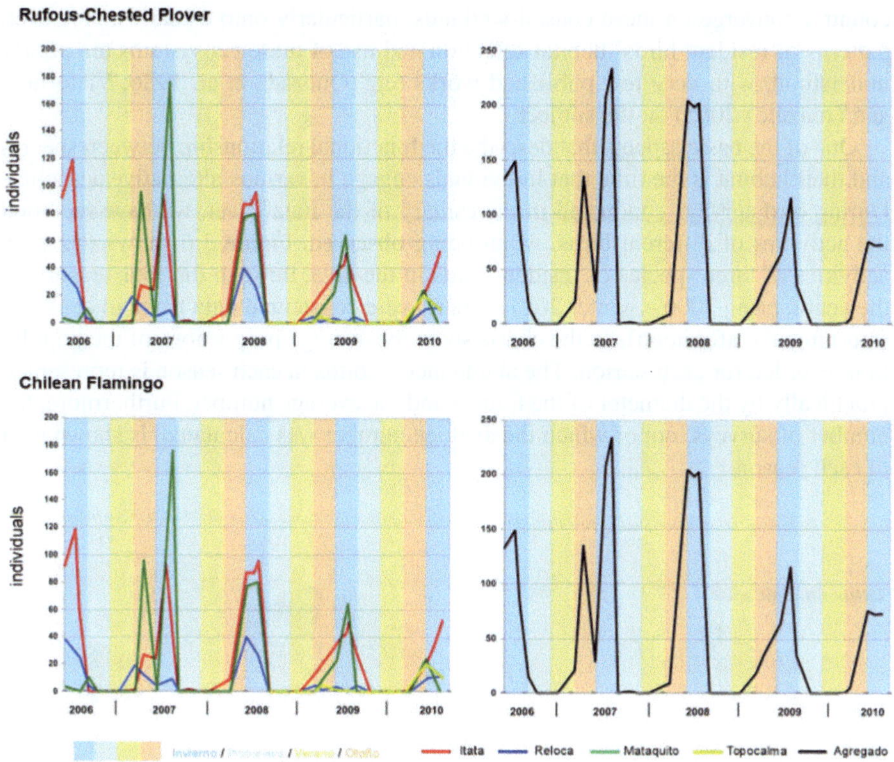

Fig. 3.5 Population trajectories for Rufus-chested Plover (*Charadrius modestus*) and *Chilean Flamingo* (*Phoenicopterus chilensis*) in four estuaries in central-south Chile (*color lines*), and these sites as a whole (*black lines*). Data shown correspond to the period of winter 2006 through winter 2010

degree of regularity, suggesting that the birds arriving to the study area are relatively defined group, but that individuals are redistributed differently between sites in different seasons. Estades et al. (2008) indicate that this "transfer" between sites is probably due to the existence of a negative relationship between the presence of the species in large estuaries and the level of rainfall of the year, probably due to the rainfall effect on habitat availability for these birds.

Role of Coastal Wetlands for Birds Residents – Selection and Habitat use

While higher concentrations of birds in coastal wetlands are migratory species, the greatest diversity of species is given by resident species (Table 3.1). Therefore, a large number of species of sea, coastal and freshwater birds that are breeding in the country, converge on these coastal wetlands, particularly onto estuaries. However, patterns of resident birds' habitat selection and use of these ecosystems are poorly understood, with very few published works (e.g. Quezada et al. 1986; Victoriano and González 2005) on the subject.

One of the basic aspects that describe the functional relationship between species and their habitat is the time that individuals engage in various alternative activities. During bird surveys conducted in the estuary of the Itata River, we have recorded the activities of different birds, when being observed. Figure 3.6 shows the main activities of three species of resident birds in the area, through the four seasons of the year (spring 2006 – winter 2007). There were considered data from all censuses (morning and afternoon), so the charts show the average proportions of the population recorded for each season. The abundance of birds in each season is represented graphically by the diameter of the figures and the average number. Furthermore, the number of surveys, out of which the average number was calculated, is shown next to each season.

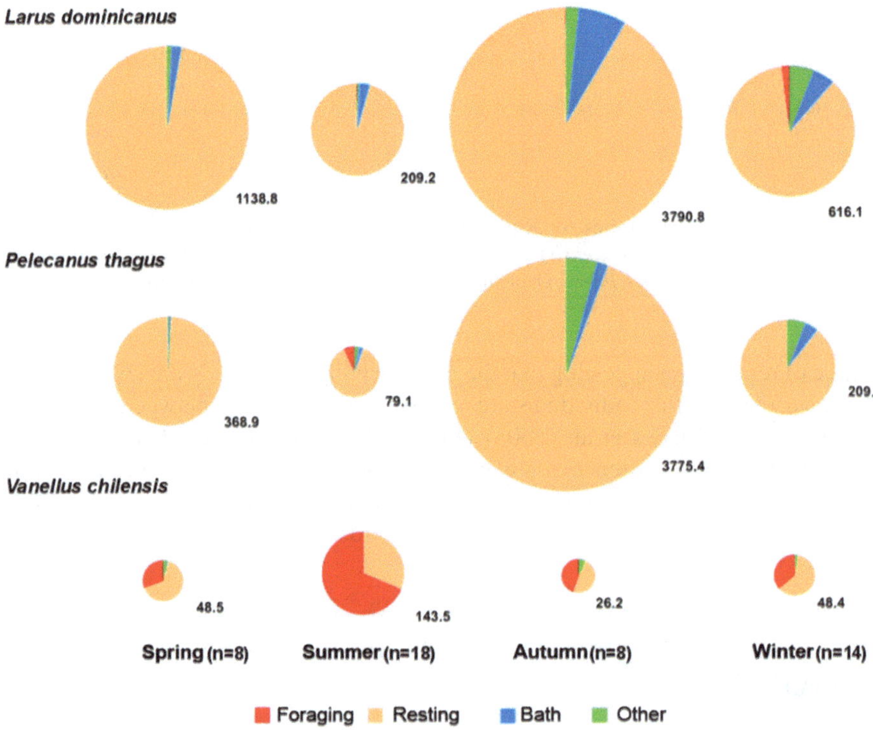

Fig. 3.6 Frequency of activities performed by three bird species in the Itata river estuary between spring 2006 and winter 2007. The dimensions of the *circles* represent the total number of individuals based on which the proportions were calculated. The category "other" includes aggression and flight

It is noted that in the case of the two marine species, the Kelp Gull (Larus dominicanus) and the Peruvian Pelican (Pelecanus Thagus), the largest proportion (87–99%) of individuals was recorded at rest. Second in importance was the bath (0.5–7%), frequent activity among these birds, which apparently use fresh water to remove salt from their plumage. The percentage of birds foraging was low (0–7%) as well as other activities (flight, aggression, etc.). No breeding activities of these species were recorded in the area.

The Southern Lapwing (Vanellus chilensis), a common species in all types of wetlands and agricultural areas, had a higher ratio of foraging dedication than the previous species (30–69%), although the number of individuals resting was also high (31–66%). Dedication to activities such as bathing and other was minimal. Unlike the two marine species, Southern Lapwing itself had some degree of reproductive activity in the area, although it was not registered during the census.

Population Dynamics

Although pretty significant number of ornithologists, both professional and amateur, conducts periodic surveys of water birds in numerous wetlands of central Chile (Espinoza 2009), the analysis of these data are almost nonexistent, so it does not yet have a reasonable characterization of the waterfowl residents' dynamics in this region.

Data derived from surveys conducted by us (CFE and MAV) in the IRMT estuaries indicate that even for most waterfowl species considered residents in central-southern Chile, there is evidence of displacement of the large numbers of the population.

Figure 3.7 shows the added population trajectories in IRMT estuaries, for six common species living in the wetlands of central Chile. It highlights the existence of a general pattern common to all species, with a minimum population observed in the spring, followed by a very sharp increase in population during the summer-autumn season. The low presence during the subsequent breeding season and high growth rate and small number of offspring observed, suggests that the dynamics of these populations is not governed by births and deaths, but mainly by the migration of individuals.

To determine the possible origin of birds, evidence of inverse population trends were sought throughout other sites in central Chile, which could account for seasonal re-distribution of individuals. For this census, data obtained by JA in environmental lagoon Farfana (33o28 'S, 70o47' W, Maipu, Metropolitan Region) were used. This lagoon, specially designed to house waterfowl, is probably the only interior wetland of the central region of Chile that has a concentration of adequate data for comparison purposes.

Figure 3.8a contrasts the population trajectory of White-tuft Grebe (Rollandia Rolland) in IRMT estuaries and Farfana. Here we see a clear reverse trend, where birds were concentrating in Farfana during reproductive periods (with the exception

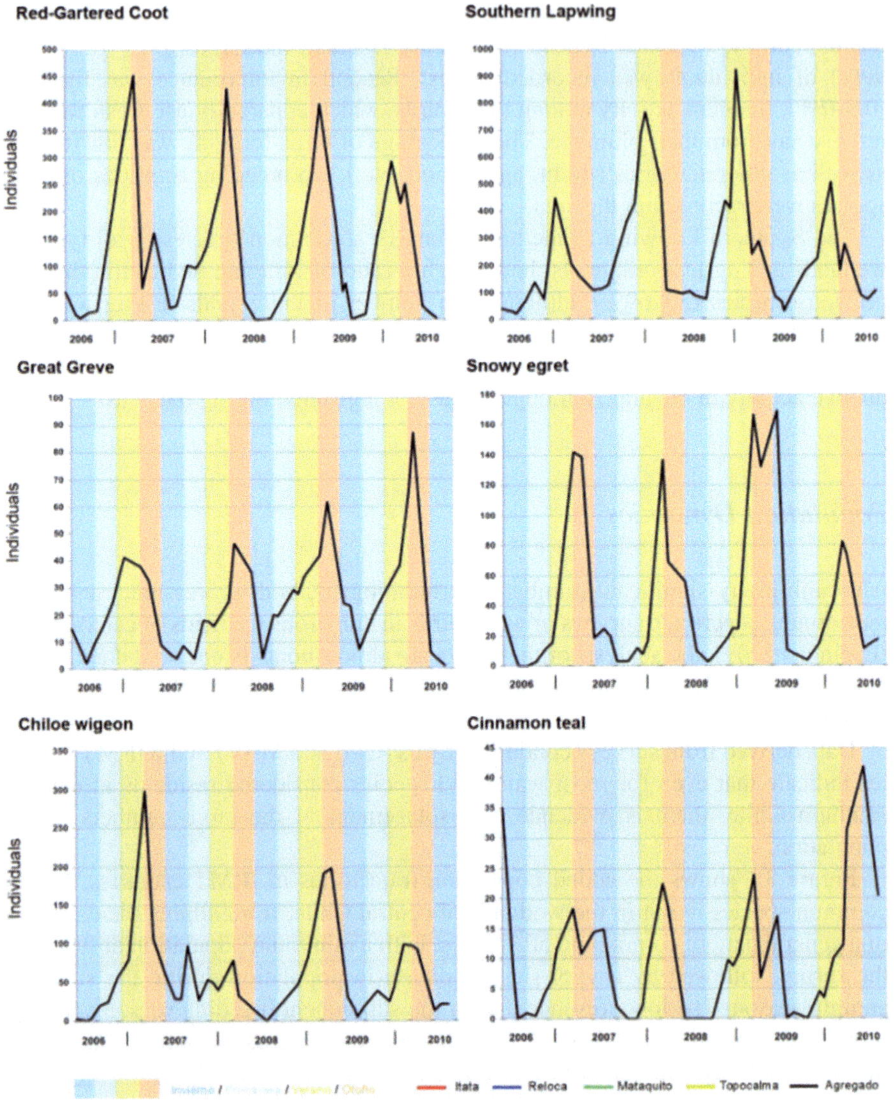

Fig. 3.7 Aggregated population trajectories for estuaries of six bird species that reside in central Chile: Red-gartered Coot (*Fulica armillata*), Southern Lapwing (*Vanellus chilensis*), Great Grebe (*Podiceps major*), Snowy Egret (*Egretta thula*), Chiloe Wigeon (*Anas sybilatrix*) and Cinnamon Teal (*Anas cyanoptera*). Data shown correspond to the period of winter 2006 through winter 2010

of spring 2007), then they would leave the area in the summer, the same period in which the species began arriving to the IRMT estuaries.

Although it is not possible to argue that there is a movement of birds from the Farfana to the IRMT estuaries (located 150–350 kilometers SW), the trend seems to suggest that at least some of the species that are concentrated in the estuaries of the

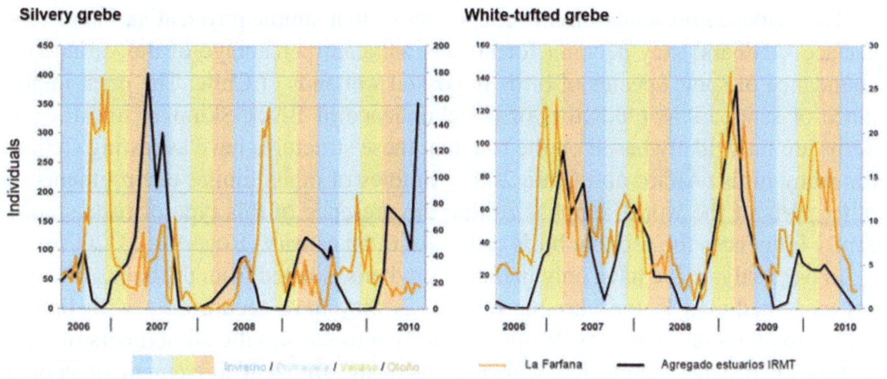

Fig. 3.8 Population trajectories of Silvery Grebe (*Podiceps occipitalis*) and White-tuft Grebe (*Rollandia rolland*) observed in the IRMT estuaries and the Farfana (Metropolitan Region). Data shown correspond to the period of winter 2006 through winter 2010

central-southern area during the post-reproductive period, are from the country-s central valley wetlands.

Far from showing a reverse trend, Fig. 3.8b reveals that the population trajectory of White-tufted Grebe (Rollandia Rolland), a relative of the previous species, virtually replicates the pattern observed in the IRMT estuaries. This suggests that in both systems, the reproduction of this species is low and that the significant population increase observed during the summer is due to individuals from other wetlands with appropriate characteristics for reproduction of these birds or, as suggested by the results of a study of the Chacabuco basin (Lopez et al., in preparation), corresponds to birds that come from the surrounding landscape.

Lopez et al. (In preparation) have observed that many water birds are dispersed in the landscape during the breeding season, so they are not registered in the regular censuses of large wetlands. Most likely, these birds breed in low extension wet environments, such as canals and small irrigation dams, wet meadows, marshes, etc., and then return to large wetlands during the post-reproductive period, during which a large portion of small scattered wetlands has dried out.

Therefore, while it is true that coastal wetlands in the central-southern Chile area have little importance as breeding sites, they play an important role in the population dynamics of many resident waterfowl, acting as post reproductive aggregation sites.

Discussion and Conclusions

The factors that determine the presence and abundance of different species of birds in coastal wetlands, at some point, include aspects of the biology of the species (e.g. resident and migratory), and the physical and biological characteristics of these environments.

The evidence presented in this chapter show that simple physical variables (e.g. surface sandbars) may account for not insignificant percentage of the variance in abundance of some species of birds in coastal wetlands of Chile. The great importance of sandbars as predictors of the abundance of Black Skimmer and Elegant Terns are most likely based on the role that these structures have as resting stations for many birds (Alfaro and Clara 2007). Studies of other similar environments, in other parts of the world, emphasize that the presence of this type of rests significantly influences the use that birds can make of an estuary (Rogers et al. 2006).

In our analysis, we have only included sandbars isolated from the banks (the others were included in the variable beach), as they have secure sites with limited access to predators like dogs. While it is normal to see significant numbers of birds resting on beaches connected with mainland, the frequent movement of people, vehicles and animals (domestic and livestock) tend to repeatedly scare the birds, so presumably, in isolated areas there should not have a higher rate of turnover. Contrary, the existence of isolated bars could ensure the permanence of flocks longer. This suggests that our model should better predict the average abundance than the maximum (more variable). In fact, although there was a dramatic difference, in all of the cases there was a better adjustment with the average abundance than with the maximum (Table 3.2).

There are many sources of error in estimating the area of sand bars in estuaries from Google Earth. Thus, the existence of images from different years, times, tide levels, etc., coupled with possible errors in photo interpretation, and the variability in the quality of data abundance of birds, suggest that in our analysis there is a high noise level. It is possible that with better information, the effect of sand bars on the abundance of birds, such as those studied, is even greater than that found by us. Nevertheless, it is possible to analyze the mechanisms by which other factors may affect the presence of birds in coastal wetlands of our country.

The uneven distribution of wetlands on the Chilean coast poses restrictions on migratory birds when it comes to the possibility of stopover sites during their movements. This is reflected in the presence of a small number of sites of concentration of these birds (Figs. 3.1 and 3.2).

For the Elegant Terns, that forages mainly in salt water (Schaffner 1986), it is possible to make intermediate stops in places without wetlands (e.g. beaches, rocks, etc.), accessing food in the sea. However, the Black Skimmer has difficulty feeding in waters with some degree of wave surf, so, with the exception of some very calm bays, this species is not foraging at sea. This implies that, notwithstanding that these birds can rest on many points between wetlands described in Fig. 3.1, they can only feeding lagoons and estuaries. Hence, the distance of over 1100 kilometers between the Lluta River and next estuary of relatively significant dimensions (Huasco, 5 ha of water) should imply greater restriction on the Black Skimmer, since it is likely that there are no important refueling places in between these points. Being so, the records of the species in Huasco are practically nil, and the wetland of Punta Teatinos (2 ha of water, La Serena, 1300 kilometers south Of Lluta) is the first place of some importance, as feeding site for the species on their journey south.

Along the Chilean migratory coastal route, the estuaries form a chain of sites that include not only centers of temporary resupply, but also wintering sites. This pattern, previously described by Skagen (1997), is exemplified by the cases of Black Skimmer and Baird's Sandpiper. So while the first species have established themselves in the IRMT estuaries throughout the season (Fig. 3.3), the second one uses these sites to stock and rest during their journey to and from more southern sites (Fig. 3.4). Aparicio (2006) observed that Sandpiper's Baird was present only in their area of study (Valdivia) during migration south, indicating that this species uses different places of detention and resupply during their migrations from and to south and north.

Water bird communities (migratory and resident) present in the estuaries on the coast of Chile are highly dynamic, albeit with a very regular and predictable structure. Although this situation does not necessarily replicate in wetlands farther south or farther north, in the IRMT estuaries the existence of a high degree of coincidence in population trends of most species between sites, even among non-migratory ones, suggests that these dynamics are largely due to regional trends rather than individual behavior of each estuary.

Thus, in many cases, the aggregate data of the population trajectories, in different sites analyzed, showed more consistency than that of each individual site, indicating the existence of a non-independent behavior of bird populations studied. The opposing trends in populations at some sites strongly suggest the existence of a redistribution of birds among sites in different seasons (e.g. Elegant Terns, Fig. 3.3, Chilean Flamingo, Fig. 3.5).

Existing information on the use that waterfowl makes of the IRMT estuaries (exemplified in this work with the case of the Itata River), clearly indicates that the main role of these environments is to serve as a resting place for many birds. Data on the activities of Kelp Gull and Peruvian Pelicans (Fig. 3.6) are representative of most species, both resident and migratory (data not shown). In particular, many sea birds foraging at sea, sleep on sandbars and beaches surrounding estuaries, so their concentration in these sites tends to increase early or late during the day.

In turn, many seabirds take advantage of the low salinity of estuarine waters for bathing, an activity that often is second in importance in terms of the number of individuals involved. Although the bathing of seabirds in freshwater sites is a common and normal behavior throughout the world, it is noteworthy that, to our knowledge, there is no assessment of the real importance this activity has on biological adaptation of individuals.

The percentage of time spent foraging is greater for freshwater species than marine ones (e.g. Southern Lapwing vs. Pelican, Fig. 3.6). Still, it is possible that the real importance of the IRMT estuaries as feeding sites for some species are underestimated due to factors such as differences in efficiency when foraging (e.g. in terms of calories per unit of time) or the effectiveness of our protocol sampling. For example, although our surveys in Itata covers a significant period of time of the day (8: 00-12: 00 and 14: 00-17: 00 h.), some species such as Black Skimmer tend to feed at night (Erwin 1977) so the observations of foraging individuals should be clearly under-represented in our samples.

Reproductive activity of water birds in the estuary of the Itata is marginal. This is a common situation in this type of environment in Chile (Victoriano et al. 2006), as well as other parts of the world, because the dynamics of the tides and their interaction with changes in river flows become very unstable in supply of nesting sites. This phenomenon is also associated with the scarcity of aquatic vegetation, which also restricts the abundance of herbivorous species.

In addition to the limitations noted, many estuaries have a high degree of human intervention. For example, in Itata, the presence of pets in the area, as well as cows and horses certainly affects the development of vegetation in which many birds build their nests. For this reason the most abundant nesting areas for species that build nests away from water, are neighboring fields and meadows.

While it is true that estuaries of central Chile play a minor role in the waterfowl nesting, these coastal wetlands do have an important role as post-reproductive aggregation site for many of these species. Indirect evidence suggests that there should be significant and regular bird movements between interior and coastal wetlands. However, the lack of detailed information on population dynamics of birds in the central valley of the country hinders proper understanding of this phenomenon.

Although there is a high predictability in the structure and dynamics of bird communities in coastal wetlands of central Chile, it is also true that there are significant yearly variations in the abundances of some species. One of the factors influencing these changes is the ENSO phenomenon, during which it has been observed that some species like the Black-necked Swan (Cygnus melanocoryphus) and some grebes increase in number in the wetlands of central Chile during wet years (Schlatter et al. 2002; Villina and Cofré 2000; Villina et al. 2002).

On the other hand, major changes in the abundances of some long-distance migratory species respond, most likely, to factors affecting mobilization and survival of these species in the Northern Hemisphere. For example, it has been described that the anchovy fisheries on the West Coast of the United States influence the numbers Elegant Terns (Schaffner 1986), which could certainly lead to changes in the amounts of these birds that visit the coasts of Chile every year.

Finally, coastal wetlands of Chile, as well as the world, are undergoing a major anthropic pressure. Deterioration factors associated with increasing human population, as well as increased levels of eutrophication, pollution and wildlife and water extraction, among others, clearly threaten the integrity of these ecosystems (Kennish 2002).

Regarding waterfowl, a disturbing aspect of this trend is the significant increase in human recreation in estuaries and other coastal wetlands, resulting in numerous impacts, particularly the shortage of peaceful rest sites for many species (Placyk and Harrington 2004). In Chile, a very important factor is the growing presence of stray dogs, which can severely affect the quality of a site as a waterfowl habitat. In addition, "off road" transit of vehicles causes great damage to ecosystems dunes commonly associated with estuaries, affecting some birds' nesting.

The data obtained by us in the IRMT estuaries delivers a relatively clear picture of how bird communities use them. However, this is a correlational information that does not allow meaningful inferences about causal relationships. Therefore, the conclusions in this paper are only valid for the conditions under which the data were taken. This poses a certain limitation to the use of this information to understand the response of birds to the future changes in these ecosystems.

For example, the knowledge of specific functional role of wetlands in the path of migratory species is essential to predict the implications that the disturbance of each of these sites could have on the conservation of these species. The lack of such information in Chile, coupled with the fact that clearly different species of migratory birds have different restrictions on their habitat requirements, makes such predictive capacity is still minimum in the country.

Acknowledgments The IRMT estuary census data come from a long-term study funded by Arauco S.A. Farfana data are the result of regular monitoring of the site funded by Aguas Andinas SA. Maria Paz Acuña produced distribution maps of Black Skimmer and Elegant Terns. The authors also thank AvesChile for facilitating census data made by members of this organization, and to the many ornithologists who contribute information through eBird.

References

Alfaro, M., and M. Clara. 2007. Assemblage of shorebirds on Rocha lagoon sandbar, Uruguay. *Ornitología Neotropical* 18: 421–432.

Aparicio, A. 2006. Abundance, distribution and migration chronology of shorebirds on exposed sandy beaches of south central Chile. In *Waterbirds around the world*, ed. G.C. en Boere, C.A. Galbraith, and D.A. Stroud, 188. Edinburgh, UK: The Stationery Office.

Chaparro, O.R., C.J. Segura, Y.A. Montiel, R.J. Thompson, and J.M. Navarro. 2008. Variations in the quantity and composition of seston from an estuary in southern Chile ondifferenttemporal scales. *Estuarine, Coastal and Shelf Science* 76: 845–860.

Cole, M.L., D.M. Leslie Jr., and W.L. Fisher. 2002. Habitatuse by shorebirds ata stopover site in the Southern Great Plains. *The Southwestern Naturalist* 47: 372–378.

Colwell, M.A. 1993. Shorebird community patterns in a seasonally dynamic estuary. *Condor* 95: 94–103.

Erwin, R.M. 1977. Black skimmer breeding ecology and behavior. *Auk* 94: 709–717.

Espinosa, L.A. 2009. Chile: Informe anual. Censo Neotropical de Aves Acuáticas 2008 [en línea]. ed. En Unterkofler D. A. Y D. E. Blanco (eds.): El Censo Neotropical de Aves Acuáticas 2008; Una herramienta para la conservación. Wetlands International, Buenos Aires, Argentina. http://lac.wetlands.org/.

Estades, C.F., J. Aguirre, and A. Camaño. 2008. Chilean Flamingos (*Phoenicopterus chilensis*) in estuaries of central Chile. *Flamingo* 16: 50–54.

Gochfeld, M., and J. Burger. 1994. Black Skimmer (Rynchops niger), The Birds of North America Online (A. Poole, Ed.). Ithaca: Cornell Lab of Ornithology. Obtenido de. http://bna.birds.cornell.edu/bna/species/108.

González, A., and P. Victoriano. 2005. Aves de los humedales costeros del Centro-Sur de Chile. Capítulo 35 En. In *Biodiversidad y Ecología de la Cordillera de la Costa*, ed. C. Smith, J. Armesto, and C. Valdovinos. Santiago: Editorial Universitaria.

Haedrich, R.L. 1983. Estuarine fishes. In *Ecosystems of the world: estuaries and enclosed seas*, ed. B. en Ketchum, 183–207. Amsterdam: Elsevier.

Ihaka, R., and R. Gentleman. 1996. R: A language for data analysis and graphics. *Journal of Computational and Graphical Statistics* 5 (3): 299–314.

Jaramillo, A. 2003. Birds of Chile. In *Including the Antarctic Peninsula, the Falkland Islands and South Georgia*. London: Helm Field Guides.

Kennish, M.J. 2002. Environmental threats and environmental future of estuaries. *Environmental Conservation* 29: 78–107.
Kusch, A., and M. Marín. 2004. Distribución del Chorlo Chileno, *Charadrius modestus* (Lichtenstein) (Charadriidae) en Chile. *Analesdel Institutodela Patagonia* 32: 69–78.
Morrison, R.I.G., and B.A. Harrington. 1979. Critical shorebirdresourcesin James Bay and eastern North America. Transactions of the 44th North American Wildlife and Natural Resources Conference, Wildlife Management Institute, Washington, DC.
Morrison, R.I.G., and J.P. Myers. 1987. Wader migration systems in the New World. *Wader Studies Group Bulletin* 49S: 57–69.
Myers, J.P., R.I.G. Morrison, P.Z. Antas, B.A. Harrington, T.E. Lovejoy, M. Sallaberry, S.E. Senner, and A. Tarak. 1987. Conservation strategy for migratory species. *American Scientist* 75: 19–26.
Page, G.W., E. Palacios, L. Alfaro, S. González, L.E. Stenzel, and M. Jungers. 1997. Numbers of wintering shorebirds in coastal wetlands of Baja California, Mexico. *Journal of Field Ornithology* 68: 562–574.
Placyk, J.S., Jr., and B.A. Harrington. 2004. Prey abundance and habitat use by migratory shorebirds at coastal stopover sites in Connecticut. *Journal of Field Ornithology* 75: 223–231.
Quezada, A.E., H. Oyarzo, and V.H. Ruiz. 1986. Distribución de avifauna en los distintos "hábitats" del Estuario Andalién, Bahía de Concepción, Chile. *Studies on Neotropical Fauna and Environment* 21: 197–206.
Rogers, D., P.F. Battley, T. Piersma, J.A. Van Gils, and K.G. Rogers. 2006. High-tide habitat choice: insights from modelling roost selection by shorebirds around a tropical bay. *Animal Behaviour* 72: 563–575.
Schaffner, F.C. 1986. Trends in Elegant Tern and northern anchovy populations in California. *Condor* 88: 347–354.
Schlatter, R.P., R.A. Navarro, and P. Corti. 2002. Effects of El Niño Southern Oscillation on numbers of Black-necked Swans at Rio Cruces Sanctuary, Chile. *Waterbirds* 25: 114–122.
Skagen, S.K. 1997. Stopover ecology of transitory populations: the case of migrant shorebirds. *Ecological Studies* 125: 244–269.
Stuardo, J., and C. Valdovinos. 1989. Estuarios y lagunas costeras: ecosistemas importantes del Chile central. *Ambiente y Desarrollo* 5: 107–115.
Victoriano, P.F., A.L. González, and R. Schlatter. 2006. Estado de conocimiento de las aves de aguas continentales de Chile. *Gayana* 70: 140–162.
Vilina, Y.A., and H. Cofré. 2000. El Niño effects onthe abundance and habitatassociation patterns of four grebe species in Chilean wetlands. *Waterbirds* 32: 95–101.
Vilina, Y.A., H.L. Cofré, C. Silva-garcía, M.D. García, and C. Pérez-friedenthal. 2002. Effects of El Niño on abundance and breeding of Black-necked Swans on El Yali Wetland in Chile. *Waterbirds* 25: 123–127.
Vilina, Y., and H. Cofré. 2006. Aves acuáticas continentales. P 270-279 en Biodiversidad de Chile: Patrimonio y Desafíos. In *Comisión Nacional del Medio Ambiente*, ed. P. Saball, M.K. Arroyo, J.C. Castilla, C.F. Estades, J.M. Ladrón de Guevara, S. Larraín, C. Moreno, F. Rivas, J. Rovira, A. Sánchez, and L. Sierralta. Santiago: Editorial Ocho Libros.

Chapter 4
Hydrophilic Flora and Vegetation of the Coastal Wetlands of Chile

Carlos Ramírez G. and Miguel Álvarez F.

Abstract In this chapter, based on the exhaustive bibliographical review, Chilean coastal wetlands are defined, marked off and classified, characterizing their macrophyte flora, in wetlands and marshes, starting with a brief review of morphology and simplification of the corms in these plants, and continuing with the physiology, highlighting details of hydraulic wealth, photosynthesis, respiration, as well as nutrient intake in aquatic environments. All of the factors that limit the distribution of aquatic macrophytes are exposed, such as salinity, water depth and movement, nature of the substrate, waterlogging time length, and nutrient availability, emphasizing the effect of anthropic eutrophication of coastal wetlands on the autotroph macrobiotas. The morphological and physiological adaptations that have been associated with macrophyte survival in the present conditions, in these habitats, are also presented. Botanical diversity of each of the life forms found in the 227 macrophytes living in Chilean coastal wetlands is analyzed throughout the chapter. The following groups are described: submerged hydrophytes (22 species), free submerged (4), free floating (6) and natants (11), as well as fresh water herbaceous halophytes (112 species), marshland halophytes (22), herbaceous (32) and woody (18) amphibians. The presence of these species in the principal types of Chilean coastal wetlands is compared to and commented on. The main types of coastal wetlands present in Chile are classified into: rivers, streams, freshwater ponds, lagoons, seasonal lagoons, swampy marsh forests, rice fields and sphagnum marshlands. The important species and the conservation efforts and status of the flora in each wetland are presented too.

Keywords Helophytes • Riparian • Herbaceous

C. Ramírez G. (✉)
Departamento de Ecología, Pontificia Universidad Católica de Chile,
Santiago, Región Metropolitana, Chile
e-mail: cramirez@uach.cl

M. Álvarez F.
InRES, Ecología Vegetal, Universidad de Bonn, Bonn, Germany

J.M. Fariña, A. Camaño (eds.), *The Ecology and Natural History of Chilean Saltmarshes*, DOI 10.1007/978-3-319-63877-5_4

Table 4.1 Classification of Chilean wetlands according to Ramírez et al. (2002)

Wetland classification			
Salt water	Marine		Littoral
	Estuarine		Estuarine
			Marshes
	Interior		Freshwater lagoon
			Salt Flats
Sweet water	Riparian	Loquacious	Rivers
			Streams
			Ponds
			Oasis
		Lentic	Lakes
			Lagoons
			Pools
	Submerge		Swamp
			Peat
			Marsh
Total			15

Introduction

The hydrophilic vascular flora, which forms the aquatic vegetation and marshy, colonizes flooded or waterlogged environments, either temporarily or permanently. Such environments are what we know as wetlands (Davis et al. 1996; Mitsh et al. 2009). The Ramsar Convention proposes a definition and classification of conservation-oriented and sustainable management of these wetlands; however, being too wide, the practical application is difficult. A better operational classification for the Chilean wetlands was proposed by Ramírez et al. (2002) and improved by Ramírez and San Martín (2006b) (Table 4.1).

Aquatic plants (macrophytes) grow preferably in shallow coastal environments of freshwater and brackish wetlands, forming part of the so-called vascular phytobenthos (Cook et al. 1974). By contrast, pelagic environments are used as a place of life of phytoplankton (microphytes), and coastal saline by macroalgae. In lakes and shallow lagoons, between macrophytes and microphytes there is an unstable equilibrium, resulting in an alternate states between the clear waters (dominated by macrophytes) and the murky ones (dominated by microphytes, Scheffer et al. 1993). At one stage macrophytes inhibit the growth of phytoplankton, by removing nutrients and reduce the available light, but in others, and after the death of macrophytes, the development of phytoplankton is stimulated (Meerhof and Mazzeo 2004). The factors that trigger this change are still unknown. In any case macrophyte communities are more stable over time than the phytoplankton (Parra et al. 2005).

Aquatic plants are the azonal type flora as described by Walter (1997), that is to say, its distribution is not dependent on macroclimate but is strongly influenced by soil conditions, as in the cases of waterlogging and flooding. Therefore, the aquatic and marshland flora shoe broad distribution, sometimes of cosmopolitan character (Ramírez et al. 1979). This "cosmopolitanism" contributes to the hydrochoric transport (Ramírez and Romero 1978), which can reach great distances, especially marshland plants (*Triglochin concinna*, *Atriplex chilensis* y*Paspalum distichum*), as well as the ornithological transport by birds, especially migratory birds that use wetlands as resting places on their long journeys (Araya and Millie 1989; Figuerola and Green 2002).

Normally, vascular macrophytes form large populations thanks to the prevalence of clonal or vegetative reproduction in them (Grace 1993); however the resulting communities are poor in species variety (Ramírez et al. 1982b). The low representation of aquatic flora, which generally does not reach 10 % of the Earth's vascular flora (Hutchinson 1975; Chambers et al. 2008), has been justified by the restrictions set by the aquatic environments for growth of vascular plants. This chapter introduces the structural and functional characteristics as well as diversity of aquatic flora (hydrophilic) present in the coastal wetlands of northern and central Chile, from Arica to Chiloe.

General Characteristics

The areal distribution of terrestrial flora is determined by the gradient of altitude, mainly by the temperature and humidity (Walter 1997). These factors do not constitute a limitation in wetlands, and thus the distribution of aquatic plants may be limited by the conditions such as salinity, depth, water movement, sediment availability as underwater substrate, and duration of the flooding and waterlogging (Arber 1963; Ramírez et al. 1982b). Macrophytes just need for their rooting a soft substrate, shallow depth and little current; however, their survival will be determined by other factors, to which the macrophytes respond with morphological characteristics that determine the so-called life forms (Ramírez and Stegmaier 1982).

Life forms of aquatic macrophytes can be grouped into rooted submerged, submerged free, natants, free floating on the surface and emerged/surface (Fig. 4.1). The first two groups live and photosynthesize submerged in the water absorbing dissolved CO_2 or through sodium bicarbonate. The natant plants are rooted to the substrate and have leaves that float on the water surface, which photosynthesize just as terrestrial plants, obtaining the CO_2 directly from the air through the stomata present on the upper side. Some species of the latter group have both submerged leaves as natants. Live free floating plants covering the water surface and photosynthesize just like terrestrial plants. Finally, emerged or surface plants, also called swamp or halophytes, are rooted in the mud, have a small part of the body in the water, but most live in the open air doing regular photosynthesis (Arber 1963; Ramírez and San Martín 2006a).

Aquatic plants are very sensitive to depth changes, caused by the accretion of sediments, which causes replacements and displacements between life forms. This natural process, which is known as succession, can also be caused by man by depositing or favoring the deposit of sediments with constructions that can stop the current. In addition, changes in nutrient concentration (nitrogen and phosphorus

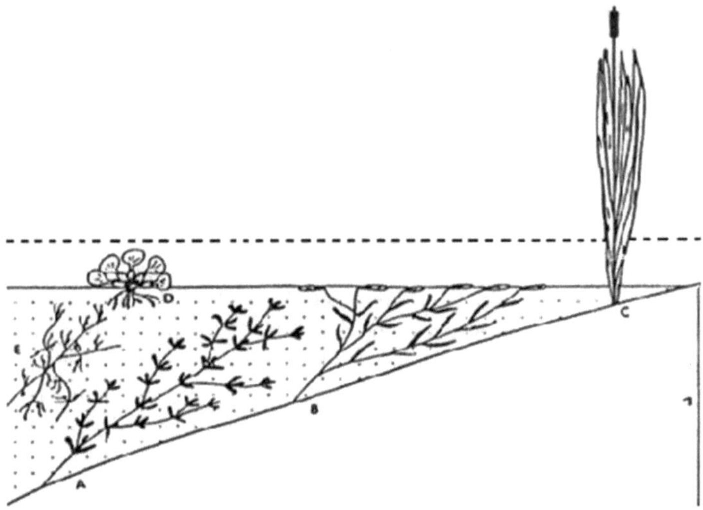

Fig. 4.1 Life forms of aquatic macrophytes. *A* = deep-rooted submerged, *B* = natant, *C* = emerged, *D* = surface free floating, *G* = submerged free floating. The dotted line indicates the highest water level (Taken from San Martín 1992)

mainly) can eliminate animal life, favoring an exaggerated development of the macrophytes, which produces an excess of organic matter, whose decomposition consumes the O2 of the water through the process of eutrophication (Keddy 2010; Ramírez et al. 2004).

Macrophytes are primary producers in wetland environments; additionally, they provide habitats for avifauna, fish, batratians and mammals (Mitsh and Gosselink 1993). They accelerate the natural sedimentation by relocating their own populations and those of the animals that depend on them. Normally they do not have much importance for the man, since, with the exception of the rice (San Martín and Ramírez 1983), they are not useful as food nor forage, because of the excess of water and minerals in their organs, especially in the submerged ones (Ramírez and San Martín 2006a) (Table 4.2). The emerged macrophytes provide material for light constructions and as of lately, aquatic plants have become important as biological indicators (Hauenstein et al. 1996; San Martín et al. 2003) and as water purifiers in artificial wetlands. In some cases, these plants acquire a negative connotation of invasive weeds that obstruct the use of water bodies, as is happening in Chile with *Egeria densa, Eichhornia crasippes, Azolla filiculoides* and *Hydrocotyle ranunculoides* (Matthei 1995; Quiroz et al. 2009).

Morphology

The colonization of fresh water by terrestrial plants is associated with the simplification of a plant body that is a corm with root, stem and leaves, which absorbs, conducts and delivers water in a regulated form. In some cases, this simplification

Table 4.2 Percentages of organic matter and ash in aquatic macrophytes' dry matter (taken from several studies by the authors). It can be seen that submerged plants and marsh halophytes present higher ash content than other species

Species	% Organic material	% Ash
Submerged		
Utricularia gibba	86.6	13.4
Myriophyllum aquaticum	82.3	17.7
Potamogeton lucens	73.1	26.9
Egeria densa	47.6	52.4
Promedio	72.4	27.6
Natant		
Nymphaea alba	93.6	6.4
Potamogeton linguatus	86.8	13.2
Ludwigia peploides	84.3	15.8
Aponogeton distachyon	83.5	16.5
Promedio	87.0	13.0
Big halophytes		
Phragmites australis	94.3	5.9
Typha angustifolia	93.6	6.4
Scirpus californicus	91.5	8.5
Promedio	93.1	6.9
Sea halophytes		
Sarcocornia fruticosa	85.7	14.4
Selliera radicans	83.5	16.5
Anagallis alternifolia	81.7	18.4
Triglochin concinna	75.1	25.0
Cotula coronopifolia	72.5	27.5
Promedio	79.7	20.4

reaches incredible extremes, such as in the case of the "water lentils" (Lemnaceas family), that consist of a simple cut of few millimeters in size. The simplification is observed in practically all organs and tissues, with the sole exception of the aerenchyma and organs that contain it (submerged stems, natant petioles, respiratory roots, etc.), which present a great development (Arber 1963) (Fig. 4.2).

The submerged leaves of aquatic plants are fine and divided, thus increasing the surface area of exchange of gases, water and minerals with the medium. For many biologists, these split leaves are comparable with the gills of the fish (Willmanns 1993). Due to the fact that water loss is not a limitation, they generally do not have a cuticle (or it is very thin) and the epidermis is transformed into a photosynthetic tissue without stomata. The natants leaves usually present stomata in the bundle and hydropores in the back, which facilitates a fast interchange with the aquatic means.

Since water and minerals can be absorbed throughout the body, the roots are either missing, are scarce or they are only fulfilling an attachment function. The conductive bundles are simple, have little lignin and are arranged centrally, which facilitates the tensile strength exerted by the current, especially in the reophyte that live in lotic environments with current, like *Glyceria multiflora* and *Agrostis stolon-*

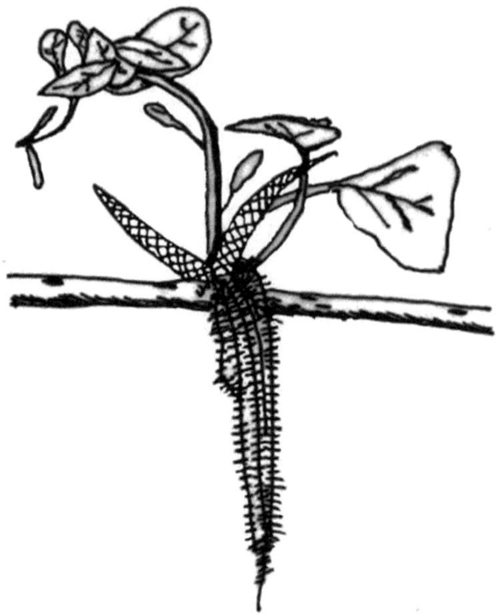

Fig. 4.2 *Ludwigia peploides* respiratory roots (growing upward) and aquatic roots (growing downward)

ifera. Since the mechanical support is provided by the same medium – water-, there is also a reduction of the sclerenchyma. The stems buried in the sediment usually form shoots that act as hibernating organs. In the free floating plants, the roots acquire a great new development, acting as an organ of balance for the stem, and this way, it can float like a buoy.

The interconnected system of intercellular spaces that runs through the entire vegetative body of aquatic plants contributes to flotation on one hand, and on the other it transports air and the oxygen produced in photosynthesis to underground organs. These intercellular spaces are found in special tissues called aerenchyma, which usually occupy more than 50% of the body volume of the plant (Fig. 4.3).

On waterfronts, where it is possible to observe different forms of life, morphologically simpler forms are found in submerged aquatic plants and do become less common as one approaches land vegetation (Fig. 4.4).

Water Wealth

Aquatic plants are hydrolabile, that is, they do not withstand drought, since outside of the water, in the terrestrial-aerial environment they cannot retain the high water content that their body needs, so they die contributing to the organic sediment at the bottom (Keddy 2010). As the bodies of water fill and diminish in their depth, the aquatic plants closest to the riverbanks die from aridity, and only those that colonize

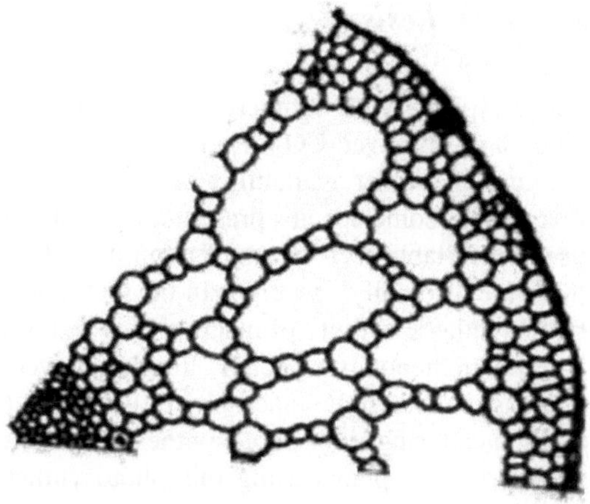

Fig. 4.3 Transversal section of an *Egeria densa* stem. The large intercellular spaces in the aerenchyma can be appreciated

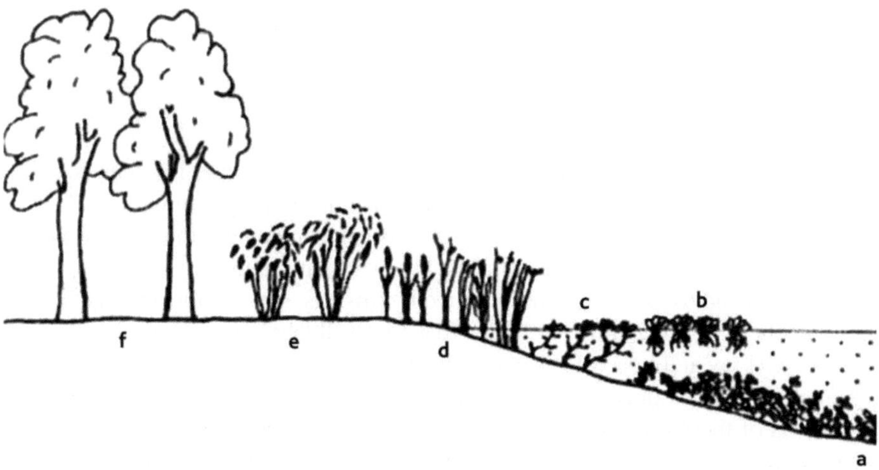

Fig. 4.4 Typical zonation of a coastal gradient in a lentic body, different zones are: *a* = submerged plants, *b* = free floating plants, *c* = natant plants, *d* = swamp with emerged plants, *e* = shrub and *f* = forest (Taken from San Martín 1992)

the deeper areas survive. We might say that they "dig their own grave" by altering the medium, whether it's by trapping the sediment or by necromass accumulation (Yarrow et al. 2009). Sedimentation is often the cause of the disappearance of submerged aquatic plants. This process of embankment is important in the lentic (without current) wetlands, which end up transforming into long-term *terra firme*; while in the lotic wetlands, the erosion dynamics determined by the current prevents this happening.

Photosynthesis

Aquatic plants with submerged green organs obtain CO_2 dissolved in water; the amount of dissolved CO_2 depends on the temperature, which is to say higher temperature and lower availability of CO_2, which happens in spring and summer. For this reason, some species present the C4 photosynthesis system, which allows the use of available CO_2 in small quantities. In addition to the seasonal variations, there are also daily variations in the CO_2 content dissolved in the water, since the photosynthetic activity of macrophytes and microphytes reduces it during the day. This phenomenon is especially limiting in wetlands with low carbonate content. In response to this, some macrophytes use CAM photosynthesis, fastening CO_2 overnight and storing it for photosynthesis. Species of the genus Isoetes represents the most primitive plants using this photosynthetic pathway (Pott and Remy 2000; Deil 2005).

Breathing

Even though air contains about 21% O_2, its dissolution in water is lower, reaching only a maximum of 5%; so the aquatic plants live in submerged environments with little oxygen (hypoxic) and their green organs can supplement the lack of oxygen with what is produced in photosynthesis; but their non-photosynthetic organs, such as rhizomes and roots, need to be supplied by this gas, since in the absence of oxygen, alcoholic fermentation processes that forms alcohol ethanol, toxic to plants, can occur (Frey and Lösch 1998). Associated with this process, there is a system of intercellular spaces in aquatic plants, through which oxygen flows to these organs, by means of thermo-osmotic processes, as the gases contained in the intercellular spaces create a pressure. Even this process of aeration or change of state of the underground organs can be done directly from the air, through dry and broken aerial stems (culms) that act almost like a "snorkel". In herbaceous swamp species, and especially in trees, the oxygen supply is made by specialized organs called pneumatophores that act as respiratory organs, for example in Ludwigia peploides. Due to the above, in the periphery of the roots of aquatic and swamp plants, a reddish oxygenated zone is formed by oxidation of iron compounds present in the sediments (Mitsh and Gosselink 1993). It should be mentioned that the low O_2 content also hinders the germination of seeds, because prior to the formation of photosynthetic organs, the seedlings are heterotrophic organisms. In the case of seasonal wetland species, many of them germinate during the terrestrial phase, but there are also species capable of using anaerobic respiration through alcoholic fermentation to be able to germinate under water.

Nutrient Absorption

Although in submerged aquatic plants the absorption of nutrients is made directly from water through the tissues, in the case of plants with aerial organs, a stream of water that conducts nutrients from the roots to the rest of the body is necessary. This current is created and maintained by hydrates (hydropores) that actively remove water in liquid form (gutting) (Mitsh and Gosselink 1993; Keddy 2010).

Aquatic and Helophytic Flora

Rooted Submerged Hydrophytes

They are herbaceous aquatic plants attached to the sediment that have their submerged photosynthetic organs, which occasionally present aerial flowers protruding from the surface (Ramírez et al. 1982a). There are two forms: one with a long stem that carries the leaves and normally occupies all of it, or a large part of the water column, and another with a short stem with a rosette of basal leaves, located at the bottom of the lagoons. Among the first is *Egeria densa* (Waterweed) and among the second is Limosella australis, a species submerged in rivers and lagoons. The greater availability of nutrients in the water allows these plants to grow significantly, which comes to 1 cm daily for waterweed plants, at the arrival of spring. Normally, pollination is carried out by air (*Myriophyllum aquaticum*) or from time to time, on and by the water surface (*Elodea potamogeton*), and underwater, by very specialized submerged hydrophytes (*Potamogeton pusillus*) (Fig. 4.5).

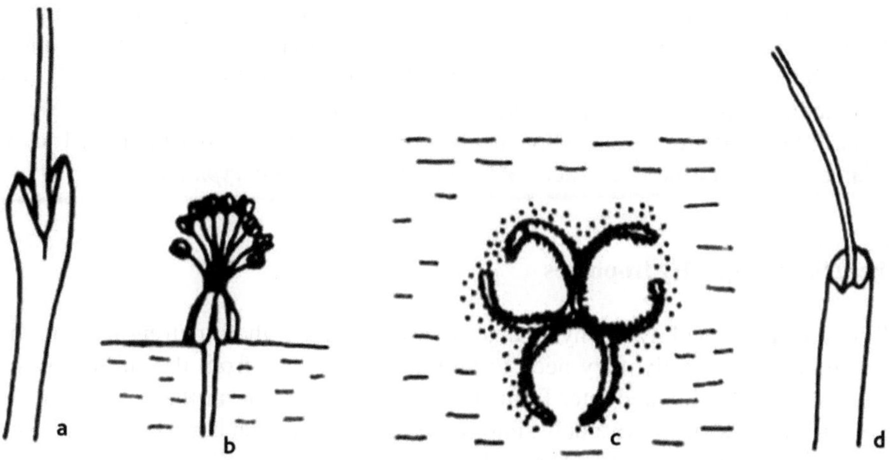

Fig. 4.5 Polinization of *Elodea potamogeton* *a* = male flower spate, *b* = male flower protruding from the water, and thus the wind blows the pollen into the water, *c* = female flower stigmas capturing floating pollen, *d* = female flower spate (Taken from Ramírez et al. 1987)

Table 4.3 Distribution of species according to macrophyte groups

Macrophyte groups	Species	Percentage
Deep submerged hydrophytes	22	9.7
Free submerged hydrophytes	4	1.8
Free floating hydrophytes	6	2.6
Natant hydrophytes	11	4.8
Glycophyte helophytes	112	49.4
Hallophilic helophytes	22	9.7
Herbaceous amphibian helophytes	32	14.1
Woody helophytes	18	7.9
Total	227	100

Submerged and rooted aquatic plants are abundant in freshwater environments, both lotic and lentic, in temperate climates. An important limitation is depth, which decreases the penetration of light, so they do not normally grow under 8 m, where they receive approximately 1% of the light present outside the water. They also grow in shallow brackish environments, while in the tropics they can live in marine environments. In Chile, only *Heterozostera chilensis* is known in this habitat, which has been collected in the Bay of La Herradura in Coquimbo (Phillips et al. 1983). In shallow coastal brackish environments appear *Zannichellia palustris*, *Potamogeton pectinatus*, *Stuckenia striata* and *Ruppia filifolia*. *Potamogeton pectinatus* also grows in sweet, running water.

This group of macrophytes presented 22 species, corresponding to 9.7% of the aquatic and marsh flora of the coastal wetlands (Table 4.3). There is a lack of taxonomic revision of the genus *Potamogeton*, in this group, which would show more species than the accepted ones at the moment. The native species of *Elatine chilensis* and *Senecio zoosteraefolius* that grow in the lower reaches of the rivers of central-southern Chile, present conservation problems due to contamination of their habitats. Egeria densa, an invasive weed of tropical South American origin, gained importance when its populations abruptly disappeared in 2004 from the "Carlos Anwandter" Nature Sanctuary on the Cruces River in Valdivia, River Region (Marín et al. 2009; Ramírez et al. 2006; San Martín et al. 2011), and which used to be source of food for the Black-necked Swan (*Cygnus melancoryphus*).

Free Submerged Hydrophytes

The free submerged hydrophytes float in in the water, without putting roots in the substrate; occasionally, they need to protrude (stand out) from the surface at the time of reproduction and later form shoots, which hibernate in the mud. They are aquatic plants with an extremely simplified body, which often lacks leaves and even corms, and measuring only a few millimeters, such as Wolffiella oblonga.

They grow in freshwater as well as brackish environments, but always in those without currents. They are plants of tropical origin, which is why, in Chile, they

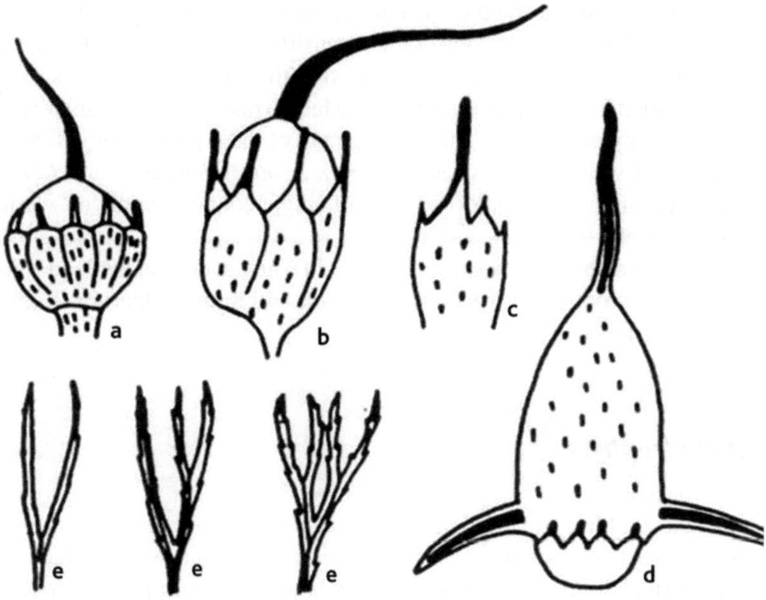

Fig. 4.6 Morphology of *Ceratophyllum chilense* *a* = female flower, *b* = fertilized female flower, *c* = involucral bract, *d* = mature fruit and *e* = different types of submerged leaves (Taken from Ramírez et al. 1987)

only appear in the warmer, central zone. Only four species are known: *Ceratophyllum chilense*, *Utricularia gibba*, *Spirodella punctata* and *Wolffiella oblonga*. *Utricularia gibba* is a carnivorous aquatic plant that captures aquatic protozoa and colonizes lentic environments up until the River (Los Ríos) Region, where it is associated with *Nymphaea alba*.

Ceratophyllum chilensis has been confused with *C. demersum*, of European origin, but its floral and reproductive characteristics make it different (Fig. 4.6).

Free Floating Hydrophytes

The free floating hydrophytes cover the surface of the water, forming dense populations that prevent the development of submerged aquatic plants and phytoplankton by interrupting the passing of light. These plants can be of great size and present a normal corm with leaves in rosette and a great root development, growing submerged (*Eichhornia crassipes*); When they are small, their body is more simplified, with a thalloid form and reduced roots (*Lemna gibba*, *L. minima* and *Azolla filiculoides*). Many of them are invasive weeds that hinder the use of bodies of water and also accelerate evaporation, reducing liquid reserves (*Eichhornia crassipes* and *Limnobium laevigatum*). These aquatic weeds, such as plants from warm climates and places, could be seen as favored by climate change, colonizing bodies of water in temperate zones, forcing us to control their transit/traffic as ornamental material.

The development of these species is limited to the current that carries them. The large ones are abundant in tropical and warm Mediterranean regions, and the smaller ones reach the temperate zones. The development of these populations is favored by eutrophication of water, such as *Lemna gibba*, *Azolla filiculoides* and *Limnobium laevigatum* colonizing bodies of water with high contamination of wastewater (Ramírez and Beck 1981).

This group presents only six species (2.6%) in Chilean coastal wetlands. *Salvinia auriculata*, was introduced as species, and was found in the Valparaiso Region, but has since disappeared.

Natant Hydrophytes

They are plants that live rooted in the substrate and only their leaves stand out, resting on the surface of the water. These natants leaves present a upper surface exposed to air with normal stomas and an underwater back with hydropores. The exposed, surfaced leaves reject the water, facilitating the flotation and also the CO_2 intake by the stomata. The flowers lie on the surface or rise a little above it.

In this group there are two distinguished types: one with short stems (rhizomes) and isomorphic leaves, and the other with long underwater stems and dimorphic leaves, that is to say, submerged and other natants, with different morphology, since the submerged ones have a more significant simplification (Fig. 4.7). The first group (*Nymphaea alba*) normally presents circular foliar leaflets and lives in environments without current, while the second group presents elongated tubular leaf-sheets (*Glyceria multiflora*) thus supporting the stream (reophyte). Some species of long stems that grow horizontally close to the surface usually present respiratory roots (*Ludwigia peploides*).

Alternanthera porrigens can grow in brackish, stream waters with current, in the Region of Valparaiso. *Aponogeton distachyon* is a neophyte of African origin, that has recently appeared in the Rivers Region (Santo Domingo lagoon, south of Valdivia), which has saponin seeds and a dispersal capacity limited by its buoyancy (San Martín et al. 1998) (Fig. 4.8, Table 4.4). *Hydrocotyle ranunculoides* abounds in littoral streams contaminated with wastewater (Steubing et al. 1980).

Eleven species of this group are presented, which contributes with 4.8% of the total aquatic flora treated here. Leaf dimorphism tends to cause confusion in the determination of species, especially in unfavorable seasons, when the leaves disappear, such as the confusion with *Potamogeton lucens* (submerged) and with *P. linguatus* (natant). The *Ranunculus aquatilis* complex should be critically reviewed, since it is a European species, not present in Chile, and because it appears to include several species. For this group there has been a reference in introduced species (*Hydrochleis nymphoides*), which has actually now disappeared from the aquatic environments of the Region of Valparaiso.

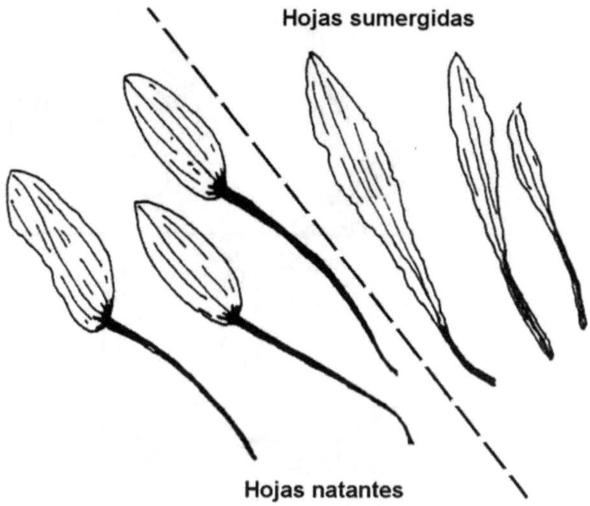

Fig. 4.7 Leaf dimorphism in *Potamogeton linguatus* (Taken from Ramírez and San Martín 2006a)

Herbaceous Glycophyte Helophytes

The helophytes, swamp plants, are grassy herbs that live in freshwater environments, distributing its presence in three different environments: the underwater substrate where the roots and other subterranean organs are; the water, where the bottom of the stem grows, and the air, where it lives and photosynthesizes most of the body of the plant. These helophytes correspond to intermediate forms between proper aquatic plants (hydrophytes) and terrestrial (terriphytess) plants. Because of its systematic origin that conditions its morphology and size, it is possible to distinguish several groups of them: ferns, plants with spores that do not present flowers or seeds *(Blechnum chilense*, *Equisetum giganteum*), erect herbs corresponding to tall but weak herbaceous plants (*Lycopus europaeus*); small creeping grass with rhizomes that grows in the underwater substrate *(Ranunculus repens*, *Hydrocotyle chamaemorus*), large and robust herbs that can present reeds, that is, stems with knots and internodes (*Phragmites australis*, *Arundo donax*); triangular cross-shaped culms sometimes with cutting blade leaves (*Scirpus californicus*, *Cyperus eragrostis*), or a stem and transverse threadlike leaves, commonly called rush (*Juncus procerus J. pallescens*) (Fig. 4.9).

Especially large-sized halophytes present large seasonal phenological changes, when much of their aerial body is dying in the winter, when the water level rises (Fig. 4.10). In these conditions, its yellow color contrasts the green grasslands of the terrestrial environment, but this is reversed in summer, when the prairies are dry and these halophytes are green (Ramírez and Añazco 1982a). They also have large storage of rhizomes buried in the mud, with several buds, which sprout in spring; portions of these rhizomes serve to breed and cultivate these plants (Fig. 4.11).

The number of this group of freshwater swamp plants is the highest in the treated wetlands, with 112 species and 49.4% of the total aquatic and marshy flora of the

Fig. 4.8 *Aponogeton distachyon a* = habit, *b* = forked inflorescence, *c* = stamen, *d* = leaf, *e* = green fruit and *f* = mature fruit (Taken from San Martín et al. 1998)

Table 4.4 Limited floatability of *Aponogeton distachyon* seeds. The seeds that float can actually disperse

Days in water	% Floating seeds	% Sunken seedshundidas
0	100	0
1	95.5	4.5
2	65.7	34.3
3	60.5	39.5
4	44.8	55.2
6	38.6	61.3
8	32.1	67.9
10	26.1	73.9
12	15.7	84.3
14	11.2	88.8
16	5.3	94.7
18	1.5	98.5

Taken from San Martín et al. (1998)

Fig. 4.9 Large glycophyte helophytes: *a* = habit and *b* = inflorescence of *Phragmites australis*, *c* = habit of *Typha angustifolia*, *d* = habit and *e* = inflorescence of *Scirpus californicus*

coastal wetlands. The group of *Pteridophytes* (Ferns, in the broad sense) presents 7 glycophyte helophytes: Blechnum chilense, B. penna-marina, Equisetum bogotense, E. giganteum, Marsillea mollis, Pilularia americana and Schizaea fistulosa. *Marsillea mollis* is a native species restricted to coastal streams of Pichidangui in the Valparaiso Region.

Calystegia sepium is a geophyte climber that ascends on *Phragmites australis* and *Scirpus californicus*. As important is the *Cyperus sculentus* species with small edible rhizomes; Gunnera tinctoria, whose very juicy and acidic petioles are eaten because of the presence of calcium oxalate (Solís 2010); *Mimulus luteus*, a native species that is consumed in salads. The rhizomes of *Scirpus californicus* have been used as a source of starch by the Andean cultures; in the same way, *Juncus procerus* was consumed by the primitive inhabitants of Monte Verde (Ramírez 1989). Finally, in the coastal foothills of the Maule Region there are rice fields (*Oriza sativa*), under the irrigation system; but when irrigation fails, the decrease of the level of flooding allows for the development of a large amount of aquatic weeds and marshes that harms the harvest (San Martín and Ramírez 1983).

Phragmites australis and *Arundo donax* might be confusing, but the first one is native and the second one is introduced and also larger. Both plants are used to build fences and musical instruments in the northern Chile. The former is used extensively to construct artificial "wetlands" for the purpose of purifying sewage with organic pollutants. The *Typha angustifolia* and *Typha dominguensis* are differentiated by the larger size of the first and by the reddish inflorescence of the second. *Typha angus-*

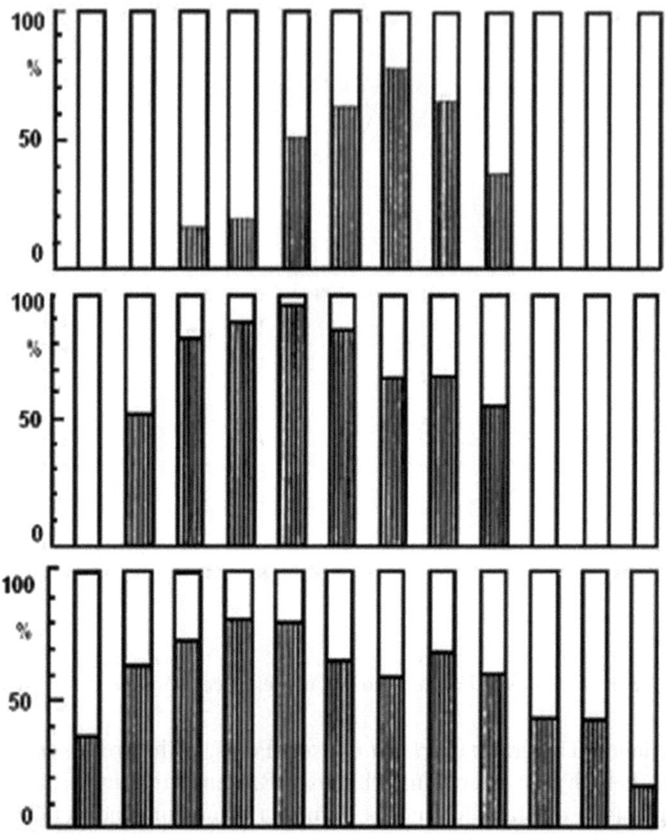

Fig. 4.10 Monthly percentage of live culms (*vertical lines*) and dead (*white*) in *Scirpus californicus* (*bottom*), *Typha angustifolia* (*center*) and *Phragmites australis* (*top*) (Taken from Ramírez and Añazco 1982)

tifolia is a thermophilic species that has increased many of its populations in central-southern Chile. *Crocosmia crocosmiflora, Iris pseudacorus, Lycopus europaeus, Lythrum salicaria and Scrophularia* umbrosa are recently introduced neophytes in central-southern Chile.

Halophile Helophytes

Marsh halophile helophytes colonize brackish swamps that form on the coast, in template regions. These marshes and moors are extreme environments where in the littoral gradient is noted a strong influence of salinity, and of the tides that cause periodic flooding. In addition, in the unfavorable winter season, salinity decreases, while in summer, it increases. These environments are usually exploited as salt

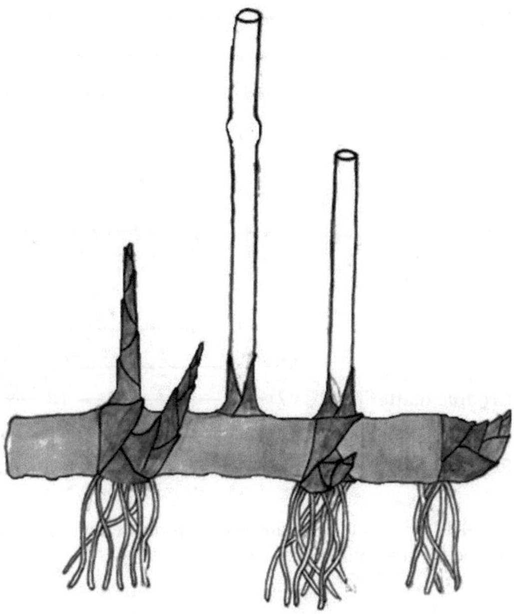

Fig. 4.11 Rhizome with *Phragmites australis* leaf buds and culms

mines or are used in pasture grazing. Both human activities cause a strong alteration of the biotope, given the low capacity of the soil caused by the fluctuating flooding (Silliman et al. 2009; Ramírez et al. 1989b) (Fig. 4.12).

Halophytes live in brackish environments as opposed to glycophytes found in freshwater environments, and the boundary between both is given by a salinity level of 0.5%. The former live in an extreme environment, because the accumulation of salts (NaCl, Na_2SO_4, $CaSO_4$) causes problems in the absorption of water and nutrients, causing a physiological drought with consequent stress for plants. An excess of ions can affect enzymes involved in photosynthesis, such as PEP-carboxylase or Rubisco. The obligated halophytes (*Sarcocornia fruticosa*) have high salt requirements, whereas the optional (*Spartina densiflora*) only tolerate the high salinity, and therefore it is possible to find them only in the upper part of the marshland (Fig. 4.12).

An excess of salt in tissues can be supported through the maintenance of organic chemicals in the vacuole of tissues, such as mannitol or proline. Excess salts can also be diluted by the accumulation of water in specific tissues, such as those associated with succulence (Keddy 2010). Most of the optional halophytes eliminate excess salt either actively, by specialized glands, or by the removal of organs that accumulate it, as in the case of leaves.

This group of halophile helophytes presents 22 species, with 9.7% of the total. Out of these, *Cotula coronopifolia* and *Spartina densiflora* are recent neophytes (introduced in historical times), but not so *Potentilla anserina*, one of the few archaeophytes (introduced in prehistoric times) of European origin that are known in Chile (Fig. 4.13).

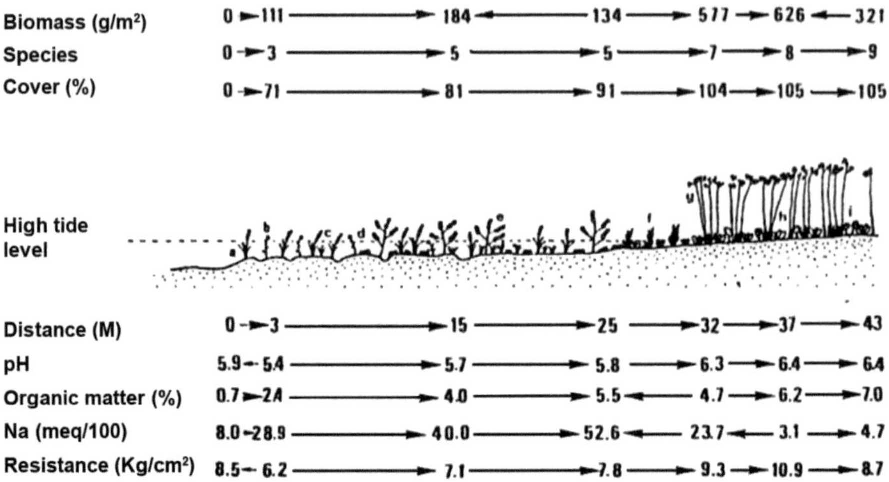

Fig. 4.12 Floristic (*top*) and soil (*bottom*) variables along a marsh gradient in central-south Chile (Taken from Ramírez et al. 1989b)

Fig. 4.13 Habit of *Potentilla anserina*

Fig. 4.14 Amphibian helophytes in seasonal lagoons of Regi6n de Araucanía: *Lythrum hissopifolia* (*left*), *Micropsis nana* (*center and top*) and *Carex fuscula* (*right*)

Amphibious Helophytes

Amphibian helophytes are small herbaceous plants with very short lifespan, since more than 50% of them complete their life cycle between spring and summer; among them are also included some perennial herbs that endure with underground reserve organs in the dry season; and finally, there are also hemicryptophytic perennial herbs that present a drastic reduction in their foliar biomass during unfavorable times. Two typical characteristics of these species are their small size (dwarfism), especially in species that germinate in the aquatic phase (of waterlogging, Fig. 4.14), and morphological flexibility, especially in plants that are established in the aquatic phase and last until the terrestrial phase, presenting changes in their way of life, from an aquatic in spring to a terrestrial in summer, allowing them to be characterized as “amphibians” (Deil 2005; Álvarez 2008; Álvarez et al. 2008; San Martín and Álvarez 2009).

Out of these dwarf amphibians there were 32 species collected, corresponding to 14.1% of the total. These plants grow in temporary natural and artificial lagoons that receive water in winter, present less waterlogging in spring and in summer dry out, going through a typical terrestrial phase. Most of them are native plants, so these environments acquire an enormous importance as a refuge for species with conservation problems (Ramírez et al. 1994).

Woody Helophytes

These are the trees that grow in river marshes of temperate regions that endure seasonal, sometimes prolonged, waterlogging. However, permanent waterlogging destroys them, as did happened with the Temo-Pitra marshy forests on the banks of the Cruces River in Valdivia, which died out after the permanent flood caused by an earthquake in 1960. Deals with native evergreen tree species belonging to the

Myrtaceae family and allochthonous deciduous trees that are members of the Salicaceae and Betulaceae families.

The swamp Myrtaceae forests are known in Chile as "Hualves" or "Pitrantos" and when the rivers bring too much water, they suffer mechanical injuries, including the removal of the soil. An extension of the flood decreases the availability of O_2 for both underground organs and lower parts of the stem, whose cambium usually dies by anoxia and causes the death of branches and individuals. Finally, with periodic flooding, they receive accretion of fine sediment (Yarrow et al. 2009).

The seeds of the Myrtaceae are of fast germination and do so in the dry season on the bare soil (Ramírez et al. 1980). The plant is anchored to the soil with tabular roots and, in addition, have branches with horizontal growth, at a certain height over the water; the submerged branches usually present a great adventitious radical development, which assures the absorption of nutrients and also the oxygen supply to the organs that live in the anoxic sediment.

There were 18 species of woody helophytes, equivalent to 7.9% of the total species. *Lepidothamnus fonckii* is a sub-bush helophyte that grows in sphagnous peatlands. *Blepharocalyx cruckshanksii, Drimys winteri and Myrceugenia exsucca* are obligatory components of the continental swamp forests (Ramírez et al. 1996), whereas *Tepualia stipularis* forms forests on the island of Chiloe, called Tepuales (Fig. 4.15). *Alnus glutinosa* is a European neophyte which arrived to the River Region after the 1960 earthquake and lately has spread, becoming a dangerous invader. *Pluchea absinthioides* is a halophylous shrub that colonizes moorlands located in the lower course of the rivers of the Norte Grande and Norte Chico of Chile, until the Region of Coquimbo. *Salix chilense* is the only native species of this

Fig. 4.15 Branches of Chilean species of the Myrtaceae: *Luma chequen* (*top*), *Myrceugeniaexsucca* (*left*), *Tepualia stipularis* (*center*) and *Blepharocalyxchruckshanksii* (*right*)

genus that colonizes riverbanks from the extreme north of Chile to the Araucania Region (Hauenstein et al. 2005).

Coastal Wetland Vegetation

Aquatic and helophytic plants form the aquatic and marshy vegetation of many watery environments (Hauenstein 2006). The following types of wetlands occur in the northern, central and central-southern coastal areas of Chile: the lower reaches of rivers, streams, lakes, brackish ponds, freshwater lagoons with marshes and wet meadows, coastal marshes, ephemeral lagoons, swamp forests, rice paddies and sphagnous peatlands. All of them present marked latitudinal variations in the presence of macrophytes (Chambers et al. 2008).

The rivers of the Norte Grande and Norte Chico have little flow rate and normally their banks and mouths are covered in marshes with high shrub vegetation. There dominate *Pluchea absinthioides, Arundo donax, Typha angustifolia and Equisetum giganteum*. There are also marshes with herbaceous vegetation of smaller size dominated by *Distichlis spicata, Sarcocornia fruticosa, Atriplex chilensis* and *Scirpus americanus*. In the waters of the Loa river mouth, there are growing submerged hydrophytes of shallow brackish water, such as *Ruppia filifolia, Zannichellia palustris* and *Potamogeton pectinatus*. The marshes of the Elqui and Limari Rivers estuaries are the richest in species.

Tidelands of the Valparaiso Region are flowing with plentiful streams in winter and display high organic pollution (eutrophication), what makes them very rich in species; great examples are: Limache, Viña and Reñaca tidelands, whose vegetation has been studied in detail by Palma *et al.* (1987), Ramírez et al. (1987) and San Martín et al. (2001). However, when in the summer eutrophication increases by the closure of the banks, very few species survive, including *Limnobium laevigatum,* capable of withstanding high contamination (San Martín et al. 2003).

In the central zone of Chile, starting at the O'Higgins Region, there are abundant lagoons, lakes and brackish lagoons, with abundant submerged, free submerged, natant, free floating and emerged vegetation. These are the most species-rich wetlands in Chile, including the Torca and Putu lagoons and Vichuquen and Budi lakes (Ramírez and San Martín 1984; Ramírez et al. 2004 and Hauenstein et al. 1992). Some of them correspond to settlements at the river mouths. They are influenced by seawater, but also by sewage pollution, which increases eutrophication, causing an exaggerated growth of macrophytes, which represent a serious problem for recreational use (Ramírez et al. 2004). There are interior freshwater lagoons with no marine influence, but also with great development of macrophytes, such as Cienaga del Name and the Aculeo and Avendaño lagoons, the latter in Quillon (Rodríguez and Dellarossa 1998). Further south from this same group there are the lakes Lanalhue and Lleulleu.

There are also abundant brackish wetlands or marshes, which are often used as salt flats. In those dominate halophytes such as: *Sarcocornia fruticosa, Atriplex chilense and Triglochin concinna.* On the basis of the information indicated above it is possible to conclude that studies of this type of vegetation are necessary to plan its conservation.

In south-central Chile, aquatic plants develop in the lower reaches of the rivers, in the so-called Potamon (Campos 1979; San Martín et al. 1998), forming extensive marshy areas such as the Carlos Anwandter Cruces River Nature Sanctuary or the Santo Domingo lagoon (Ramírez et al. 1989a, 1991; San Martín et al. 1999, 1993), in the River Region. A great development here shows that the marshes formed by *Sarcocornia fruticosa, Cotula coronopifolia, Selliera radicans and Puccinellia glaucescens* are dominant. Several of these have been colonized by *Spartina densiflora*. These marshes are very abundant on the island of Chiloe, but their state of conservation is precarious.

Throughout the coastal zone, between the regions of Valparaiso and Lakes, there are ephemeral lagoons or ponds that form in depressions where water accumulates in winter and spring, and dries out in summer. Amphibian plants are abundant in the aquatic spring phase, while they behave like terrestrial plants in summer (Deil 2005). These lagoons are places of refuge for a very special native flora; however, there is still a need to raise awareness of this fact, both at the public and private levels, in order to avoid its disappearance, a danger that is raised by drastic changes that causes land use (changes from grazing to crops, fruit or forest plantations, implementation of irrigation and drainage, etc.) that are taking place in this area.

Coastal swamps forests occur throughout central and south-central Chile, colonizing depressions of coastal terraces and valleys with prolonged flooding.

In the summer, the phreatic levels decrease, without an edaphic drought (Ramírez et al. 1983). They present a very clear latitudinal differentiation, so that their floristic composition changes from north to south, first in the northern forests of Pitra-Chequen, then to the south towards the forest of Temo-Pitra and then the swamp forest of Tepu, which has its maximum expansion on the island of Chiloe, in the so-called Tepuales. On the coast of Valdivia there is a small area called the Chin-Chin-Canelo forest (Ramírez and San Martín 2005; San Martín et al. 2008).

The swamp Myrtaceae forests are evergreen, and have a well-developed tree layer, with a closed canopy that prevents the passage of light, so there are no lower strata, which is also conditioned by prolonged waterlogging. In addition, both extreme conditions restrict the biological activity in the soil. When these forests are destroyed, Junquillo meadows are formed, which are often used for grazing (Hauenstein et al. 2002). They serve as habitat for the Huillin or River Otter (Lontra provocax), a native carnivorous mammal in danger of extinction, and Vega shrimp (Parastacus nicoletti), which feeds on roots (Muñoz 2003).

Distribution of Flora Through Wetland Types

Table 4.5 shows the distribution of the 227 species of macrophytes present in different coastal wetlands studied. Wetlands with greater species are streams and lagoons, with 105 and 103 species each (Table 4.6). Seasonal rivers and lagoons have 49 and

Table 4.5 Presence of chilean macrophytes in coastal wetlands

	Wetland type									
Specie	R	A	Al	l	lt	M	bp	Ar	te	Total
Deep submerged hydrophytes (22 especies)										
Callitriche lechleri	X	X		X	X					4
Calittriche palustris		X		X	X					3
Crassula peduncularis	X				X					2
Egeria densa	X	X		X						3
Elatine chilensis				X	X					2
Elodea canadensis		X		X						2
Elodea potamogeton		X	X	X						3
Juncus bulbosus		X		X						2
Lilaeopsis macloviana		X		X	X					3
Limosella australis		X		X						2
Myriophyllum aquaticum	X	X		X						3
Myriophyllum quitense	X	X	X	X						4
Potamogeton berteroanus	X	X	X	X						4
Potamogeton gayi		X								1
Potamogeton lucens		X								1
Potamogeton pectinatus	X	X	X							3
Potamogeton pusillus				X	X					2
Potamogeton striatus	X									1
Ruppia filifolia	X	X								2
Senecio zoosteraefolius	X									1
Verónica anagallis-aquatica	X	X		X						3
Zannichellia palustris	X	X								2
Hidrófitos sumergidos libres (4 especies)										
Ceratophyllum chilense			X	X						2
Utricularia gibba				X						1
Spirodela punctata			X							1
Wolffiella oblonga			X							1
Free natant hydrophytes (6 species)										
Azolla filiculoides		X	X	X	X					4
Eichhornia crassipes		X	X	X						3
Lemna gibba		X	X	X						3
Lemna minuscula		X		X						2
Lemna valdiviana		X		X						2
Limnobium laevigatum	X	X		X						3
Natant hydrophytes (12 species)										
Agrostis stolonifera	X	X		X						3
Alternanthera porrigens		X	X	X						3
Aponogeton distachyos				X						1
Callitriche autumnalis		X		X						2
Callitriche verna				X						1

(continued)

Table 4.5 (continued)

	Wetland type									
Specie	R	A	Al	l	lt	M	bp	Ar	te	Total
Glyceria multiflora	X	X								2
Hydrocotyle ranunculoides		X								1
Ludwigia peploides		X	X	X						3
Ludwigia octovalvis		X	X	X						3
Nymphaea alba	X			X						2
Potamogeton linguatus	X	X		X						3
Ranunculus aquatilis				X	X					2
Glycophyte helophytes (112 especies)										
Alisma lanceolatum	X	X	X							3
Alisma plantago-aquatica	X	X	X				X			4
Anmannia coccinea							X			1
Anagallis alternifolia			X	X						2
Apium nodiflorum		X								1
Arundo donax	X	X								2
Aster vahlii	X	X	X		X					4
Baccharis sagittalis	X	X	X							3
Bidens aurea	X	X								2
Bidens laevis	X	X								2
Blechnum chilense	X	X	X			X				4
Blechnum penna-marina									X	1
Calystegia sepium			X							1
Cardamine bonariensis		X								1
Cardamine alsophila		X								1
Cardamine nasturtioides		X								1
Carex brongniartii				X						1
Carex canescens				X						1
Carex fuscula					X					1
Carex magellanica									X	1
Carex pseudocyperus		X		X						2
Carex riparia				X						1
Carpha alpina									X	1
Centella asiatica				X						1
Centella triflora				X						1
Centipeda elatinoides				X						1
Chenopodium ambrosioides				X						1
Crocosmia crocosmiflora	X	X		X						3
Cyperus conceptionis		X								1
Cyperus cyperoides		X								1
Cyperus eragrostis		X		X			X			3
Cyperus rotundus				X				X		2
Cyperus sculentus								X		1

(continued)

Table 4.5 (continued)

	Wetland type									
Specie	R	A	Al	l	lt	M	bp	Ar	te	Total
Drosera uniflora									X	1
Echinochloa colona		X						X		2
Echinochloa crusgalli		X						X		2
Eleocharis bonariensis				X						1
Eleocharis macrostachya				X						1
Eleocharis melanostachya				X						1
Epilobium australe		X		X						2
Epilobium puberulum		X		X						2
Equisetum bogotense	X	X								2
Equisetum giganteum	X	X								2
Galega officinalis		X		X						2
Galium leptum				X						1
Gentiana lactea									X	1
Gunnera tinctoria	X	X		X						3
Habenaria paucifolia				X						1
Hedyotis salzmannii				X						1
Holcus lanatus		X		X						2
Hydrocotyle chamaemorus		X		X		X				3
Hypolepis rugosula						X				1
Iris pseudacorus		X								1
Juncus cyperoides				X						1
Juncus pallescens	X	X		X						3
Juncus lesueurii				X						1
Juncus microcephalus	X	X		X						3
Juncus procerus	X	X		X		X				4
Leptocarpus chilensis	X			X						2
Lilaea scilloides		X		X						2
Marsippospermum philippi									X	1
Lindernia anagallidea								X		1
Littorella australis				X						1
Lotus uliginosus		X		X						2
Lycopus europaeus		X		X						2
Lythrum album		X				X				2
Lythrum salicaria	X	X		X						3
Marsillea mollis					X					1
Mentha aquatica		X								1
Mentha rotundifolia		X		X						2
Mimulus bridgessii		X								1
Mimulus glabratus		X								1
Mimulus luteus		X								1
Myosotis palustris		X		X			X			3

(continued)

Table 4.5 (continued)

	Wetland type									
Specie	R	A	Al	l	lt	M	bp	Ar	te	Total
Nasturtium bonariensis		X								1
Oriza sativa								X		1
Perezia palustris									X	1
Phragmites australis	X	X		X						3
Pilularia americana				X	X					2
Polygonum hydropiperoides		X		X	X					3
Polygonum lapathifolium				X						1
Polypogon australis				X		X				2
Polypogon monspelliesis		X						X		2
Ranunculus bonariensis		X			X					2
Ranunculus chilensis				X						1
Ranunculus cymbalaria		X								1
Ranunculus flagelliformis		X			X					2
Ranunculus minutiflorus		X					X			2
Ranunculus monanthus						X				1
Ranunculus muricatus		X								1
Ranunculus peduncularis		X					X			2
Ranunculus repens		X		X			X			3
Rorippa nasturtium-aquaticum		X								1
Rorippa palustris		X		X	X					3
Rumex conglomeratus		X		X						2
Rumex crispus	X	X		X						3
Rumex romasa	X	X		X						3
Sagittaria montevidensis	X			X			X			3
Scirpus californicus	X	X		X		X				4
Scirpus inundatus		X								1
Scrophularia umbrosa		X								1
Schizaea fistulosa									X	1
Senecio acanthifolius									X	1
Senecio aquaticus				X						1
Senecio fistulosus	X	X								2
Sisyrinchium junceum				X	X					2
Tetroncion magellanicum									X	1
Triglochin palustris						X				1
Triglochin striata	X	X								2
Typha angustifolia	X	X		X						3
Typha dominguensis	X			X						2
Verbena corymbosa		X		X						2
Halophile helophytes (22 especies)										
Alternanthera halimifolia						X				1
Apium australe						X				1

(continued)

Table 4.5 (continued)

	Wetland type									
Specie	R	A	Al	l	lt	M	bp	Ar	te	Total
Atriplex chilense						X				1
Cotula coronopifolia						X				1
Distichlis spicata	X					X				2
Juncus arcticus						X				1
Lobelia alata		X				X				2
Lythrum maritimum						X				1
Paspalum distichum						X				1
Potentilla anserina						X				1
Puccinellia glaucescens						X				1
Ranunculus apiifolius						X				1
Ranunculus uniflorus						X				1
Rumex cuneifolius						X				1
Salsola kali						X				1
Samolus repens						X				1
Sarcocornia fruticosa						X				1
Scirpus americanus						X				1
Selliera radicans						X				1
Spartina densiflora						X				1
Spergularia rubra						X				1
Triglochin concinna						X				1
Amphibian helophytes (32 especies)										
Anagallis minima					X					1
Blennosperma chilensis					X					1
Callitriche terrestris					X					1
Cardamine valdiviana					X					1
Crassula paludosa					X					1
Cyperus reflexus					X					1
Eleocharis pachycarpa				X	X					2
Eryngium humifusum					X					1
Eryngium pseudojunceum					X					1
Glynus radiatus					X					1
Gnaphalium cymatioides					X					1
Gratiola peruviana				X	X					2
Hydrocotyle cryptocarpa					X					1
Juncus bufonius					X					1
Juncus capillaceus					X					1
Juncus planifolius		X			X					2
Juncus stipulatus					X					1
Juncus tenuis					X					1
Lasthenia kunthii					X					1
Legenere valdiviana					X					1

(continued)

Table 4.5 (continued)

	Wetland type									
Specie	R	A	Al	l	lt	M	bp	Ar	te	Total
Lythrum hyssopifolia					X					1
Lythrum portula					X					1
Mentha pulegium				X	X					2
Micropsis nana					X					1
Montia fontana					X					1
Navarretia involucrata					X					1
Nierembergia repens					X					1
Nothoscordum striatellum		X		X	X					3
Phyla nodiflora		X		X	X					3
Plagiobothrys fulvus					X					1
Plagiobothrys pratensis					X					1
Scirpus cernuus					X					1
Woody helophytes (19)										
Alnus glutinosa	X	X		X						3
Blepharocalyx cruckschanksii							X			1
Lepidothamnus fonckii									X	1
Crinodendron patagua							X			1
Drimys winteri							X			1
Escallonia resoluta							X			1
Escallonia virgata							X			2
Luma chequen							X			1
Luma gayana							X			1
Myrceugenia exsucca							X			1
Myrceugenia parvifolia							X			1
Pluchea absinthioides	X					X				2
Salix babilonica	X	X		X						3
Salix caprea				X						1
Salix humboldtiana	X	X								2
Salix viminalis	X	X								2
Sambucus nigra		X		X						2
Tepualia stipularis							X			1
Total species per wetlandl	49	105	13	103	49	29	19	10	11	388/299

Abbreviations of the types of wetlands are shown in Table 4.6

47 species, that is, about half of the previous ones. The rest of them have few species, except the marshes, which have 29 species, which is equivalent to 7.51% of the total. These proportions are repeated when analyzing the aquatic bodies of other regions of Chile, as indicated by San Martín et al. (2010).

There are species that are repeated and common in different wetlands, but the maximum frequency found only corresponds to 4, which means less than half of the wetlands present, what indicates a great diversity of habitats in different environ-

Table 4.6 Species richness of macrophytes in each type of Chilean coastal wetland

Wetland type	Abbreviation	Species	Percentage
Rivers	R	49	12.7
Streams	A	105	27.2
Freshwater lagoons	Al	13	3.4
Lagoons	L	103	26.7
Temporary lagoons	Lt	47	12.2
Marshes	M	29	7.5
Swamp forests	Bp	19	4.9
Rice fields	Ar	10	2.6
Sphagnous swamps	Te	11	2.9

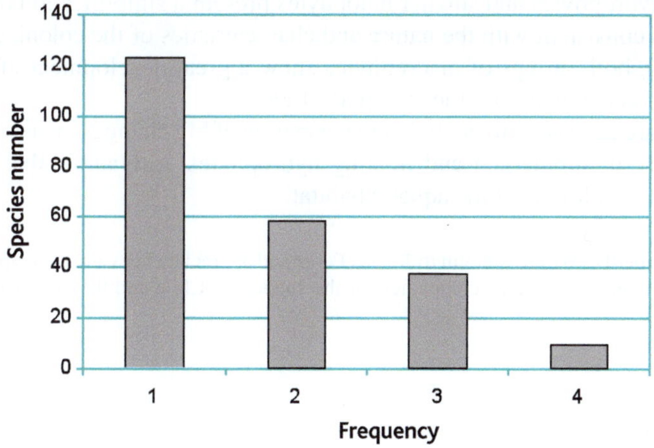

Fig. 4.16 Frequency of macrophyte species in Chilean coastal wetlands

ments, but also low floristic affinity. Frequency 3 further increases the number of species in a gradual way (Fig. 4.16).

Discussion and Conclusions

In this chapter, Chilean coastal wetlands have been defined by characterizing their flora. All the factors that limit the distribution of aquatic macrophytes, and the morphological and physiological modifications that allow the survival of the macrophytes are presented. Also, the floristic richness of each of the life forms present in the coastal wetlands of Chile is analyzed in detail.

- The coastal wetlands of the Chilean Pacific coast present a great diversity of environments, conditioned by natural and artificial causes.

- The flora (floristic) richness of these wetlands is high, even though taxonomic revisions are needed that will surely broaden the flora spectrum. These wetlands serve as a refuge for a native aquatic flora and marshes, whose habitats are being reduced dangerously.
- The flora of vascular macrophytes, despite being of the azonal type, presents differentiation in the latitudinal gradient, which has to do with the temperature, but mainly with the physical conditions of the water and sediments, as well as with eutrophication of anthropic origin. These same conditions allow for the development and expansion of certain forms of life in them.
- Morphological differentiation in life forms is greater in hydrophytes than in halophytes, which corresponds to the greater differentiation of habitats in the lower portion of the coastal flooding gradient.
- Larger hydrophytes and smaller helophytes present a simplification of the corm, which is consistent with the nature and characteristics of the colonized wetland. However, both groups of macrophytes show a great development of the aerenchyma, tissue that allows the transport of gases.
- Studies are necessary to get to know the causes of big changes in population size, especially in submerged and floating hydrophytes, a dynamic that drastically alters the conditions of the aquatic habitat.

Acknowledgments We are grateful to Jessica Perez and appreciate her valuable help in the preparation of the figures, to Arauco companies for the support in Chile and the Schimper-Stiftung in Germany.

References

Álvarez, M. 2008. Diasporenbank und Ökologie der Vegetation temporärer Gewässer in Cholchol-Gebiet (9. Región, Chile). Dissertationes Botanicae 407:1–87.

Álvarez, M., C. Ramírez, and U. Deil. 2008. Ecología y distribución de *Hydrocotyle cryptocarpa* Speg. En Sudamérica. *Gayana Botánica* 65 (2): 139–144.

Araya, B., and G. Millie. 1989. *Guía de campo de las aves de Chile*, 405 pp. Santiago: Editorial universitaria.

Arber, A. 1963. *Water plants: A study of aquatic angiosperms*, 436 pp. Cambridge: Cambridge University Press.

Campos, H. 1979. El recurso de agua dulce en Chile. *Boletín Informativo Limnológico* 3: 6–17.

Chambers, P.A., P. Lacoul, K.J. Murphy, and S.M. Thomaz. 2008. Global diversity of aquatic macrophytes in freshwater. *Hydrobiologia* 595: 9–26.

Cook, C., B. Gut, E.M. Rix, E. Schneller, and M. Seita. 1974. *Water plants of the world – A manual for the identification of the genera of freshwater macrophytes*, 425 pp. La Haya: Dr. W. Junk.

Davis, J., D. Blascoym, and M. Carbonell. 1996. *Manualdela Convención Ramsar– Unaguía A la convención sobre los humedales de importancia internacional*, 211 pp. Gland, Switzerland

Deil, U. 2005. A review on habitats, plant traits and vegetation of ephemeral wetlands – A global perspective. *Phytocoenologia* 35: 533–705.

Figuerola, J., and A.J. Green. 2002. Dispersal of aquatic organisms by waterbirds: A review of past research and priorities for future studies. *Freshwater Biology* 47: 483–494.

Frey, W., and R. Lösch. 1998. *Lehrbuch der Geobotanik – Pflanze und Vegetation in Raum und Zeit*, 436 pp. Stuttgart: Gustav Fischer.

Grace, J.B. 1993. The adaptive significance of clonal reproduction in angiosperms: An aquatic perspective. *Aquatic Botany* 44: 159–180.

Hauenstein, E. 2006. Visión sinóptica de los macrófitos dulceacuícolas de Chile. *Gayana Botánica* 70 (1): 16–23.

Hauenstein, E., C. Ramírez, M. González, and C. San Martín. 1992. Comparación de la flora macrofítica de tres lagos del Centro-Sur de Chile (Budi, Llanquihue y Cayutue). *Revista Geográfica de Valparaíso* 22 (23): 175–193.

Hauenstein, E., C. Ramírez, M. González, L. Leiva, and C. San Martín. 1996. Flora hidrófila del lago Villarrica (IX Región, Chile) y su importancia como elemento indicador de contaminación. *Medio Ambiente* 13 (1): 88–96.

Hauenstein, E., M. González, F. Peñaya, and A. Muñoz-pedreros. 2002. Clasificación y caracterización de la flora y vegetación de los humedales de la costa de Toltén (IX Región, Chile). *Gayana Botánica* 59 (2): 87–100.

Hauenstein, E., F. Peña-Cortés, M. Gonzálezyr, and R. Schlatter. 2005. Nuevos límites para la distribución de *Salix humboldtiana* Willd. Salicaceae, en Chile. *Gayana Botánica* 62 (1): 44–46.

Hutchinson, G.E. 1975. *A treatise on Limnology*. Vol. 3, 660 pp. New York: Limnological botany.

Keddy, P.A. 2010. *Wetland ecology: Principles and conservation*, 478 pp. Cambridge, MA: Cambridge University Press.

Marín, V., A. Tironi, L.E. Delgado, M. Contreras, F. Novoa, M. Torres-gómez, R. Garreaud, I. Vila, and I. Serey. 2009. On the suden disappearance of *Egeria densa* from a Ramsar wetland site of southern Chile: A climatic event trigger model. *Ecological Modelling* 220: 1752–1763.

Matthei, O. 1995. *Manual de las malezas que crecen en Chile*, 545 pp. Concepción: Alfabeta Impresores.

Meerhof, M., and N. Mazzeo. 2004. Importancia de las plantas flotantes libres de gran porte en la conservación y rehabilitación de lagos someros de Sudamérica. *Ecosistemas* 13: 13–22.

Mitsh, W.J., and J.G. Gosselink. 1993. *Wetlands*, 722 pp. New York: Van Nostrand.

Mitsh, W.J., J.G. Gosselink, L. Zhan, and C.J. Anderson. 2009. *Wetland ecosystems*, 290 pp. Hoboken: Wiley.

Muñoz, A. 2003. *Guía de los Humedales del Río Cruces*, 143 pp. Valdivia: CEA ediciones.

Palma, B., C. San Martín, M. Rosales, L. Zúñiga, and C. Ramírez. 1987. Distribución espacial de la flora y la vegetación acuática y palustre del estero Marga-Marga en Chile Central. *Anales del Instituto de Ciencias del Mar y Limnología de la Universidad Nacional Autónoma de México* 14 (2): 125–132.

Parra, O., C. Valdovinos, S. Basualto, and R. Urrutia. 2005. Diversidad fitoplanctónica de los lagos nahuelbutanos (Chile Centra). In *Historia, biodiversidad y ecología de los bosques costeros de Chile*, ed. C. Smith-Ramírez, J. Armesto, and C. Valdovinos, 146–157. Santiago: Editorial Universitaria.

Phillips, H.C., B. Santelices, R. Bravoy, and C.P. Macroy. 1983. Heterozostera tasmanica (Martens ex Ascher). Den Hartog in Chile. *Aquatic Botany* 15: 195–200.

Pott, R., and D. Remy. 2000. *Gewässer des Binnenlandes*, 225 pp. Stuttgart: Eugen Ulmer.

Quiroz, C., A. Pauchard, A. Marticorena, and L. Cavieres. 2009. *Manual de plantas invasoras del centro-sur de Chile*, 45 pp. Concepción: Universidad de Concepción.

Ramírez, C. 1989. The macrobotanical remains. In *Monte Verde: A late pleistocene settlement in Chile. Vol. I. Paleoenvironment and site context*, ed. T. Dillehay, 147–170. Washington/London: Smithsonian Institution Press.

Ramírez, C., and N. Añazco. 1982a. Variaciones estacionales en el desarrollo de *Scirpus californicus*, *Typha angustifolia* y *Phragmites communis* en pantanos valdivianos, Chile. *Agro Sur* 10 (2): 111–123.

Ramírez, C., and S. Beck. 1981. Makrophytische Vegetation und Flora in Gewässern der Umgebung von La Paz, Bolivien. *Archiv für Hydrobiologie* 91 (1): 82–100.

Ramírez, C., and M. Romero. 1978. El Pacífico como agente diseminante en el litoral chileno. *Ecología (Argentina)* 3 (1): 19–30.
Ramírez, C., and J. San Martín. 1984. Hydrophylous vegetation of a coastal lagoon in Central Chile. *International Journal Ecological Environmental Science* 10: 93–100.
Ramírez, C., and C. San Martín. 2005. Asociaciones vegetales de la Cordillera de la Costa de la Región de Los Lagos. In *Historia, biodiversidad y ecología de los bosques costeros de Chile*, ed. C. Smith-Ramírez, J. Armesto, and J. Valdovinos, 206–224. Santiago: Editorial Universitaria.
Ramírez, C and C. San Martín. 2006a. Diversidad de macrófitos chilenos. In *Macrófitas y vertebrados de los sistemas límnicos de Chile*, ed. I. Vila, A. Veloso, R. Schlatter, and C. Ramírez, 21–61. Santiago: Editorial Universitaria.
Ramírez, C. and C. San Martín. 2006b. Ecosistemas dulceacuícolas. In *Biodiversidad de Chile. Patrimonio y Desafíos*, ed. P. Saball, M. Arroyo, J.C. Castilla, C. Estades, J.M. Ladrón de Guevara, S. Larraín, C. Moreno, F. Rivas, J. Rovira, A. Sánchez, and L. Sierralta, 112–124. Santiago: Comisión Nacional del Medio Ambiente.
Ramírez, C., and E. Stegmaier. 1982. Formas de vida en hidrófitos chilenos. *Medio Ambiente* 6 (1): 43–54.
Ramírez, C., M. Romeroy, and M. Riveros. 1979. Habit, habitat, origin and geographical distribution of Chilean vascular hydrophytes. *Aquatic Botany* 7: 241–253.
Ramírez, C., M. Romeroyo, and O. Henríquez. 1980. Estudios de germinaciónensemillas de Mirtáceas chilenas. *Bosque* 3 (2): 106–114.
Ramírez, C., R. Godoy, and E. Hauenstein. 1982a. Las especies de "Luchecillos" (*Hydrocharitaceae*) que prosperan en Chile. *Anales del Museo de Historia Natural de Valparaíso* 14: 47–55.
Ramírez, C., R. Godoy, D. Contreras, and E. Stegmaier. 1982b. *Guía de plantas acuáticas y palustres Valdivianas*, 52 pp. Valdivia: Instituto de Botánica, Universidad Austral de Chile.
Ramírez, C., F. Ferriereyh, and H. Figueroa. 1983. Estudio fitosociológico de los bosques pantanosos templados del Sur de Chile. *Revista Chilena de Historia Natural* 56 (1): 11–26.
Ramírez, C., J. San Martín, C. San Martín, and D. Contreras. 1987. Estudio florístico Y vegetacional de la laguna El Peral, Quinta Región de Chile. *Revista Geográfica de Valparaíso* 18: 105–120.
Ramírez, C., H. Figueroa, E. Hauenstein, and D. Contreras. 1989a. Distribution of benthic flora in the lower course of the Valdivia River, Chile. *Estuaries* 12 (2): 111–118.
Ramírez, C., C. San Martín, D. Contreras, and J. San Martín. 1989b. Flora de las marismas del centro-sur de Chile. *Medio Ambiente* 10 (2): 11–24.
Ramírez, C., C. San Martín, R. Medina, and D. Contreras. 1991. Estudio de la flora hidrófila del Santuario de la Naturaleza "Río Cruces" (Valdivia, Chile). *Gayana Botanica* 48 (1/4): 67–80.
Ramírez, C., C. San Martín, D. Contreras, and J. San Martín. 1994. Estudio fitosociológico de la vegetación pratense del valle del río Chol-Chol (Cautín, Chile). *Agro Sur* 22 (1): 41–56.
Ramírez, C., C. San Martín, and J. San Martín. 1996. Estructura florística de los bosques pantanosos de Chile central. In *Ecología del bosque nativo de Chile*, ed. J. Armesto, M.T. Kalin-Arroyo, and C. Villagrán, 215–234. Santiago: Editorial Universitaria.
Ramírez, C., C. San Martín, and H. Rubilar. 2002. Una propuesta para la clasificación de los humedales chilenos. *Revista Geográfica de Valparaíso* 32 (33): 265–273.
Ramírez, C., C. San Martín, and J. San Martín. 2004. Colmatación por macrófitos del complejo lacustre Vichuquén (VII Región, Chile) y clave de determinación. *Revista Geográfica de Chile Terra Australis* 49: 179–196.
Ramírez, C., E. Carrasco, S. Mariani, and N. Pacheco. 2006. La desaparición del Luchecillo (*Egeria densa*) del Santuario del Río Cruces (Valdivia, Chile): Una hipótesis plausible. *Ciencia & Trabajo* 20: 79–86.
Rodríguez, R., and V. Dellarossa. 1998. *Plantas vasculares acuáticas en la Región del Bio-Bío*, 38 pp. Concepción: Universidad de Concepción.
San Martín, C. 1992. *Flora, vegetación y dinámica vegetacional de la laguna Santo Domingo (Valdivia, Chile)*, 192 pp. Valdivia: Tesis, Escuela de Graduados, Universidad Austral de Chile,

San Martín, C., and M. Álvarez. 2009. Floristic composition of anthropogenic seasonal wetlands in the coastal mountain range of Cautín, Chile. *Agro Sur* 37 (1): 9–25.

San Martín, J., and C. Ramírez. 1983. Flora de malezas en arrozales de Chile Central. *Ciencia e Investigación Agraria* 10 (3): 207–222.

San Martín, C., R. Medina, P. Ojeda, and C. Ramírez. 1993. La biodiversidad vegetacional del santuario de la naturaleza "Río Cruces" (Valdivia, Chile). *Acta Botanica Malacitana* 18: 259–279.

San Martín, C., J. Barrera, and C. Ramírez. 1998. Dispersal mechanism and germinationof *Aponogetondistachyon* L. F. In Valdivia, Chile. *Phyton* 63 (1/2): 31–38.

San Martín, C., C. Ramírez, and P. Ojeda. 1999. Distribución de macrófitos y patrones de zonación ribereña en la cuenca del río Valdivia, Chile. *Revista Geográfica de Valparaíso* 30: 117–126.

San Martín, C., C. Ramírez, J. San Martín, and R. Villaseñor. 2001. Flora y vegetación del estero Reñaca (V Región, Chile). *Gayana Botánica* 58 (1): 31–46.

San Martín, C., C. Ramírez, and M. Álvarez. 2003. Macrófitos como bioindicadores: Una propuesta metodológica para caracterizar ambientes dulceacuícolas. *Revista Geográfica de Valparaíso* 34: 243–253.

San Martín, C., C. Ramírez, and J. San Martín. 2008. Distribución geográfica de los bosques pantanosos de Mirtáceas en Chile. *Revista Geográfica de Chile Terra Australis* 51 (52): 49–64.

San Martín, C., Y. Pérez, D. Montenegro, and M. Álvarez. 2010. Diversidad, hábito y hábitat de macrófitos acuáticos en la Patagonia occidental (Región de Aysén, Chile). *Anales Instituto Patagonia (Chile)* 39 (1): 23–41.

San Martín, C., C. Ramírez, E. Carrasco, O. Vidal, and G. Toledo. 2011. La hipótesis natural del aumento de radiación UV en la desaparición de *Egeria densa* (Hydrocharitaceae) del humedal del río Cruces (Valdivia, Chile). *Agro Ciencia. Rev. Chil. Cs Agropec.* 26 (1): 15–24.

Scheffer, M., S.H. Hosper, M.L. Meijer, B. Moss, and E. Jeppesen. 1993. Alternative equilibria in schallow lakes. *Trends in Ecology and Evolution* 8: 275–279.

Silliman, B.R., E.D. Grosholz, and M.D. Bertness. 2009. *Human impacts of Saltmarshes*, 412 pp. Berkeley: University of California Press.

Solís, JL. 2010. *Biomineralización en plantas: Análisis cuantitativo y cualitativo de los cristales de oxalato de calcio en Gunneráceas chilenas (Rosopsida)*, 82 pp. Valdivia: Tesis, Escuela de Ciencias, Universidad Austral de Chile

Steubing, L., C. Ramírezy, and M. Alberdi. 1980. Energy content of water- and bog-plant associations in the region of Valdivia (Chile). *Vegetatio* 43: 153–161.

Walter, H. 1997. *Zonas de vegetación y clima*, 245 pp. Barcelona: Edic. Omega A.A.

Willmanns, O. 1993. *Ökologische Pflanzensoziologie*, 479 pp. Heidelberg/Wiesbaden: Quelle & Meyer.

Yarrow, M., V. Marín, M. Finlayson, A. Tironi, L. Delgadoy, and F. Fischer. 2009. The ecology of *Egeria densa* (Liliopsida: Alismatales): A wetland ecosystem engineer? *Revista Chilena de Historia Natural* 82: 299–313.

Chapter 5
Coastal Wetlands of Northern Chile

Walter Sielfeld, Ronny Peredo, Rosa Fuentes, Vinko Malinarich, and Flavio Olivares

Abstract The coastal wetlands of the northern zone, being the focus of water presence in an extremely arid area, are particular ecosystems that shelter and serve as a refuge for exclusive flora and fauna species, and at the same time have been strongly associated with the history of the population and use of the territory by humans. This chapter analyzes the composition, diversity and abundance of flora and avifauna species of the coastal wetlands associated with the mouths of the rivers and streams of the northern zone of Chile and discusses their conservation status. The vegetation associated with these ecosystems tends to follow an order in fringes, presenting itself from the littoral zone to the submerged, natants, emerged/surfaced, and free floating species, marshy ones and finally shrubs. In terms of species, a total of 69 species are recognized in 4 classes. Out of these species, 80% are native and 14% introduced, with only three species being endangered. In the case of birds, there is a presence of 194 species, and the Lluta wetland is the one with the greatest representation of this group (154 spp.). Most of these species are residents of wetlands (140 spp.), but there are also visiting species (26 spp.), as well as accidental ones (28 spp.). Out of all of the avifauna recorded in these wetlands, 32 species are endangered. In general terms, the development of mining and agricultural activities, urbanization and modification of the water flows associated with the coastal

W. Sielfeld (✉) • R. Fuentes
Department of Marine Sciences, Universidad Arturo Prat, Iquique, Chile

Departamento Ciencias del Mar, Universidad Arturo Prat, Iquique, Chile
e-mail: walter.sielfeld@unap.cl

R. Peredo
Arica, XV Regi6n de Arica, Parinacota, Chile

V. Malinarich
Agricultural and Livestock Service, Ministry of Agriculture, Tarapacá Region, Iquique, Chile

Servicio Agrícola y Ganadero I, Ministerio de Agricultura, Regi6n de Tarapacá, Iquique, Chile

F. Olivares
Tierra del Sol – Research and Development consultores, Region of Atacama, Vallenar, Chile

Tierra del Sol – Investigaci6n y Desarrollo consultores. Regi6n de Atacama, Vallenar, Chile

J.M. Fariña, A. Camaño (eds.), *The Ecology and Natural History of Chilean Saltmarshes*, DOI 10.1007/978-3-319-63877-5_5

wetlands of northern Chile are the main threats to this type of ecosystems, and should be considered in the regional conservation strategies for avifauna, flora and biodiversity in general.

Keywords Birds • Basin • Aridity

Introduction

The coastal wetlands of northern Chile (Regions I, II, III and XV) have a series of characteristics that transform them into unique ecosystems (Aguirre and Torres 2005). They stand out primarily as places with presence of water, in an area marked by its extreme aridity, which is why these were the centers of development of pre-Columbian cultures typical of the area (Chinchorro culture, Camarones, Loa, etc.). Later, with the arrival of the Spanish conquerors, the modern settlements were centered in same places, so the current population growth puts practically all the wetlands of this region at risk (Cortez 2010). Among the actions that affect them the most are: urban sewage, industrial waste, tracking, habitat destruction due to the exploitation of aggregates (mines), the advancement of agricultural and urban borders with wetlands and surrounding lands, reduction of water flows because of urban developments and mining, the effect of introduced species (especially dogs and rats) and irrigation ditches; modification of watercourses by canalization and dam construction, urbanization, destruction of waterfront vegetation, lack of legal and practical mechanisms to ensure the future viability of the few currently protected wetlands, and lack of marine parks (considering that a large part of the coastal wetlands are located within 80 m of the high tide line). Many wetlands also function as clandestine landfills, and as camping and picnic sites, activities that heavily disturb wildlife (obs. personal, W. Sielfeld).

From the biological point of view, the coastal wetlands of northern Chile contain exclusive desert areas species such as the Northern River Shrimp (Cryphiops caementarius) and the Silverside Northern River Shrimp (Basilichthys semotilus), the latter endemic, and today largely extinguished from its original range. Except for the fishing ban, there are no effective protection measures for these species.

From the biogeographic point of view, northern Chile, generally recognized as Norte Grande and Norte Chico, as well as and under various equivalent nomenclatures, has been characterized as an independent unit, marked by its endemism, aridity and specific composition, both coastal marine and terrestrial and its inland waters. At the neotropics level, this zone is identified as a biogeographic unit by Jeannel (1942, 1967), Monrós (1958), Kuschel (1964), Schwabe (1969) and Cabrera and Willink (1980). Cabrera (1957, 1971) and Cabrera and Willink (1980) who base their classification on aspects of phytogeographic type, and Fittkau (1969) and Müller (1973), in vertebrates. Studies related to the distribution of arthropod fauna are scarcer, being an exception Peña (1966) for Chile, Roig Juñent (1994) and Morrone (1996) for the neotropical regions in general. In the case of Chile, a zoogeographic scheme was proposed by Mann (1960) and a zonation scheme of the vegetation was proposed by Pisano (1956) and Gajardo (1995). Di Castri (1969) elaborated a bioclimatic scheme, based on the country's climatic types and the characteristics that they imprint onto the flora and fauna. Luebert and Pliscoff (2006) have spatially delimited the vegetative units of Chile, based on the geographic variation of the composition of the species and the life forms of the plant communities.

The northern zone of Chile, and especially the coastal desert of the regions of Arica and Parinacota, Tarapaca and Antofagasta, are included in the "coastal desert zone" by Di Castri (op. Cit.), and the "coastal desert" by Mann (1960). Gajardo (1995) defines this area as the "sub-region of coastal desert/coastal desert of Tocopilla", that extends along the coast of Chile from the regions of Tarapaca to Coquimbo and, consequently, includes the wetlands treated here.

Several coastal wetlands in the former zone have been recognized as the fauna concentration centers, which is why they have been proposed to be protection areas (CONAMA 2007). The Lluta River is currently "Sanctuary of the Lluta River Wetland" as well as the Shorebird Reserve, named for the Hemispheric Shorebird Network (WHSRN). The Loa River mouth is a priority conservation site in the Regional Biodiversity Strategy, Antofagasta Region lll. The Copiapo River mouth is protected under the name of "Protected Marine and Coastal Area for Multiple Uses", and the Huasco River mouth is a priority site for conservation and protection in the Regional Biodiversity Strategy, Region III of Atacama.

This contribution analyzes the composition, diversity and abundance of flora and avifauna species of the coastal wetlands, associated with the river mouths and gullies of northern Chile, in order to show their importance as main biodiversity concentration centers in this area and to lay the foundations for a protection and conservation program that also considers its importance as biological corridor for the numerous migratory species that visit the Chilean coast.

Study Area

The study was focused on the coastal wetlands associated with the river mouths in northern Chile, from the northern end (Lluta River) and the Atacama Region III (Huasco River). Wetlands of the Lluta, San Jose, Vitor, Camarones, Tana, Loa, Salado, and Copiapo and Huasco rivers are included (Fig. 5.1).

Data Origin

The section corresponding to the wetland flora is based on information from several bibliographic sources, mainly highlighting: Silob (2001) for the Lluta River; Squeo et al. (2006) and Andrade (2007) for the Copiapo River; and CADE-IDEPE (2004) for the Huasco River. Original information obtained by the authors for the

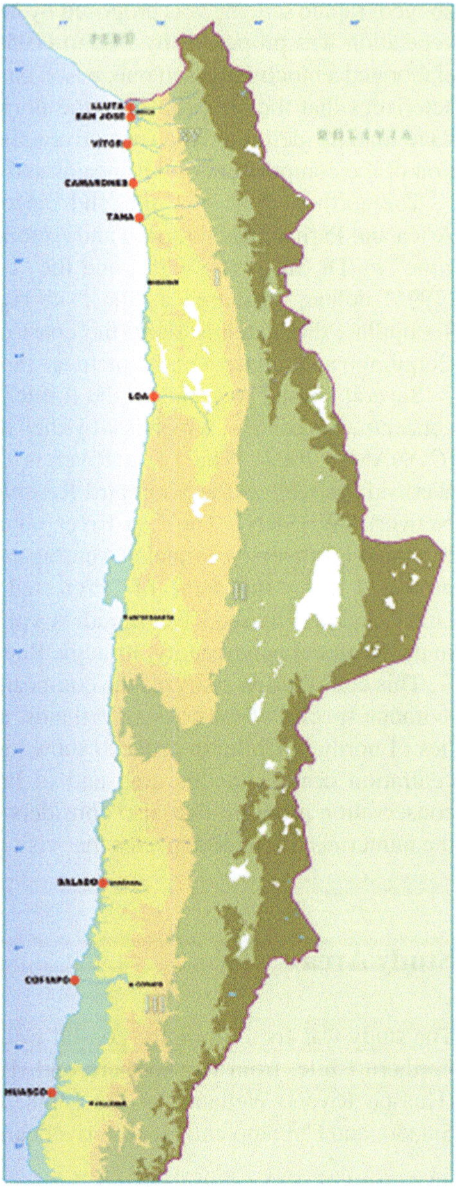

Fig. 5.1 Location of the study sites

river mouths of the Lluta (Fuentes 2000–2009, ESSAT/UNAP Environmental Monitoring), Camarones, Tana, and Loa rivers is also included. The classification of flora, according to the growth development, was done on the basis of the Raunkiaer system (1934), considering as phanerophytes those plants that are usually woody, growing erect, and whose replacement buds are above the ground more than 25 cm; camaephytes as either erect or climbing/creeping qualities, whose overwintering buds are near the ground, at no more than 25 cm; hemicryptophytes as those with buds at the soil surface; cryptophytes, those with rhizomes modified stems, which are under ground or under water (geophytes, helophytes, and hydrophilics); and finally, as therophytes we have considered those annual species that develop in the favorable period and survive harsh conditions as seeds.

In the case of bird species, the study was based on diverse bibliographical references, mainly on Peredo and Miranda (2000a, b, 2001), Peredo and Amado (1995a, b) and Peredo and Miranda (2000a, b) to Lluta River; Malinarich (2005) for Loa River; Luna and Cortés (2007) and Squeo et al. (2006), Copiapo River; and Olivares (2006), Huasco River .

Original information obtained by the authors in the Lluta river mouths (Sielfeld & Ledesma 2000–2009, ESSAT / UNAP Environmental Monitoring) is also included. For these organisms, the census system used by the authors in different surveys was carried out generally through the detection, recording and counting of birds in each of the study sectors, considering the first 5 min for the offset of the birds behavior (see Reynolds et al. 1980), from which the counts were done for up to 5 min (see Jarvinen 1978, Fuller and Lanslow 1984). The counts are always double and in the case of differences greater than 5%, they are repeated until an adequate adjustment is achieved. The value recorded is the average. In order to compare the overlapping of groups of bird species, the CLUSTER routine analysis was used for the conglomerates of the PRIMER 6 program (Clarke and Gorley 2006) based on the Bray-Curtis similarity measure for abundances and the Sorensen coefficient for presence/absence data. With the CLUSTER routine, the SIMPROF permutation test was used (at 5% level) to determine the significance of the internal structure of the conglomerates. In the figure, these are highlighted by thick gray lines. For each wetland, the diversity was calculated using the Shannon-Wiener index (H') where Pi is the gravimetric ratio of the i-th taxon of prey and S is the number of prey taxons and the Simpson index. The information obtained was also analyzed according to species richness (number of species and Margalef index), ecological diversity (Shannon-Wiener index: Lloyd et al. 1968), and relative abundance (numerical %), frequency (% of occurrence) and species classification by dominance and constancy scale proposal by Bodenheimer (1955, in Avendaño and Sáiz 1977) where the following categories are considered:

- Dominance:
  - Dominant Species >5.0%
  - Accessory Species 2.5–5.0%
  - Accidental Species <2.5%
- Constance:
  - Constant Species >50.0%
  - Accessory Species 25.0–50.0%
  - Accidental Species <25.0%

General Background of the Northern Chilean Wetlands

The basins of the northern regions of Chile generally have little pollution, although in many cases the poor quality of water limits the possibility of use for this element (as in the Lluta, Camarones and Loa rivers). In the case of the Loa river, chemical contamination is predominant over bacteriological contamination, a situation that is reversed in the Region III rivers (Copiapo and Huasco), where bacteriological contamination predominates (Table 5.1). The studied basins have very low average surplus flows (Table 5.2), a situation that is aggravated in periods of great drought, where surplus flows are practically zero (Salazar 2003). During very rainy years, especially those associated with ENSO periods, their conditions can change violently at outflows of up to 4.5 m^3/s, as happened with the Lluta River (Estrella de Arica, February 10, 2004) and 1.46 m^3/s with the San Jose River (Estrella de Arica, February 17, 2004).

The following is a detail of the water courses associated with the coastal wetlands of northern Chile. The information collected comes from different reports by government agencies (SAG, DGA) that have focused on the flora or fauna of these bodies of water (example Malinarich 2005).

Table 5.1 Level and kind of contamination of the principal basin associated with the sites study (from: mop, Salazar 2003)

	Contamination grade contaminación		Contamination type	
Region/basin	Actual	Potential	Actual	Potential
I Region, San Jose River	Nil to low	Nil to low	–	–
II Region, Loa River	Medium	High	Chemical and Bacteriological	Chemical and Bacteriological
III Región, Copiapó River	Low	Mid	Bacteriological	Chemical and Bacteriological
III Región, Huasco River	Low	Mid	Chemical and Bacteriological	Chemical and Bacteriological

Table 5.2 Averaged flow rates of the hydrological systems associated with the study sites (from: Peña 1993, Salazar 2003)

Region	Basin	Avg. flow entry (m^3/s)	Avg. flow exit (m^3/s)	Avg.daily min exit (m^3/s)
I	Lluta river	2.3	1.4	0.09
I	San Jose river	1.5	0.0	0.0
II	Loa River	2.8	0.6	0.04
III	Copiapo river	2.9	0.1	0.04
III	Huasco river	3.5	1.7	0.30

Lluta River

Exoreic river that flows through homonymous valley and exhibits its mayor floods during the summer months as a result of the "highland winter" rains. It corresponds to a pre-Andean basin that rises southwest of Putre in Humapalca, and ends in the sector of Chacalluta, to the northwest of the city of Arica, near the border with Peru. It has a length of 147 km and covers an area of 3378 km^2; its main tributaries are the Azufre River and the Gorges of Caracarani, Colpitas and Socoroma. The quality of its waters is characterized by high content of boron, presenting the element as a growing trend along the course of the river (Campos et al. 2007), thus limiting agriculture. In the months of water surplus, the underground beds are recharged by feeding vegas and coastal lagoons in the lower section of the channel. It has scarse vegetation in the upper part of the basin, which increases in the middle and lower part, consisting mainly of communities of horsetail, chilca shrubs, salt grass and reed. Its mouth forms a wetland of approximately 300 hectares, emphasizing the avifauna by its high diversity. This wetland is threatened by the urban activities and construction of the city of Arica, and the limitation of water resources for human consumption and agriculture.

San Jose River

Exoreic river of pre-Andean origin that runs through the gorge of Azapa and flows into the center of the city of Arica. It exhibits permanent flow only in its upper course, which manages to reach the sea only in the form of an alluvial torrent in the summer months of the "highland/altiplano" winter of the rainy years. There is also a recharge of the underground lagoons that fill the slopes of the river mouth, and which have now been intervened, which also feeds vegas and coastal lagoons, for irrigation of green areas. It covers an area of 3060 km^2 and a length of 128 km. In the widest, medium and low parts of the basin, it presents suitable soils for agriculture with waters less brackish than those of the Lluta River (Campos 2005). Its mouth has an area of approximately 10 hectares, with scarse vegetation, and was formed by a rock and sand beach. It stands out for its avifauna that is conditioned by the descent of the river, except for the one of marine origin. This wetland is threatened by the scarcity of water, aggravated by the increasing agricultural exploitation of the Azapa valley, and the intervention of the groundwater sources, to which are added channeling works and urbanization.

Camarones River

The Camarones River is exoreic throughout the year, it ends in the form of a very narrow gorge that gives origin to this narrow wetland, which does not represent an important center of concentration of avifauna. Much of the lower course of the river (sector from Cuya to the mouth) is altered by agricultural activities and facilities, which also makes the sector less important.

Tana River

Endorheic river that reaches the coast only during extremely rainy years of the highlands. Consequently, its mouth, located immediately to the north of the city of Pisagua is a dry course, devoid of vegetation, which is why it is mainly used as a rest area for marine birds only.

Loa River

The river presents a regime related to the development of the altiplanic (highland winter) winter, which determines the amount of floods between January and March of each year (DGA 2004 a, b). The current river flow at the mouth is very low (Table 5.2) as a result of the upstream extraction of an important part of its flow, mainly because of the mining activities. On the other hand, the violent contamination with xanthates, isopropanol, detergents and heavy metals that has occurred in 1997, generated a massive destruction of shrimp and silver back shrimp (pejerreyes), among others, an event the system has not yet recovered from. This situation has marked the middle and lower reaches of the river, including the lagoon at its mouth.

Salado River

The contamination of the Chañaral Bay and the Salado River mouth dates back to 1938, when Potrerillos vessels, owned by the American Andes Copper Mining Company, were emptied into the sea through the Salado River channel, passing through Llanta, Diego de Almagro and El Salado. Once the Potrerillos copper deposit was exhausted in 1959, El Salvador mine was starting to be exploited, continuing the discharge of tailings, a process that continues under Codelco Chile, once the copper (extraction) was nationalized, during the Popular Unity government. At the beginning of the '70s, the presence of black and viscous mud in the bay of Chañaral and its detriment to the mechanized port of the Santa Fe Mine, located south of Chañaral, has motivated, in 1975, the Division of Salvador de Codelco to move the waste dump site 9 km north of Chañaral. Subsequently, the Ministry of Public Works of the military regime ordered the diversion of the Salado River to the north, which, due to the coastal dynamics, has spread the contamination over an area of at least 20 miles, reaching the area of the National Park Pan de Azucar.

Copiapo River

The use of the Copiapo river water has intensified notably in the 1980s and 1990s, mainly for agricultural and mining use. Currently, 31% is used in mining and 53% in agriculture (DGA/CADE-IDEPE 2004). The aquifer naturally replenishes about 4500 l/s, out of which almost double the amount is extracted, equivalent to 7000 l/s, and the rights granted to date quadruple the recharge (Godoy 2010). Consequently, the aquifer is not able to recharge and recover, and is on its way to disappear. The river feeds on tributaries that descend from the glaciers of the Andes, draining them for 160 kilometers until reaching the Pacific Ocean, even though, today, that no longer happens. The last time the Copiapo River was seen to reach the Marfue was in 2015, after a great flood. Currently, the sector that was its river mouth is part of the "Punta Morro Protected Marine and Coastal Area - Copiapo River mouth, Atacama Region" of the Marine and Coastal Protected Areas of Multiple Uses system (AMCPMU), implemented by CONAMA.

Huasco River

In the specific case of the Huasco River, its flow is mainly regulated by the "Santa Juana" reservoir, located 22 km upstream from Vallenar, which suffers overflows mainly in the winter period, generating floods in the lower part of the river. In its mouth there is a wetland with several lagoons, which are related with extensive bulrushes.

Flora Characteristics of the Northern Chilean Wetlands

Littoral Region

The vegetation associated with these limnic systems tends to follow a fringe arrangement; the aquatic species are distributed in the coastal zone, with very few of them located in deep areas. The order of the littoral zone is towards land: first come submerged species, then natants, emerged and free floating ones. On land, the first strip are marshy species, such as reeds, totora and reeds, and then as the moisture gradient decreases, there is also a change in vegetation structure, with shrubs such as chilca and cachiyuyos dominating.

Flora and Vegetation of the Wetlands

The systematic list (Table 5.3) of the northern Chilean vegetation species from the Lluta to Huasco River indicates the presence of 69 species, divided in 17 groups and 28 families. This group order corresponds to: Poales (22%), Asterales (13%),

Table 5.3 Plant species occurring in the saltmarshes of the northerns Chilean coast

Order	Family	Specie	Common name
Lycopodiopsida			
Isoetales	Isoetaceae	*Isoëtes savatieri* Franch.	Isete
Equisetopsida			
Equisetales	Equisetaceae	*Equisetum bogotense* Kunth	Limpiaplata
Equisetales	Equisetaceae	*Equisetum giganteum* L.	Cola de caballo
Rosopsida (eudicotiledóneas)			
Apiales	Apiaceae	*Sium latifolium* L.	Berro brasileño
Apiales	Araliaceae	*Hydrocotyle ranunculoides* L. F.	Hierba de la plata Tangue
Asterales	Asteraceae	*Baccharis juncea* (Lehm.) Desf.	Suncho
Asterales	Asteraceae	*Baccharis pingraea* DC. Var. *Pingraea*	Chilquilla
Asterales	Asteraceae	*Baccharis salicifolia* (Ruiz & Pav.) Pers.	Chilca/suncho
Asterales	Asteraceae	*Baccharis scandens* (Ruiz & Pav.) Pers.	Chilca
Asterales	Asteraceae	*Baccharis spartioides* (Hook. & Arn. Ex DC.) J. Remy	Pichana
Asterales	Asteraceae	*Grindelia glutinosa* (Cav.) Mart.	Chiñe
Asterales	Asteraceae	*Hypochaeris sp.*	–
Asterales	Asteraceae	*Pluchea chingoyo* (Kunth) DC.	Chingoyo
Asterales	Asteraceae	*Tessaria absinthioides* (Hook. & Arn.) DC.	Brea/Sorona
Caryophyllales	Aizoaceae	*Carpobrotus chilensis* (Molina) n.E. Br.	Doca/Dedo de agua
Caryophyllales	Aizoaceae	*Tetragonia macrocarpa* Phil.	Escarcha/pasto guanoso
Caryophyllales	Aizoaceae	*Tetragonia maritima* Barnéoud	Aguanosa
Caryophyllales	Amaranthaceae	*Atriplex atacamensis* Phil.	Cachiyuyo
Caryophyllales	Amaranthaceae	*Atriplex deserticola* Phil.	Cachiyuyo
Caryophyllales	Amaranthaceae	*Atriplex mucronata* Phil.	Cachiyuyo
Caryophyllales	Amaranthaceae	*Sarcocornia fruticosa* (L.) A.J. Scott	Hierba sosa
Caryophyllales	Amaranthaceae	*Suaeda foliosa* Moq.	–
Caryophyllales	Frankeniaceae	*Frankenia chilensis* C. Presl	Hierba del salitre
Caryophyllales	Tamaricaceae	*Tamarix sp.*	Tamarindo

(continued)

Table 5.3 (continued)

Order	Family	Specie	Common name
Fabales	Fabaceae	*Adesmia littoralis* Burkart	Añahua/Pircún
Fabales	Fabaceae	*Astragalus chamissonis* (Vogel) Reiche	–
Fabales	Fabaceae	*Caesalpinia angulata* (Hook. & Arn.) Baill.	Retamilla
Fabales	Fabaceae	*Geoffroea decorticans* (Gillies ex Hook. & Arn.) Burkart	Chañar
Fabales	Fabaceae	*Melilotus indicus* (L.) All.	Alfalfilla
Fabales	Fabaceae	*Prosopis alba* Griseb. Var. *Alba*	Algarrobo blanco
Fabales	Fabaceae	*Prosopis chilensis* (Molina) Stuntz emend. Burkart var. *Chilensis*	Algarrobo
Gentianales	Apocynaceae	*Skytanthus acutus* Meyen	Cuerno de cabra
Lamiales	Boraginaceae	*Heliotropium curassavicum* L. Var. *Curassavicum*	Cola de mico Alacrancillo
Lamiales	Boraginaceae	*Heliotropium floridum* (A. DC.) Clos	Heliotropo/Palo negro
Lamiales	Plantaginaceae	*Bacopa monnieri* (L.) Pennell	Bacopa enana
Lamiales	Plantaginaceae	*Plantago major* L.	Llantén
Lamiales	Verbenaceae	*Phyla reptans* (Kunth) Greene	Tiqui-tiqui Hierba de la virgen
Malpighiales	Salicaceae	*Salix humboldtiana* Willd. Var. *Humboldtiana*	Sauce amargo
Malvales	Malvaceae	*Cristaria aspera* Gay var. *Aspera*	Malvita
Malvales	Malvaceae	*Cristaria viridiluteola* Gay var. *Pinnata* (Phil.) M. Muñoz	Malvita
Myrtales	Onagraceae	*Ludwigia peploides* (Kunth) P.H. Raven ssp. *Montevidensis* (Spreng.) P.H. Raven	Clavito de agua
Myrtales	Onagraceae	*Oenothera coquimbensis* Gay	Don Diego de la noche
Sapindales	Anacardiaceae	*Schinus polygamus* (Cav.) Cabrera	Huingán
Saxifragales	Haloragaceae	*Myriophyllum* sp.	Pino de agua
Solanales	Convolvulaceae	*Cressa truxillensis* Kunth	–

(continued)

Table 5.3 (continued)

Order	Family	Specie	Common name
Solanales	Nolanaceae	*Nolana albescens* (Phil.) I.M. Johnst.	Suspiro
Solanales	Nolanaceae	*Nolana carnosa* (Lindl.) Miers ex Dunal	Suspiro
Solanales	Nolanaceae	*Nolana divaricata* (Lindl.) I.M. Johnst.	Suspiro
Solanales	Nolanaceae	*Nolana salsoloides* (Lindl.) I.M. Johnst.	Suspiro
Solanales	Nolanaceae	*Nolana sedifolia* Poepp.	Suspiro
Solanales	Solanaceae	*Solanum chilense* (Dunal) Reiche	Tomatillo
Solanales	Solanaceae	*Solanum peruvianum* L.	–
Liliopsida (Monocotiledóneas)			
Alismatales	Juncaginaceae	*Triglochin palustris* L.	Hierba de la paloma
Najadales	Ruppiaceae	*Ruppia maritima* L.	Pelo de marisma
Poales	Cyperaceae	*Carex* sp.	–
Poales	Cyperaceae	*Eleocharis* sp.	–
Poales	Cyperaceae	*Schoenoplectus americanus* (Pers.) Volkart ex Schinz & R. Keller	Totora azul
Poales	Cyperaceae	*Schoenoplectus californicus* (C.A. Mey.) Soják var. *Californicus*	Totora, Tromé
Poales	Juncaceae	*Juncus acutus* L.	Junco espinoso
Poales	Juncaceae	*Juncus cyperoides* Laharpe	Ihua-ihua
Poales	Juncaceae	*Juncus procerus* E. Mey.	Junquillo grande
Poales	Poaceae	*Arundo donax* L.	Caña común
Poales	Poaceae	*Cortaderia speciosa* (nees & Meyen) Stapf	Cola de zorro
Poales	Poaceae	*Distichlis spicata* (L.) Greene.	Grama salada
Poales	Poaceae	*Leptochloa fusca* (L.) Kunth ssp. *Uninervia* (J. Presl) n. Snow	–
Poales	Poaceae	*Paspalum vaginatum* Sw.	Chepica blanca
Poales	Poaceae	*Phragmites australis* (Cav.) Trin. Ex Steud.	Carrizo
Poales	Poaceae	*Polypogon monspeliensis* (L.) Desf.	–
Poales	Typhaceae	*Typha angustifolia* L.	Totora/Vatro

Table 5.4 Systematic ordination of the plant species

Class	n° Species	Percentage (%)
Lycopodiopsida	1	1.4
Equisetopsida	2	2.9
Rosopsida	49	71.0
Liliopsida	17	24.6

Solanales (12%), Fabales (10%) y Caryophyllales (10%), being Asteraceae (9sp.), Fabaceae (7sp.) and Poaceae (7sp.) as mostly represented families.

At the level of the class distribution, Table 5.4 indicates the presence of four classes, two representatives of Pteridophyta family and two of Magnoliophyta one, the latter group being the dominant, with about 95% of the species present in the coastal wetlands between Lluta (Arica-Parinacota) and Huasco (Atacama).

According to the Catalog of Vascular Plants of the Southern Cone, 80% of the species recorded in the northern coastal wetlands are native, and 14% introduced (Table 5.5). According to the Raunkiaer System growth habit, 31% are cryptophytes, 29% phanerophytes, 16% camphorites, 18% hemicryptophytes and 6% therophytes.

Regarding the distribution of species, it should be mentioned that the richness exhibiting in the Lluta river, as well as the presence of some species only in this sector, is mainly due to the topography of the same, since the wetland of the Lluta river presents a greater extension, which allows for riparian shrub species of gorges and oasis, such as *Equisetum giganteum, Grindelia glutinosa* and *Geoffroea decorticans*, to name a few, to be found only in Lluta and not at the mouth of the other rivers, areas in which they are found only in the valley.

The analysis of similarity (Fig. 5.2) in relation to the information contained in Table 5.6, indicates a great closeness between Camarones, Tana and Loa (about 0.88); and the rest of the wetlands, which are presented as independent units, since Lluta has a similarity of only 0.56% with previous ones (generally considered low), and the other two (Huasco and Copiapo) less than 0.48%. Both Huasco and Copiapo have a greater species richness, 22 and 40 respectively, which separates them from the wetlands of Arica-Parinacota and Tarapaca, which, with the exception of Lluta (24 sp.), do not present more than 15 species, 12 of which are common to Camarones, Tana and Loa, which explains their similarity.

State of Flora Conservation in the Northern Chilean Wetlands

Out of all the species, only three are endangered according to the different classification studies that were considered, *Equisetum giganteum* and *Isoëtes savatieri* as Vulnerable (VU) according to the National Museum of Natural History Bulletin 47 (1998) and *Prosopis alba* Vulnerable (VU) according to CONAF.

Table 5.5 Origin and life style of the plant species

Specie	Origin	Habit
Adesmia littoralis	Native	Camephyts
Arundo donax	Introduced	Cryptophytes
Astragalus chamissonis	Introduced	Camephyts
Atriplex atacamensis	Native	Phanerophytes
Atriplex deserticola	Native	Phanerophytes Faneróĺita
Atriplex mucronata	Native	Camephyts
Baccharis juncea	Native	Cryptophytes
Baccharis pingraea var. *pingraea*	Native	Cryptophytes
Baccharis salicifolia	Native	Phanerophytes
Baccharis scandens	Native	Phanerophytes
Baccharis spartioides	Native	Phanerophytes
Bacopa monnieri	Introduced	Cryptophytes Criptófita
Caesalpinia angulata	Native	Phanerophytes
Carex sp.	–	Cryptophytes Criptófita
Carpobrotuschilensis	Native	Hemicryptophytesfita
Cortaderia speciosa	Native	Cryptophytes HemicryptophytesCriptófita
Cressa truxillensis	Introduced	Hemicryptophytes
Cristaria aspera var. *aspera*	Native	Theophytes
Cristaria viridiluteola var. *pinnata*	Native	Theophytes
Distichlis spicata	Native	Cryptophytes Criptófita
Eleocharis sp.	–	Cryptophytes Criptófita
Equisetum bogotense	Native	Cryptophytes Criptófita
Equisetum giganteum	Native	Cryptophytes Criptófita
Frankenia chilensis	Native	Camephyts
Geoffroeadecorticans	Native	Phanerophytes
Grindelia glutinosa	Native	Phanerophytes
*Heliotropium curassavicum*var. *curassavicum*	Native	Hemicryptophytesiptófita
Heliotropiumfloridum	Native	Phanerophytes
Hydrocotyleranunculoides	Native	Cryptophyte
Hypochaeris sp.	–	–
Isoëtes savatieri	Native	Cryptophytes Criptófita
Juncus acutus	Introduced	Hemicryptophytes
Juncus cyperoides	Native	Hemicryptophytes
Juncus procerus	Native	Hemicryptophytes
Leptochloa fusca ssp. *uninervia*	Native	Hemicryptophytes
*Ludwigia peploides*ssp. *montevidensis*	Native	Criptophytes
Melilotus indicus	Introduced	Camephyts
Myriophyllum sp.	–	Criptophytes
Nolana albescens	Native	Phanerophytes
Nolana carnosa	Native	Camephyts

(continued)

Table 5.5 (continued)

Specie	Origin	Habit
Nolana divaricata	Native	Camephyts
Nolana salsoloides	Native	Phanerophytes
Nolana sedifolia	Native	Phanerophytes
Oenotheracoquimbensis	Native	Therophytes
Paspalum vaginatum	Native	Cryptophytes
Phragmites australis	Native	Cryptophytes
Phyla reptans	Native	Hemicryptophytes
Plantago major	Introduced	Hemicryptophytes
Pluchea chingoyo	Native	Phanerophytes
Polypogon monspeliensis	Introduced	Hemicryptophytes
Prosopis alba	Native	Phanerophytes
Prosopis chilensis var. *chilensis*	Native	Phanerophytes
Ruppia maritima	Native	Cryptophytes
Salix humboldtiana var. *humboldtiana*	Native	Phanerophytes
Sarcocornia fruticosa	Native	Criptophytes
Schinus polygamus	Native	Phanerophytes
Schoenoplectus americanus	Native	Cryptophytes
Schoenoplectus californicus	Native	Cryptophytes
Sium latifolium	Introduced	Hemicryptophytes
Skytanthus acutus	Native	Camephyts
Solanum chilense	Native	Camephyts
Solanum peruvianum	Native	Camephyts
Suaeda foliosa	Native	Phanerophytes
Tamarix sp.	Introduced	Phanerophytes
Tessaria absinthioides	Native	Phanerophytes
Tetragonia macrocarpa	Native	Therophytes
Tetragonia maritima	Native	Camephyts
Triglochin palustris	Native	Hemicryptophyte
Typha angustifolia	Native	Criptophyte

Characteristics of Avifauna of the Northern Chilean Wetlands

Systematic Composition of the Bird Assembly

The wetlands north of the Huasco River are used by 194 bird species, with the following details by wetland (Table 5.7): Lluta 154, San Jose 73, Vitor 84, Camarones 83, Tana 30, Loa 37, Salado 33, Copiapo 56, and Huasco 84 species. These species are contained in 16 orders and 41 families. The most numerous orders are Charadriiformes with 60 species (30.93%) and Passeriformes with 54 species (27.84%), both with 8 families. The most prominent families are Scolopacidae with 24 species (12.37%), Anatidae with 16 species (8.25%), Tyrannidae 15 species

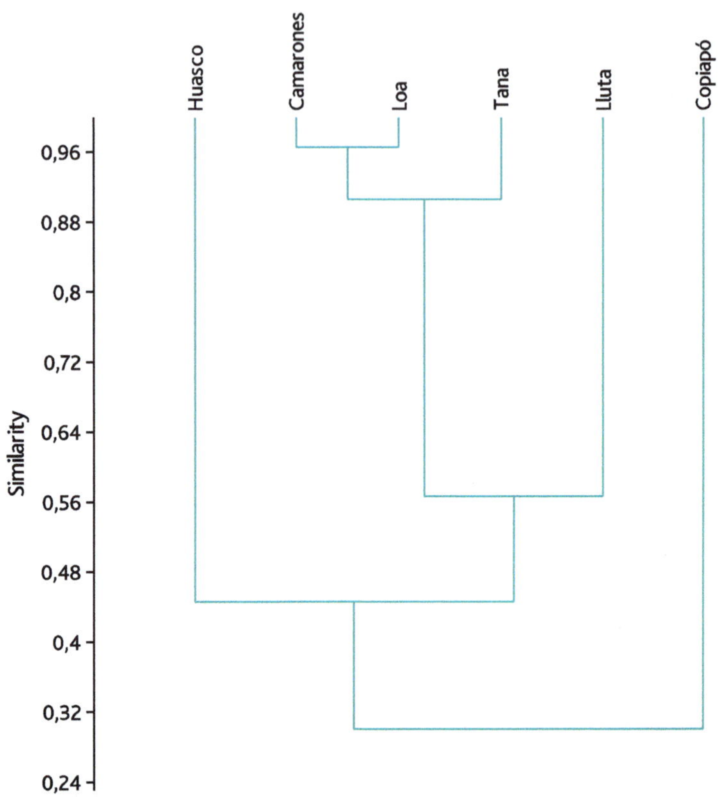

Fig. 5.2 Similarity dendogram of the plant species associations recorded on the coastal saltmarshes of the northern coast of Chile

(7.73%), Charadriidae 11 species (5.67%), and Ardeidae 10 species (5.15%). The most common ones are Calidris (beach) with 9 species (4.64%), and Charadrius (chorlos) and Anas (ducks), both with 7 species (3.61%).

Conservation State of the Bird Species

From the studied set, 32 species present conservation problems to some degree, according to the different classification systems consulted, the most threatened species are the Arica hummingbird (In danger, IUCN, In danger and Rare, CONAMA, Vulnerable, SAG and CONAF), the Small Seagull (Endangered, IUCN, CONAMA and CONAF, In Danger of Extinction, SAG) and the Nun Seagull (Near Threatened, IUCN, Vulnerable, SAG and CONAF). Other species at risk are the tamarugal Sierra Finches (Vulnerable, IUCN); the Red Knot (In Danger, CONAMA); the Black-faced Ibis (In Danger of Extinction, SAG, Vulnerable, CONAF); the Grey Gull (Vulnerable, SAG and CONAF); the Andean gull (Vulnerable, SAG); The Giant

Table 5.6 Geographical location of the plant species

Specie	Lluta	Camarones	Tana	Loa	Copiapo	Huasco
Adesmia littoralis					x	x
Arundo donax	x	x		x		
Astragalus chamissonis	x					
Atriplex atacamensis		x	x	x		
Atriplex deserticola					x	x
Atriplex mucronata					x	
Baccharis spartioides						x
Baccharis juncea		x		x		
Baccharis pingraea var. *pingraea*					x	
Baccharis salicifolia	x					x
Baccharis scandens	x	x	x	x		
Bacopa monnieri	x					x
Caesalpinia angulata						x
Carex sp.					x	
Carpobrotuschilensis						x
Cortaderia speciosa	x	x	x	x	x	x
Cressa truxillensis					x	
Cristaria aspera var. *aspera*					x	
Cristaria viridiluteola var. *pinnata*						x
Distichlis spicata	x	x	x	x	x	x
Eleocharis sp.					x	
Equisetum bogotense					x	
Equisetum giganteum	x					
Frankenia chilensis					x	
Geoffroeadecorticans	x					
Grindelia glutinosa	x					
*Heliotropium curassavicum*var. *curassavicum*	x	x	x	x		
Heliotropiumfloridum					x	x
Hydrocotyleranunculoides					x	
Hypochaeris sp.					x	
Isoëtes savatieri					x	
Juncus acutus	x	x	x	x	x	x
Juncus cyperoides					x	
Juncus procerus					x	
Leptochloa fusca ssp. *uninervia*	x					
*Ludwigia peploides*ssp. *montevidensis*					x	
Melilotus indicus	x					
Myriophyllum sp.	x					
Nolana albescens					x	
Nolana carnosa					x	
Nolana divaricata					x	

(continued)

Table 5.6 (continued)

Specie	Lluta	Camarones	Tana	Loa	Copiapo	Huasco
Nolana salsoloides						x
Nolana sedifolia					x	
Oenotheracoquimbensis					x	
Paspalum vaginatum					x	
Phragmites australis					x	
Phyla reptans	x					
Plantago major					x	
Pluchea chingoyo	x					
Polypogon monspeliensis					x	
Prosopis alba				x		
Prosopis chilensis var. *chilensis*					x	
Ruppia maritima	x	x	x	x	x	x
Salix humboldtiana var. *humboldtiana*						x
Sarcocornia fruticosa	x	x	x	x	x	x
Schinus polygamus						x
Schoenoplectus americanus	x	x	x	x	x	x
Schoenoplectus californicus					x	
Sium latifolium					x	
Skytanthus acutus						x
Solanum chilense		x	x	x		
Solanum peruvianum	x					
Suaeda foliosa					x	
Tamarix sp.	x					
Tessaria absinthioides	x	x	x	x	x	x
Tetragonia macrocarpa	x					
Tetragonia maritima					x	
Triglochin palustris					x	
Typha angustifolia	x	x	x	x	x	x

Coot (Vulnerable, SAG and CONAF); The Peregrine Falcon (Vulnerable, SAG; Endangered, CONAF); The Osprey (Vulnerable, SAG and CONAF); the Puna Ibis (Vulnerable, CONAF); the Guanay cormorant (Vulnerable, CONAF); and Chilean Flamenco (Vulnerable, CONAF). It should be noted that Arica's hummingbird is the only species that also possesses the Natural Monument Status (Decree No. 2 of January 9, 2006, Ministry of Agriculture) (Table 5.8).

Bird Continuity and Seasonal Intermittence

Based on the bird residence in Chile established by Araya et al. (1995), and Martinez and González (2004), the study was composed of 140 resident species (72.2%), 26 visitor species (13.4%) and 28 accidental species (14.4%).

Table 5.7 Bird species recorded on the study sites

Order	Family	Common name	Scientific name
Anseriformes	Anatidae	Speckled teal	*Anas flavirostris*
		White-cheeked pintail	*Anas bahamensis*
		Cinnamon teal	*Anas cyanoptera*
		Blue-winged teal	*Anas discors*
		Yellow-billed pintail	*Anas georgica*
		Puna teal	*Anas puna*
		Chiloé wigeon	*Anas sibilatrix*
		Red shoveler	*Anas platalea*
		Silver teal	*Anas versicolor*
		Crested duck	*Lophonettaspecularioides*
		Southern pochard	*Nettaerythrophthalma*
		Lake duck	*Oxyura vitatta*
		Ruddy duck	*Oxyura jamaicensis*
		Fulvous whistling duck	*Dendrocygnabicolor*
		Black-bellied whistling duck	*Dendrocygnaautumnales*
		Black-necked swan	*Cygnusmelanocoryphus*
Podiciformes	Podicipedidae	Silvery grebe	*Podiceps occipitalis*
		White-tufted grebe	*Podiceps rolland*
		Great grebe	*Podiceps major*
		Pied-billed grebe	*Podilymbus podiceps*
Phoenicopteriformes	Phoenicopteridae	Chilean flamingo	*Phoenicopteruschilensis*
Pelecaniformes	Pelecanidae	Peruvian pelican	*Pelecanus thagus*
		Brown pelican	*Pelecanusoccidentales*
	Sulidae	Peruvian booby	*Sula variegata*
		Blue-footed booby	*Sula nebouxii*
	Phalacrocoracidae	Neotropic cormorant	*Phalacrocorax brasilanus*

(continued)

Table 5.7 (continued)

Order	Family	Common name	Scientific name
		Red-legged cormorant	*Phalacrocorax gaimardi*
		Guanay cormorant	*Phalacrocorax bougainvillii*
Ciconiformes	Ardeidae	Great egret	*Cosmerodius albus*
		Snowy egret	*Egretta thula*
		Little blue heron	*Egretta caerulea*
		Cattle egret	*Bubulcus ibis*
		Cocoi heron	*Ardea cocoi*
		Tricolored heron	*Egretta tricolor*
		Striated heron	*Butorides striata*
		Black-crowned night heron	*Nycticorax nycticorax*
		Stripe-backed bittern	*Ixobrichus involucris*
		Least bittern	*Ixobrychus exilis*
	Thereskiornithidae	Puna ibis	*Plegadis ridgwayi*
		Black-faced ibis	*Theristicus melanosis*
		Roseate spoonbill	*Platalea ajaja*
Cathartiformes	Ciconidae	Wood stork	*Mycteria americana*
	Cathartidae	Turkey vulture	*Cathartes aura*
		Black vulture	*Coragyps atratus*
Accipitrifomes	Pandionidae	Osprey	*Pandion haliaetus*
	Accipitridae	Cinereous harrier	*Circus cinereus*
		Harris's hawk	*Parabuteo unicinctus*
		White-tailed kite	*Elanus leucurus*
		Variable hawk	*Buteo polyosoma*
		Black-chested buzzard-eagle	*Geranoaetusmelanoleucus*
Falconiformes	Falconidae	American kestrel	*Falco sparverius*
		Peregrine falcon	*Falco peregrinus*
		Aplomado falcon	*Falco femoralis*

(continued)

Table 5.7 (continued)

Order	Family	Common name	Scientific name
		Southern crested caracara	*Caracara plancus*
		Chimango caracara	*Milvago chimango*
Gruiformes	Rallidae	Common gallinule	*Gallinula galatea*
		Spot-flanked gallinule	*Gallinula melanops*
		Andean coot	*Fulica ardesiaca*
		White-winged coot	*Fulica leucoptera*
		Red-gartered coot	*Fulica armillata*
		Red-fronted coot	*Fulica rufifrons*
		Giant coot	*Fulica gigantea*
		Plumbeous rail	*Pardirallus sanguinolentus*
		Black rail	*Laterallus jamaicensis*
Charadriiformes	Charadriidae	Andean lapwing	*Vanellus resplendens*
		Southern lapwing	*Vanellus chilensis*
		Semipalmated plover	*Charadrius semipalmatus*
		Kentish plover	*Charadrius alexandrinus*
Charadriiformes		Two-banded plover	*Charadrius falklandicus*
		Wilson's plover	*Charadrius wilsonia*
		Collared plover	*Charadrius collares*
		Killdeer	*Charadrius vociferus*
		Rufous-chested plover	*Charadrius modestus*
		Grey plover	*Pluvialis squatarola*
		American golden plover	*Pluviales dominica*
	Haematopidae	American oystercatcher	*Haematopus palliatus*
		Blackish oystercatcher	*Haematopus ater*
	Recurvirosteridae	Black-necked stilt	*Himantopus mexicanus*

(continued)

Table 5.7 (continued)

Order	Family	Common name	Scientific name
		Andean avocet	*Recurvirostra andina*
	Burhinidae	Peruvian thick-knee	*Burhinus superciliaris*
	Scolopacidae	Lesser yellowlegs	*Tringa flavipes*
		Greater yellowlegs	*Tringa melanoleuca*
		Solitary sandpiper	*Tringa solitaria*
		Wandering tattler	*Tringa incana*
		Willet	*Tringa semipalmata*
		Spotted sandpiper	*Actitis macularius*
		Ruddy turnstone	*Arenaria interpres*
		Surfbird	*Aphriza virgata*
		Least sandpiper	*Calidris minutilla*
		Semipalmated sandpiper	*Calidris pusilla*
		Baird's sandpiper	*Calidris bairdii*
		Sanderling	*Calidris alba*
		Pectoral sandpiper	*Calidris melanotos*
		Stilt sandpiper	*Calidris himantopus*
		Red knot	*Calidris canutus*
		White-rumped sandpiper	*Calidris fuscicollis*
		Western sandpiper	*Calidris mauri*
		Buff-breasted sandpiper	*Tryngites subruficollis*
		Whimbrel	*Numenius phaeopus*
		Hudsonian godwit	*Limosa haemastica*
		Short-billed dowitcher	*Limnodromus griseus*
		Wilson's phalarope	*Phalaropus tricolor*
		Red-necked phalarope	*Phalaropus lobatus*

(continued)

Table 5.7 (continued)

Order	Family	Common name	Scientific name
		Red phalarope	*Phalaropus fulicaria*
Charadriiformes	Thinocoridae	Least seedsnipe	*Thinocorusrumicivorus*
	Jacanidae	Wattled jacana	*Jacana jacana*
	Laridae	Grey-headed gull	*Chroicocephalus cirrocephalus*
		Andean Gull	*Chroicocephalus serranus*
		Brown-hooded Gull	*Chroicocephalus maculipennis*
		Grey gull	*Leucophaeus modestus*
		Laughing gull	*Leucophaeus atricilla*
		Franklin's gull	*Leucophaeus pipixcan*
		Belcher's gull	*Larus belcheri*
		Kelp gull	*Larus dominicanus*
		Inca tern	*Larosterna inca*
		Peruvian tern	*Sternula lorata*
		Arctic tern	*Sterna paradisaea*
		Common tern	*Sterna hirundo*
		South American tern	*Sterna hirundinacea*
		Gull-billed tern	*Gelochelidon nilotica*
		Elegant tern	*Thalasseus elegans*
		Royal tern	*Thalasseus maximus*
		Sandwich tern	*Thalasseus sandvicensis*
		Black skimmer	*Rhynchops niger*
Columbiformes	Columbidae	Croaking ground dove	*Columbina cruziana*
		West Peruvian dove	*Zenaida meloda*
		Eared dove	*Zenaida auriculata*
		Picui ground dove	*Columbina picui*
Strigiformes	Tytonidae	Barn owl	*Tyto alba*
	Strigidae	Short-eared owl	*Asio flammeus*
		Burrowing owl	*Athene cunicularia*
Caprimulgiformes	Caprimulgidae	Band-winged nightjar	*Caprimulgus longirostris*
Apodiformes	Apodidae	Andean swift	*Aeronautes andecolus*
		Chimney swift	*Chaetura pelágica*

(continued)

Table 5.7 (continued)

Order	Family	Common name	Scientific name
	Trochilidae	Giant hummingbird	*Patagona gigas*
		Green-backed firecrown	*Sephanoidessephanoides*
		Oasis hummingbird	*Rhodopis vesper*
		Peruvian sheartail	*Thaumastura cora*
		Chilean woodstar	*Eulidia yarrellii*
Coraciformes	Alcedinidae	Green kingfisher	*Chloroceryleamericana*
Passeriformes	Furnaridae	Buff-winged cinclodes	*Cinclodes fuscus*
		Dark-bellied cinclodes	*Cinclodes oustaleti*
		Chilean seaside cinclodes	*Cinclodesnigrofumosus*
		Greyish miner	*Geositta maritime*
		Common miner	*Geossitta cunicularia*
		Scale-throated earthcreeper	*Upucerthia dumetaria*
		Plain-mantled tit-spinetail	*Leptasthenura aegithaloides*
		Wren-like rushbird	*Phleocrypthesmelanops*
	Tyrannidae	Dark-faced ground tyrant	*Muscisaxicola macloviana*
		Spot-billed ground tyrant	*Muscisaxicola maculirostris*
		Rufous-naped ground tyrant	*Muscisaxicola rufivertex*
		Black-billed shrike-tyrant	*Agriornis* montana
		White-crested elaenia	*Elaenia albiceps*
		Yellow-billed tit-tyrant	*Anairetes flavirostris*
		Pied-crested tit-tyrant	*Anairetes reguloides*
		Vermilion flycatcher	*Pyrocephalus rubinus*

(continued)

Table 5.7 (continued)

Order	Family	Common name	Scientific name
		Many-coloured rush tyrant	*Tachuris rubigastra*
		Bran-colored flycatcher	*Myiophobus fasciatus*
		Austral negrito	*Lessonia rufa*
		Andean negrito	*Lessonia oreas*
		Short-tailed field tyrant	*Muscigralla brevicauda*
		Cazamoscas tijereta	*Tyrannus savana*
		Spectacled tyrant	*Hymenops perspicillatus*
	Hirundinidae	Blue-and-white swallow	*Pygochelidon cyanoleuca*
		Southern martin	*Progne elegans*
		American cliff swallow	*Petrochelidonpyrrhonota*
		Sand martin	*Riparia riparia*
		Barn swallow	*Hirundo rustica*
		Chilean swallow	*Tachycineta mellen*
		Tawny-headed swallow	*Alopochelidon fucata*
	Motacillidae	Yellowish pipit	*Anthus lutescens*
		Correndera pipit	*Anthus correndera*
	Troglodytidae	House wren	*Troglodytes aedon*
Passeriformes		Sedge Wren	*Cistophorisplatenses*
	Cotingidae	Rufous-taile Plantcutter	*Phytotoma rara*
	Thaupidae	Cinereus Conebill	*Conirostrum cinereum*
		Tamarugo Conebill	*Conirostrum tamarugense*
	Emberizidae	Rufous-collared Sparrow	*Zonotrichiacapensis*
		Slender-billed Finch	*Xenospingus concolor*
		Band-tailed Seedeater	*Catamenia anales*

(continued)

Table 5.7 (continued)

Order	Family	Common name	Scientific name
		Blue-black grassquit	*Volatinia jacarina*
		Black-tawny Seedeater	*Sporophila Velasco*
		Mourning Sierra Finch	*Phrygilus fruticetti*
		Band-tailed Sierra Finch	*Phrygilus alaudinus*
		Common Diuca Finch	*Diuca diuca*
	Cardinalidae	Black-backed Grosbeak	*Pheucticus aureoventris*
	Icteridae	Peruvian Medowlark	*Sturnella bellicosa*
		Long-tailed Medowlark	*Sturnella loyca*
		Shiny Cowbird	*Molothrus bonariensis*
		Austral Blackbird	*Curaeus curaeus*
		Yellow-backed Blackbird	*Agelasticus thilius*
	Fringillidae	Hooded Siskin	*Carduelis magellanica*
		Black-chinned Siskin	*Carduelis barbata*
	Passeridae	House Sparrow	*Passer domesticus*

By the status and wetland, the highest numbers were for resident species, with Lluta presenting 104 (74.3%) and Huasco 76 (39.2%) breeds. For visiting species, Lluta has displayed 25 (96.2%), San Jose 21 (80.8%), Camarones 20 (76.9%) and Vitor 19 (73.1%) of species.

In the case of accidental species, Lluta presented 25 species (89.3%), San Jose 4 (14.3%), Vitor and Camarones 3 species (10.7%) each. In all of the studied wetlands, the resident birds were the most numerous ones, while the visiting birds, in their totality, corresponded to boreal migrants, as curlew, Franklin gull, and the accidental birds come mostly from North America, and the neotropical region (the tropical America), with normal ranges of distribution to the north or east of Chile. Among them are the Western Sandpiper and the Wood Stork. Table 5.9 presents a summary of the state for each one of the wetlands studied.

Table 5.8 Conservation status of the bird species

	Residence in Chile			State of conservation			
Common name	Visitor	Resident	Accidental	IUCN	CONAMA	SAG	CONAF
Pato jergón chico		X		LC			
Pato gargantillo		X		LC	R	R	
Pato colorado		X		LC			
Pato alas azules		X		LC			
Pato jergón grande		X		LC			
Pato puna		X		LC			
Pato real		X		LC			
Pato cuchara		X		LC	I	IC	
Pato capuchino		X		LC			
Pato juarjual		X		LC			
Pato castaño		X		LC			
Pato rana de pico delgado		X		LC			
Pato rana pico ancho		X		LC			
Pato silbón		X		LC			
Pato silbón de ala blanca		X		LC			
Cisne de cuello negro		X		LC			
Blanquillo		X		LC			
Pimpollo		X		LC			
Huala		X		LC			
Picurio		X		LC			
Flamenco chileno		X		nT		V	
Pelícano		X		nT			
Pelícano pardo		X		LC			
Piquero		X		LC	I	IC	
Piquero de patas azules		X		LC			
Yeco		X		LC			
Lile		X		nT		IC	
Guanay		X		nT		V	
Garza grande		X		LC			
Garza chica		X		LC			
Garza azul		X		LC			
Garza boyera		X		LC			
Garza cuca		X		LC		R	
Garza tricolor		X		LC			
Garcita azulada		X		LC			
Huairavo		X		LC			

(continued)

Table 5.8 (continued)

Common name	Residence in Chile			State of conservation			
	Visitor	Resident	Accidental	IUCN	CONAMA	SAG	CONAF
Huairavillo		X		LC		R	
Huairavillo menor		X		LC			
Cuervo de pantano de la puna		X		LC		V	
Bandurria		X		LC	P	V	
Espátula		X		LC			
Cigüeña de cabeza pelada		X		LC			
Jote cabeza colorada		X		LC			
Jote cabeza negra		X		LC			
Águila pescadora		X		LC	V	V	
Vari		X		LC			
Peuco		X		LC			
Bailarín		X		LC			
Aguilucho		X		LC			
Águila		X		LC			
Cernícalo del norte		X		LC			
Halcón peregrino		X		X	LC	V	EP
Halcón perdiguero		X		LC			
Traro		X		LC			
Tiuque		X		LC			
Tagüita del norte		X		LC			
Tagüita		X		LC			
Tagua andina		X		LC			
Tagua chica		X		LC			
Tagua común		X		LC			
Tagua frente roja		X		LC			
Tagua gigante		X		LC	V	V	
Pidén		X		LC			
Pidencito		X		LC	I	IC	
Queltehue de la puna		X		LC			
Queltehue		X		LC			
Chorlo semipalmado		X		LC			
Chorlo nevado		X		LC			
Chorlo de pico grueso		X		LC			
Chorlo de doble collar		X		LC			
Chorlo de collar		X		LC			

(continued)

Table 5.8 (continued)

Common name	Residence in Chile			State of conservation			
	Visitor	Resident	Accidental	IUCN	CONAMA	SAG	CONAF
Chorlo gritón		X		LC			
Chorlo chileno		X		LC			
Chorlo ártico		X		LC			
Chorlo dorado		X		LC			
Pilpilén		X		LC			
Pilpilén negro		X		LC			
Perrito		X		LC			
Caití		X		LC			
Chorlo cabezón		X		LC			
Pitotoy chico		X		LC			
Pitotoy grande		X		LC			
Pitotoy solitario		X		LC			
Playero gris		X		LC			
Playero grande		X		LC			
Playero manchado		X		LC			
Playero vuelvepiedra		X		LC			
Playero de las rompientes		X		LC			
Playero enano		X		LC			
Playero semipalmado		X		LC			
Playero de Baird		X		LC			
Playero blanco		X		LC			
Playero pectoral		X		LC			
Playero de patas largas		X		LC			
Playero ártico		X		LC	EP		
Playero de lomo blanco		X		LC			
Playero occidental		X		LC			
Playero canela		X		nT			
Zarapito		X		LC			
Zarapito de pico recto		X		LC			
Becacina chica		X		LC			
Pollito de mar tricolor		X		LC			
Pollito de mar boreal		X		LC			
Pollito de mar rojizo		X		LC			

(continued)

Table 5.8 (continued)

	Residence in Chile			State of conservation			
Common name	Visitor	Resident	Accidental	IUCN	CONAMA	SAG	CONAF
Perdicita chica		X		LC			
Jacana		X		LC			
Gaviota de capucho gris		X		LC			
Gaviota andina		X		LC	V	R	
Gaviota cáhuil		X		LC			
Gaviota garuma		X		LC	V	V	
Gaviota reidora		X		LC			
Gaviota de Franklin		X		LC			
Gaviota peruana		X		LC			
Gaviota dominicana		X		LC			
Gaviotín monja		X		nT	V	V	
Gaviotín chico		X		En	EP	P	EP
Gaviotín ártico		X		LC			
Gaviotín boreal		X		LC			
Gaviotín sudamericano		X		LC			
Gaviotín de pico grueso		X		LC			
Gaviotín elegante		X		nT			
Gaviotín real		X		LC			
Gaviotín de Sandwich		X		LC			
Rayador		X		LC			
Quiguagua		X		LC			
Paloma de alas blancas		X		LC			
Tórtola		X		LC			
Tortolita cuyana		X		LC			
Lechuza		X		LC			
Nuco		X		LC	I	IC	
Pequén		X		LC			
Gallinita ciega		X		LC			
Vencejo chico		X		LC			
Vencejo chimenea		X		nT			
Picaflor gigante		X		LC			
Picaflor		X		LC			
Picaflor del norte		X		LC			
Picaflor de Cora		X		LC			
Picaflor de Arica		X		En	EP	V	V

(continued)

Table 5.8 (continued)

Common name	Residence in Chile			State of conservation			
	Visitor	Resident	Accidental	IUCN	CONAMA	SAG	CONAF
y R							
Martín pescador chico		X		LC			
Churrete acanelado		X		LC			
Churrete chico		X		LC			
Churrete costero		X		LC			
Minero chico		X		LC			
Minero común		X		LC			
Bandurilla		X		LC			
Tijeral		X		LC			
Trabajador		X		LC			
Dormilona tontita		X		LC			
Dormilona chica		X		LC			
Dormilona nuca rojiza		X		LC			
Mero gaucho		x		LC			
Fío-fío		X		LC			
Cachudito del norte		X		LC			
Cachudito de cresta blanca		X		LC			
Saca-tu-real		X		LC			
Siete colores		X		LC			
Cazamosca pico chato		X		LC			
Colegial		X		LC			
Colegial del norte		X		LC			
Cazamoscas de cola corta		X		LC			
Cazamoscas tijereta		X		LC			
Run-rún		X		LC			
Golondrina de dorso negro		X		LC			
Golondrina negra		X		LC			
Golondrina grande		X		LC			
Golondrina barranquera		X		LC			
Golondrina bermeja		X		LC			
Golondrina chilena		X		LC			

(continued)

Table 5.8 (continued)

Common name	Residence in Chile			State of conservation			
	Visitor	Resident	Accidental	IUCN	CONAMA	SAG	CONAF
Golondrina de cabeza rojiza		X		LC			
Bailarín chico peruano		X		LC			
Bailarín chico		X		LC			
Chercán		X		LC			
Chercán de las vegas		X		LC			
Rara		X		LC			
Comesebo chico		X		LC			
Comesebo de los tamarugales		X		VU	IC		
Chincol		X		LC			
Pizarrita		X		nT			
Semillero		X		LC			
Negrillo		X		LC			
Corbatita		X		LC			
Yal		X		LC			
Platero		X		LC			
Diuca		X		LC			
Rey del bosque		X		LC			
Loica peruana		X		LC			
Loica		X		LC			
Mirlo		X		LC			
Tordo		X		LC			
Trile		X		LC			
Jilguero peruano		X		LC			
Jilguero		X		LC			
Gorrión		X		LC			

IUCN (Lista Roja de la UICn de Especies Amenazadas 2010 V.3:): *En* endangered, *VU* vulnerable, *nT* almost endangered, *LC* minor worry

CONAMA (Primera y Segunda Clasificación de Especies Silvestres según su Estado de Conservación, 2007–2008): *EP* endangered, *R* rare

SAG (Ley de Caza y su Reglamento, 2009): *R* rare, *P* in danger of extinction, *I* scarcely or inadequately known, *V* vulnerable

CONAF (Libro Rojo de los Vertebrados de Chile, 2da. Edición, 1993): *EP* endangered, *V* vulnerable, *R* rare, *IC* insufficiently known

Use of Bird Habitat

Out of all the habitats evaluated, the most visited by the birds were the lagoon shores (95 sp.), river mouths and associated seacoast (94 sp.), and lagoons (83 sp.), for feeding and resting. The most frequent use by habitat was: on tidal strip: feeding

(60 sp.); on the beach: resting (77 sp.); on the lagoon shore: feeding (95 sp.); in the lagoons: feeding (83 sp.); in the river mouth: resting (94 sp.); in the humid vegetation zone: the resting; in the area of low vegetation or without vegetation: feeding (58 sp.); and in the area of high vegetation: resting (69 sp.). The habitat most frequented by use, was: resting at the river mouth (94 sp.); feeding on the lagoon shores (95 sp.), and reproduction in the high vegetation zone (25 sp.) (Table 5.10).

Latitud Analysis of Assembly Structure

Specific Assembly Structure

The analysis of similarity (dendrogram), according to information in Table 5.11, indicates a remarkable separation (similarity 40.8%) between the assemblies of birds of the Copiapo and Huasco rivers, with those north of Chañaral (Lluta, San Jose, Vitor, Camarones, Tana and Loa) (Fig. 5.3).

Table 5.9 Residence condition of the birds occurring on the coastal saltmarshes of northern Chile

Wetland	Total species(= 194)	Visitors	Residents	Accidental
Lluta	154 (79.38%)	25	104	25
San Jose	73 (37.62%)	21	48	4
Vitor	84 (43,30%)	19	61	3
Camarones	83 (42.78%)	20	60	3
Tana	30 (15.46%)	3	28	0
Loa	37 (19.10%)	5	31	1
Salado	33 (17.01%)	3	30	0
Copiapo	56 (28.87%)	6	50	0
Huasco	84 (43.30%)	8	76	1

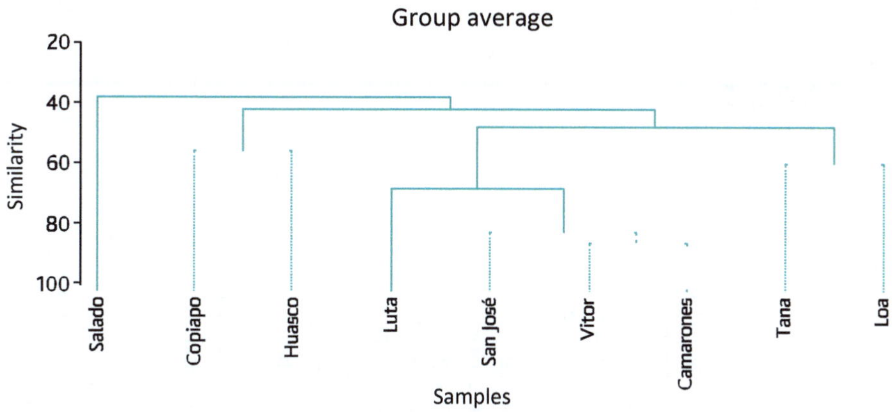

Fig. 5.3 Species composition similarity between the bird assemblies occurring in the saltmarshes of the northern coast of Chile

Table 5.10 Habitat use of the birds within each wetland of the northern Chilean coast

Common name	Tidal strip ((Sand and/or rocks)	Beach (Sand and/or rocks)	Lagoon shore (with vegetation or scarce)	lagoons (interior)	River mouth	Humid vegetation zone (totoral, reeds)	Low vegetation zone (pastiure) or without	High vegetation zone(bushes, shrubs)	Bird type
Pato jergón chico			D	C	C	D	D		A
Pato gargantillo			D	C	C	D	D		A
Pato colorado			D	C	C	D	D		A
Pato alas azules			D	C	C	D	D		A
Pato jergón grande			D	C	C	D	D		A
Pato puna			D	C	C	D	D		A
Pato real			D	C	C	D	D		A
Pato cuchara			D	C	C	D	D		A
Pato capuchino			D	C	C	D	D		A
Pato juarjual			D	C	C	D	D		A
Pato castaño			D	C	C	D	D		A
Pato rana de pico delgado			D	C	C	D	D		A
Pato rana pico ancho			D	C	C	D	D		A
Pato silbón			D	C	C	D	D		A
Pato silbón de ala blanca			D	C	C	D	D		A
Cisne de cuello negro			D	C	C	D	D		A
Blanquillo			D	C	C	D	D		A
Pimpollo			D	C	C	D	D		A
Huala	C		D	C	C	D	D		A
Picurio			D	C	C	D	D		A
Flamenco chileno	C	D	D	C	C				A
Pelícano	C	D							M

Pelícano pardo	C	D							M
Piquero	C	D							M
Piquero de patas azules	C	D							M
Yeco	C	D	D	C	C				M
Lile	C	D							M
Guanay	C	D							M
Garza grande		D	D	C	C	D	D		A
Garza chica	C	D	D	C	C	D	D		A
Garza azul		D	D	C	C	D	D	D	A
Garza boyera			C			D	C-D	D	A
Garza cuca			D	C	C	D	C-D		A
Garza tricolor			D	C	C	D	D		A
Garcita azulada						C-D			A
Huairavo	C	D	D	C	C	D	D	D	A
Huairavillo						C-D			A
Huairavillo menor						C-D			A
Cuervo de pantano de la puna		D	D	C	C				A
Bandurria			C-D		C		C-D		A
Espátula			D	C	C	D	D		A
Cigüeña de cabeza pelada			C-D	C	C	D	D		A
Jote cabeza colorada		C-D	C-D				R-C-D	R-D	T
Jote cabeza negra		C-D	C-D				R-C-D	R-D	T
Águila pescadora	C			C	C			D	T
Vari		C	C			C	C	C-D	T
Peuco		C	C			C	C	C-D	T

(continued)

Table 5.10 (continued)

Common name	Tidal strip ((Sand and/or rocks)	Beach (Sand and/or rocks)	Lagoon shore (with vegetation or scarce)	lagoons (interior)	River mouth	Humid vegetation zone (totoral, reeds)	Low vegetation zone (pastiure) or without	High vegetation zone(bushes, shrubs)	Bird type
Bailarín		C	C			C	C	C-D	T
Aguilucho		C	C			C	C	C-D	T
Águila		C	C			C	C	C-D	T
Cernícalo		C	C			C	C	R-C-D	T
Halcón peregrino	C	C-D	C	C	C	C	C	C-D	T
Halcón perdiguero	C	C	C	C	C	C	S	C-D	T
Traro		C-D	C-D				C-D	D	T
Tiuque	C	C-D	C	C	C	C	C	D	T
Tagüita del norte			C	C	C	R-D	D		T
Tagüita			C	C	C	D	D		A
Tagua andina			C	C	C	D	D		A
Tagua chica			C	C	C	D	D		A
Tagua común			C	C	C	D	D		A
Tagua frente roja			C	C	C	D	D		A
Tagua gigante			C	C	C	D	D		A
Pidén			C	C	C	D	D		A
Pidencito			C	C	C	D	D		A
Queltehue de la puna		D	C				C-D		A
Queltehue		D	C				C-D		A
Chorlo semipalmado		D	C		C				A
Chorlo nevado	C	R-D	C						A
Chorlo de pico grueso	C	D	C						A
Chorlo de doble collar		D	C						A
Chorlo de collar		D	C						A

Chorlo gritón		D	C		C		R-C-D		A
Chorlo chileno	C	D	C-D		C		C-D		A
Chorlo ártico	C	D	C-D		C		C-D		A
Chorlo dorado	C	D	C-D		C		C-D		A
Pilpilén	C	R-D	C		C		R-D		A
Pilpilén negro	C	D							A
Perrito		D	C	C	C	D			A
Caití		D	C	C	C	D			A
Chorlo cabezón							C-D		T
Pitotoy chico	C	D	C	C	C	D	D		A
Pitotoy grande	C	D	C	C	C	D	D		A
Pitotoy solitario	C	D	C	C	C	D	D		A
Playero gris	C	D	C	C	C	D	D		A
Playero grande	C	D	C						A
Playero manchado		D	C		C	C-D			A
Playero vuelvepiedra	C	D	C		C				A
Playero de las rompientes	C	D	C		C				A
Playero enano	C	D	C		C				A
Playero semipalmado	C	D	C		C				A
Playero de Baird	C	D	C		C				A
Playero blanco	C	D	C		C				A
Playero pectoral	C	D	C		C				A
Playero de patas largas	C	D	C		C				A
Playero ártico	C	D	C		C				A
Playero de lomo blanco	C	D	C		C				A

(continued)

Table 5.10 (continued)

Common name	Tidal strip ((Sand and/or rocks)	Beach (Sand and/or rocks)	Lagoon shore (with vegetation or scarce)	lagoons (interior)	River mouth	Humid vegetation zone (totoral, reeds)	Low vegetation zone (pastiure) or without	High vegetation zone(bushes, shrubs)	Bird type
Playero occidental	C	D	C		C				A
Playero canela	C	D	C		C				A
Zarapito	C	D	C		C				A
Zarapito de pico recto	C	D	C		C				A
Becacina chica			C	C	C	C-D			A
Pollito de mar tricolor		D	C	C	C	C			A
Pollito de mar boreal		D	C	C	C	C			A
Pollito de mar rojizo		D	C	C	C	C			A
Perdicita chica							C-D		T
Jacana			C			C-D			A
Gaviota de capucho gris	C	D	C	C	C				M
Gaviota andina	C	D	C	C	C		D		M
Gaviota cáhuil	C	D	C	C	C				M
Gaviota garuma	C	D	C	C	C				M
Gaviota reidora	C	D	C	C	C		D		M
Gaviota de Franklin	C	D	C	C	C				M
Gaviota peruana	C	D	C	C	C				M
Gaviota dominicana	C	D	C	C	C		D		M
Gaviotín monja	C	D	C	C	C				M
Gaviotín chico	C	D	C	C	C				M
Gaviotín ártico	C	D	C	C	C				M
Gaviotín boreal	C	D	C	C	C				M
Gaviotín sudamericano	C	D	C	C	C				M
Gaviotín de pico grueso	C	D	C	C	C				M

Gaviotín elegante	C	D	C	C	C				M
Gaviotín real	C	D	C	C	C				M
Gaviotín de Sandwich	C	D	C	C	C				M
Rayador	C	D	D	C	C				M
Quiguagua							C	R-D	T
Paloma de alas blancas							C	R-D	T
Tórtola							C	R-D	T
Tortolita cuyana							C	R-D	T
Lechuza							C	D	T
nuco							C	D	T
Pequén							R-D-C		T
Gallinita ciega							R-D-C		T
Vencejo chico			C	C	C	C	C	C	T
Vencejo chimenea			C	C	C	C	C	C	T
Picaflor gigante								C-R-D	T
Picaflor								C-R-D	T
Picaflor del norte								C-R-D	T
Picaflor de Cora								C-R-D	T
Picaflor de Arica								C-R-D	T
Martín pescador chico				C	C	C-D			T
Churrete acanelado									T
Churrete			C		C	C-D	C	D	T
Churrete costero	C								T
Minero chico							C-D		T
Minero común							C-D		T

(continued)

Table 5.10 (continued)

Common name	Tidal strip ((Sand and/or rocks)	Beach (Sand and/or rocks)	Lagoon shore (with vegetation or scarce)	lagoons (interior)	River mouth	Humid vegetation zone (totoral, reeds)	Low vegetation zone (pastiure) or without	High vegetation zone(bushes, shrubs)	Bird type
Bandurrilla							C-D		
Tijeral								C-D	T
Trabajador						C-R-D		C	T
Dormilona tontita		C-D	C				C-D	C-D	T
Dormilona chica		C-D	C				C-D	C-D	T
Dormilona nuca rojiza		C-D	C				C-D	C-T	T
Mero gaucho							C-D	D	T
Fío-fío								C-R-D	T
Cachudito del norte								C-R-D	T
Cachudito de cresta blanca								C-R-D	T
Saca-tu-real				C		C-D	C	C-R-D	T
Siete colores						C-R-D			T
Cazamosca pico chato						C-D	C	C-R-D	T
Colegial		C	C		C		C-D		T
Colegial del norte		C	C		C		C-D		T
Cazamoscas de cola corta							C-D	C-D	T
Cazamoscas tijereta							C	C-D	T
Run-rún						C-R-D	C	C-D	T
Golondrina de dorso negro			C	C	C	C	C	C	T
Golondrina negra			C	C	C	C	C	C-D	T
Golondrina grande			C	C	C	C	C	C-D	T
Golondrinabarranquera			C	C	C	C	C	C-D	T
Golondrina bermeja			C	C	C	C	C	C-D	T
Golondrina chilena			C	C	C	C	C	C-D	T

Golondrina de cabeza rojiza			C	C	C	C	C	C-D	T
Bailarín chico peruano							C-D		T
Bailarín chico							C-D		T
Chercán						C	C-R-D	C-R-D	T
Chercán de las vegas						C	C-R-D	C-R-D	T
Rara							C	C-R-D	T
Comesebo chico							C	C-R-D	T
Comesebo de los tamarugales							C	C-R-D	T
Chincol						C	C	C-R-D	T
Pizarrita						C	C	C-R-D	T
Semillero							C	C-D	T
negrillo							C	C-D	T
Corbatita							C	C-R-D	T
Yal							C	C-D	T
Platero							C	C-D	T
Diuca							C	C-D	T
Rey del bosque							C	C-D	T
Loica peruana							C-R-C	C-D	T
Loica							C	C-D	T
Mirlo							C	C-D	T
Tordo							C	C-D	T
Trile			C		C	C-D	C-D	C-D	T
Jilguero peruano							C	C-D	T
Jilguero							C	C-D	T
Gorrión							C	C-D	T

R Rest, *M* Marine, *F* Feeding, *A* Aquatic, *B* Breeding, *T* Terrestrial, *S* Flying predator

Table 5.11 Geographical distribution of the birds species on the saltmarshes

Common name	specie	lluta	san Jose	Vitor	camarones	tana	loa	salado	copiapo	Huasco
Pato jergón chico	*Anas flavirostris*	X		X	X		X		O	X
Pato gargantillo	*Anas bahamensis*	X		X	X		X		X	
Pato colorado	*Anas cyanoptera*	X	X	X	X		X		X	X
Pato alas azules	*Anas discors*	X								
Pato jergón grande	*Anas geórgica*	X		X	X		X		X	X
Pato puna	*Anas puna*	X	X		X					
Pato real	*Anas sibilatrix*								X	X
Pato cuchara	*Anas platalea*	X								X
Pato capuchino	*Anas versicolor*	X								
Pato juarjual	*Lophonettaspecularioides*	X								
Pato castaño	*Nettaerythrophthalma*	X								
Pato rana de pico delgado	*Oxyura vitatta*									X
Pato rana pico ancho	*Oxyura jamaicensis*	X								
Pato silbón	*Dendrocygnabicolor*	X								
Pato silbón de ala blanca	*Dendrocygnaautumnales*	X								
Cisne de cuello negro	*Cygnusmelanocoryphus*									X
Blanquillo	*Podiceps occipìtalis*	X							X	
Pimpollo	*Podiceps rolland*									X
Huala	*Podiceps major*	X	X							X
Picurio	*Podilymbus podiceps*								X	X
Flamenco chileno	*Phoenicopteruschilensis*	X	X	X	X	X				X
Pelícano	*Pelecanus thagus*	X	X	X	X	X	X	X		X
Pelícano pardo	*PelecanusoccidentalIs*	X	X							
Piquero	*Sula variegata*	X	X	X	X	X	X	X		
Piquero de patas azules	*Sula nebouxii*	X	X	X	X					

Yeco	*Phalacrocorax brasilanus*	X	X	X	X	X	X	X	X	X
Lile	*Phalacrocorax gaimardi*	X	X	X	X	X	X		X	
Guanay	*Phalacrocorax bougainvillii*	X	X	X	X	X	X	X		
Garza grande	*Cosmerodius albus*	X	X	X	X			X	X	X
Garza chica	*Egretta thula*	X	X	X	X	X	X	X		X
Garza azul	*Egretta caerulea*	X	X	X						X
Garza boyera	*Bubulcus ibis*	X	X		X			X		X
Garza cuca	*Ardea cocoi*	X								X
Garza tricolor	*Egretta tricolor*	X								
Garcita azulada	*Butorides striata*	X					X			
Huairavo	*Nycticorax nycticorax*	X	X	X	X	X	X	X		X
Huairavillo	*Ixobrichus involucris*									X
Huairavillo menor	*Ixobrychus exilis*									
Cuervo de pantano de la puna	*Plegadis ridgwayi*	X								
Bandurria	*Theristicus melanosis*	X			X			X	X	X
Espátula	*Platalea ajaja*	X								
Cigüeña de cabeza pelada	*Mycteria americana*	X								
Jote cabeza colorada	*Cathartes aura*	X	X	X	X	X	X	X		X
Jote cabeza negra	*Coragyps atratus*	X								X
Águila pescadora	*Pandion haliaetus*	X	X	X	X					
Vari	*Circus cinereus*	X							X	X
Peuco	*Parabuteo unicinctus*	X								X
Bailarín	*Elanus leucurus*									X
Aguilucho	*Buteo polyosoma*							X	O	X
Águila	*Geranoaetus melanoleucus*	X	X	X	X		X			

(continued)

Table 5.11 (continued)

Common name	specie	lluta	san Jose	Vitor	camarones	tana	loa	salado	copiapo	Huasco
Cernícalo del norte	*Falco sparverius*	X	X	X	X	X			O	
Halcón peregrino	*Falco peregrinus*	X	X	X	X					
Halcón perdiguero	*Falco femoralis*	X	X	X	X					
Traro	*Caracara plancus*	X					X		X	X
Tiuque	*Milvago chimango*							X	X	X
Tagüita del norte	*Gallinula galatea*	X	X	X	X	X	X		X	X
Tagüita	*Gallinula melanops*	X								X
Tagua andina	*Fulica ardesiaca*	X			X		X			X
Tagua chica	*Fulica leucoptera*	X					X		X	X
Tagua común	*Fulica armillata*								X	X
Tagua frente roja	*Fulica rufifrons*	X							O	X
Tagua gigante	*Fulica gigantea*	X								
Pidén	*Pardirallus sanguinolentus*	X		X		X			O	X
Pidencito	*Laterallus jamaicensis*									X
Queltehue de la puna	*Vanellus resplendens*	X								
Queltehue	*Vanellus chilensis*							X	X	X
Chorlo semipalmado	*Charadrius semipalmatus*	X	X	X						
Chorlo nevado	*Charadrius alexandrinus*	X	X		X				X	X
Chorlo de pico grueso	*Charadrius wilsonia*		X							
Chorlo de doble collar	*Charadrius falklandicus*	X								X
Chorlo de collar	*Charadrius collares*	X								X
Chorlo gritón	*Charadrius vociferus*	X		X	X					
Chorlo chileno	*Charadrius modestus*	X						X	X	
Chorlo ártico	*Pluvialis squatarola*	X	X	X	X				X	
Chorlo dorado	*Pluviales dominica*	X	X	X	X					
Pilpilén	*Haematopus palliatus*	X	X	X	X	X	X	X	X	X
Pilpilén negro	*Haematopus ater*	X	X	X	X	X		X	X	X

Perrito	*Himantopus mexicanus*	X								X
Caití	*Recurvirostra andina*	X								
Chorlo cabezón	*Burhinus superciliaris*	X		X						
Pitotoy chico	*Tringa flavipes*	X	X	X	X		X			X
Pitotoy grande	*Tringa melanoleuca*	X	X	X	X				X	X
Pitotoy solitario	*Tringa solitaria*	X			X					
Playero gris	*Tringa incana*	X								
Playero grande	*Tringa semipalmata*	X	X	X	X					X
Playero manchado	*Actitis macularius*	X	X	X	X					
Playero vuelvepiedra	*Arenaria interpres*	X	X	X	X			X	X	X
Playero de las rompientes	*Aphriza virgata*	X	X	X	X					X
Playero enano	*Calidris minutilla*	X	X	X	X					
Playero semipalmado	*Calidris pusilla*	X	X	X	X					
Playero de Baird	*Calidris bairdii*	X	X	X	X		X		O	X
Playero blanco	*Calidris alba*	X	X	X	X				X	X
Playero pectoral	*Calidris melanotos*	X	X							
Playero de patas largas	*Calidris himantopus*	X			X					
Playero ártico	*Calidris canutus*	X		X						
Playero de lomo blanco	*Calidris fuscicollis*	X								
Playero occidental	*Calidris mauri*	X								
Playero canela	*Tryngites subruficollis*	X								
Zarapito	*Numenius phaeopus*	X	X	X	X	X	X	X	X	X
Zarapito de pico recto	*Limosa haemastica*	X								
Becacina chica	*Limnodromus griseus*	X								
Pollito de mar tricolor	*Phalaropus tricolor*	X	X	X	X		X			

(continued)

Table 5.11 (continued)

Common name	specie	lluta	san Jose	Vitor	camarones	tana	loa	salado	copiapo	Huasco
Pollito de mar boreal	*Phalaropus lobatus*	X			X					
Pollito de mar rojizo	*Phalaropus fulicaria*	X								
Perdicita chica	*Thinocorusrumicivorus*	X						X	O	
Jacana	*Jacana jacana*	X								
Gaviota de capucho gris	*Chroicocephalus cirrocephalus*	X	X							
Gaviota andina	*Chroicocephalus serranus*	X	X	X	X					X
Gaviota cáhuil	*Chroicocephalus maculipennis*									X
Gaviota garuma	*Leucophaeus modestus*	X	X	X	X	X	X		X	X
Gaviota reidora	*Leucophaeus atricilla*	X								
Gaviota de Franklin	*Leucophaeus pipixcan*	X	X	X	X	X	X	X	X	X
Gaviota peruana	*Larus belcheri*	X	X	X	X	X	X			X
Gaviota dominicana	*Larus dominicanus*	X	X	X	X	X	X	X	X	X
Gaviotín monja	*Larosterna inca*	X	X	X	X		X			
Gaviotín chico	*Sternula lorata*	X			X		X			
Gaviotín ártico	*Sterna paradisaea*	X	X		x					X
Gaviotín boreal	*Sterna hirundo*	X	X	X	X					
Gaviotín sudamericano	*Sterna hirundinacea*	X	X	X	X					
Gaviotín de pico grueso	*Gelochelidon nilotica*	X								
Gaviotín elegante	*Thalasseus elegans*	X	X	X	X	X	X			
Gaviotín real	*Thalasseus maximus*	X								
Gaviotín de Sandwich	*Thalasseus sandvicensis*	X								
Rayador	*Rhynchops niger*	X	X	X	X					
Quiguagua	*Columbina cruziana*	X	X	X	X	X				
Paloma de alas blancas	*Zenaida meloda*	X	X	X	X	X				X
Tórtola	*Zenaida auriculata*	X	X	X	X					X
Tortolita cuyana	*Colombina picus*								O	

Lechuza	*Tyto alba*	X		X						
nuco	*Asio flammeus*	X							O	
Pequén	*Athene cunicularia*	X	X	X	X	X			O	
Gallinita ciega	*Caprimulgus longirostris*	X		X	X					
Vencejo chico	*Aeronautes andecolus*	X		X	X	X				
Vencejo chimenea	*Chaetura pelágica*	X								
Picaflor gigante	*Patagona gigas*									X
Picaflor	*Sephanoidessephanoides*							X		X
Picaflor del norte	*Rhodopis vesper*	X	X	X	X	X				X
Picaflor de Cora	*Thaumastura cora*	X	X	X	X					
Picaflor de Arica	*Eulidia yarrellii*	X	X	X						
Martín pescador chico	*Chloroceryleamericana*	X								
Churrete acanelado	*Cinclodes fuscus*							X		X
Churrete chico	*Cinclodes oustaleti*							x		X
Churrete costero	*Cinclodesnigrofumosus*			X	X	X				
Minero chico	*Geositta maritime*	X							O	
Minero común	*Geossitta cunicularia*	X			X			X	O	
Bandurrilla	*Upucerthia dumentaria*	X								
Tijeral	*Leptasthenura aegithaloides*			X	X			X	X	
Trabajador	*Phleocrypthesmelanops*	X			X	X	X		O	X
Dormilona tontita	*Muscisaxicola macloviana*	X	X	X	X	X	X	X		
Dormilona chica	*Muscisaxicola maculirostris*	X					X			
Dormilona nuca rojiza	*Muscisaxicola rufivertex*								X	
Mero gaucho	*Agriornis montana*							X		
Fío-fío	*Elaenia albiceps*	X		X	X					

(continued)

Table 5.11 (continued)

Common name	specie	lluta	san Jose	Vitor	camarones	tana	loa	salado	copiapo	Huasco
Cachudito del norte	*Anairetes flavirostris*	X								
Cachudito de cresta blanca	*Anairetes reguloides*				X					
Saca-tu-real	*Pyrocephalus rubinus*	X	X	X	X					
Siete colores	*Tachuris rubigastra*						X		O	X
Cazamosca pico chato	*Myiophobus fasciatus*	X		X						
Colegial	*Lessonia rufa*	X	X	X	X				X	X
Colegial del norte	*Lessonia oreas*	X	X	X	X					
Cazamoscas de cola corta	*Muscigralla brevicauda*	X								
Cazamoscas tijereta	*Tyrannus savana*	X								
Run-rún	*Hymenops perspicillatus*									X
Golondrina de dorso negro	*Pygochelidon cyanoleuca*	X	X	X	X				X	
Golondrina negra	*Progne elegans*	X								
Golondrina grande	*Petrochelidonpyrrhonota*	X		X						
Golondrinabarranquera	*Riparia riparia*	X								
Golondrina bermeja	*Hirundo rustica*	X	X	X	X	X	X		O	X
Golondrina chilena	*Tachycineta mellen*							X	O	X
Golondrina de cabeza rojiza	*Alopochelidon fucata*									X
Bailarín chico peruano	*Anthus lutescens*	X								
Bailarín chico	*Anthus correndera*								X	
Chercán	*Troglodytes aedon*	X	X	X	X			X	X	X
Chercán de las vegas	*Cistophorisplatenses*								X	X
Rara	*Phytotoma rara*									X
Comesebo chico	*Conirostrum cinereum*	X	X	X	X					
Comesebo de los tamarugales	*Conirostrum tamarugense*	X		X	X	X				
Chincol	*Zonotrichiacapensis*	X	X	X	X	X	X	X		X
Pizarrita	*Xenospingus concolor*	X	X	X	X	X				

Semillero	*Catamenia anales*	X								
negrillo	*Volatinia jacarina*	X								
Corbatita	*Sporophila telasco*	X		X						
Yal	*Phrygilus fruticetti*									X
Platero	*Phrygilus alaudinus*								X	
Diuca	*Diuca diuca*	X							X	
Rey del bosque	*Pheucticus aureoventris*	X								
Loica peruana	*Sturnella bellicosa*	X		X						X
Loica	*Sturnella loyca*									X
Mirlo	*Molothrus bonariensis*	X	X	X	X				X	X
Tordo	*Curaeus curaeus*									
Trile	*Agelasticus thilius*								X	X
Jilguero peruano	*Carduelis magellanica*	X		X	X					
Jilguero	*Carduelis barbata*									X
Gorrión	*Passer domesticus*	X	X	X			X	X		

O: datos de presencia de Squeo et al. (2006)

This segregation derives mainly from the fact that the wetlands of the Copiapo/Huasco region are on the northern border limit of the distribution of several species of the typical Mediterranean-type communities in Chile, a point previously highlighted by Myers (1994), WCMC (1992), Orians 1994). The species associated with wetlands are as it follows:

Copiapo Wetland: Picurio, vari, tiuque, tagua girl, common tagua, trile, bandurria, queltehue. The area of Copiapo is in turn, the southern limit of distribution of the northern hummingbird.

Vallenar River Wetland: Huala (*), Crane Heron, Black-Headed Pigeon, Mallard Duck, Black-necked Swan, Puppy, Cahuilian Gull, Dancer, Common Hummingbird, Rare, Thrush, Chilean Loon, Queltehue .

Most of the terrestrial species, such as the diuca, yal, tapaculo, diucón, cachudito, tjeral, giant hummingbird, have their northern limit of coastal distribution in Paposo and surroundings (Aramayo et al. 2005; Sielfeld obs. -2004).

It should also be considered that in the case of several species common to the two assemblies (Lluta/Loa with Salado/Huasco), there are subspecies between both sectors, an aspect that surely derives from the geographic isolation, and emphasizing for example the cases of the species: Worker, Kestrel, seven colors, chincol.

On the other hand, among the non-common species highlighted ecologically, such as collegiate/collegiate to the north, Peruvian loica/loica, male dancer/peruvian boy dancer, queltehue/queltehue de la puna, tagüita/tagüita del norte, among others.

A special case today is the Salado River, characterized by the intense contamination of its reservoirs from Potrero, which started in 1939. Its future recovery is practically impossible if one considers both the intensity of pollution, the extent of the contaminated area and the increasing expansion of the urban edge of Chañaral, which in the future should cover part of the former wetland.

Due to its location and close relationship with the Caldera region, in the past it must have been part of an important biological corridor for migratory waterfowl, hosting a bird ensemble similar to the one that today characterizes the mouth of the Copiapo River.

The wetlands at the river mouths of rivers and gorges north of Chañaral form a fairly homogeneous group (similarity 46.75%) in which three specie assemblies can be recognized: San Jose/Vitor/Camarones, Tana/Loa and Lluta.

Out of these, the most important one is undoubtedly the Lluta river wetland, which stands out for its high number of species and concentration of individuals (131 species, concentrations of about 3000 individuals). This is because this wetland is the satellite of the extensive ITE wetland (Tacna, Peru), and from which the Lluta is visited temporarily, without being able to sustain the populations of many of the larger species that arrive in the area by itself. Therefore, it represents an "inn" type wetland, which by its geographic location also constitutes the southernmost distributional limit of tropical, accidental species and unique records for Chile.

Of the remaining wetlands, the San Jose river one only reaches the sea during extremely rainy years of the "highland"/altiplano region, being normally endorheic. The Vitor and Camarones River mouths, both of exoreic type, are in general narrow, and therefore do not form extensive wetlands. Vitor is affected by mining pollution and Camarones by agriculture. They harbor species characteristic of the extreme north of Chile, but their concentrations are not comparable with those of the Lluta River.

The Tana ravine, normally endorheic type, empties north of the port of Pisagua, and does not generate a wetland, therefore it is a site mainly used by the typical seacoast birds. A similar situation occurs at the Loa river mouth, which is exoreic and presents a terminal lagoon, which, due to the increasing salinization of the river and the pollution derived from the different industries and the urbanization upstream, determines an assembly dominated by fundamentally marine species. The structure of the previous assemblies is also related to the number of taxa present per wetland, especially the wetland associated with the Lluta river, which contains more than 130 species, corresponding to 26.7% of all species, and 59% of all the families known to Chile (Table 5.8). The remaining wetlands have less than 80 species, Salado river being the most poorly represented, with only 34 species (Table 5.12).

Regarding the residence of the species, the Lluta and Loa rivers stand out with a very low percentage of resident species (10–11%), which, in the case of the Lluta river wetland, is related to its previously mentioned reason of being a satellite of the ITE (Tacna) wetland, from which short-term migrations to Lluta are generated. In the case of the Loa river mouth, its present condition is only of passage for several birds in transit. Only the Passeriformes and Gruiformes species are stable (Table 5.13).

Community Organization of the Assemblies

For this case, only the information derived from monthly registration programs obtained in the wetlands of the Lluta, Loa and Huasco rivers was available (Table 5.14).

In the wetland of the Lluta river, the numerically more representative species are the pelican (5.5%), the Dominican seagull (10.5%), the Franklin seagull (26.3%) and the grey gull (30.5%).

The most frequent aquatic species (= constants) are the pelican (58.8%), the great egret and the snowy (small) heron (both 82.4%), the red headed jay (70.6%), common gallinule (58.8%), the wader (64.7%), Baird's sandpiper (88.3%), Belcher's gull (76.5%), Dominican seagull (70.59%) and Rufous-colored sparrow (76.5%).

Seasonal and occasional and/or rare birds are all either an accessory or accidental, but they contribute to the high diversity that characterizes the wetland of the Lluta River, and are its main attraction and the reason for its protection.

Table 5.12 Systematic ordination of the bird assemblies in the saltmarshes

	Chile	Lluta		Loa		Salado		Copiapo		Huasco	
Classification	n	n	%	n	%	n	%	n	%	n	%
Order	22	14	63.6	10	45.5	7	31.8	12	54.5	11	50.0
Families	61	36	59.0	24	39.3	20	32.8	24	39.3	26	42.6
Species	487	131	26.9	54	11.9	33	6.8	56	11.5	79	16.2

Table 5.13 Residence status of the bird species

	Lluta		Loa		Salado		Copiapo		Huasco	
Residence status	n	%	n	%	n	%	n	%	n	%
Residents	16	12.0	6	11.1	30	90.9	50	89.3	46	58.2
Nesting	23	18.0	27	50.0	?	?	?	?	52	65.8
Visitants or migrants	46	35.0	14	25.9	3	9.1	6	10.7	16	20.3
Accidentals/occasional	40	35.0	34	62.9	–	–	–	–	17	21.5

In the case of the Loa river wetland, the species that are numerically more important (= dominant) are the red headed jot (11.4%), gargo gull (15.0%), Dominican gull (13.1%), Chercán (8.3%), The most frequent species are piquero (58.0%), pilpilen (3.3%), gull gull (89.5%), Dominican gull (79.0%), worker (68.4%), Tontito (73.7%), chercán (100.0%) and chincol (73.7%).

In the case of the Huasco river wetland, the common tagua (24.3%) and the Dominican seagull (28.7%) are numerically important (= dominant).

Regarding the occurrence frequency of species, it is generally noted that 37 species (46.8%) have a presence equal to or greater than 50% (= constant species), highlighting the continuous presence of the color duck, duck jergón grande, duck frog Slender, yeco, jote redhead, common tagua, black pilpilen, puppy, curlew, Dominican gull, collegiate and chincol, all with a frequency of 100%; Duck (92.0%), pimpollo (50.0%), huala (83.0%), picurio (75.0%), great heron. This shows that this system, unlike previous wetlands, is highly stable and with little replacement of species.

Diversities (Shannon and Simpson indexes; Table 5.15) calculated for the three wetlands generally indicate very close values (Shannon 2.3–2.6; Simpson 0.8–0.9). The Loa river wetland stands out for exhibiting the highest values, which is a result of generally low dominance values (all less than 15%, Table 5.8) in the main components. The lowest values correspond to the Lluta river wetland as a result of the high dominance of various species of marine type, among them Dominican seagull, Franklin gull and garuma gull.

Finally, it should be noted that under consideration of the general quantity of species per wetland, the number of seasonal visitors, the presence of rare and/or accidental species, either from highland environments or from the Peruvian area, it stands out without a doubt the Lluta river wetland. This site is uniquely situated in Chile, characterized by the presence of 10 species of Ardeidae, 2 species of Ciconidae, 10 species of Rallidae, 11 Charadriidae, 24 Scolapacidae, 16 Laridae, along with the terrestrial avifauna characteristic of the southern gorges of Peru, which includes negrillos, Peruvian seedlings, charlatans, corbatita, saca-tu-real, Peruvian loica, cachudito of the north and white crest, among other species.

Table 5.14 Frequency of abundance of the bird species recorded on Lluta, Loa, Copiapo and Huasco marshes

Name	Lluta[a]				Loa[b]				Huasco[c]			
Indicators	n%	D	f%	c	n%	D	f%	c	n%	D	f%	C
Yellow-billed teal	0.06	Acc	17.65	Acc					0.36	Acc	8.0	Acc
White-cheeked pintail	0.02	Acc	5.9	Acc	0.09	Acc	5.26	Acc				
Cinnamon teal	0.16	Acc	17.65	Acc	0.92	Acc	26.32	Acs	2.46	Acs	100.0	C
Blue-winged teal					0.09	Acc	10.53	Acc				
Yellow-billed pintail	0.04	Acc	29.41	Acs	0.13	Acc	15.79	Acc	4.56	Acs	100.0	C
Puna teal	0.01	Acc	5.9	Acc								
Chiloé wigeon									6.23	D	92.0	C
Red shoveler									0.08	Acc	25.0	Acs
Lake duck									0.64	Acc	100.0	C
Black-necked swan									0.01	Acc	8.0	Acc
Silvery grebe					0.04	Acc	5.26	Acc				
White-tufted Grebe									0.10	Acc	50.0	C
Great grebe									0.36	Acc	83.0	C
Pied-billed grebe									0.30	Acc	75.0	C
Chilean flamingo	0.01	Acc	5.9	Acc					0.03	Acc	17.0	Acc
Peruvian pelican	5.54	D	58.8	C	1.80	Acc	47.37	Acs	0.30	Acc	50.0	C
Peruvian booby					2.72	Acs	57.89	C				
Neotropic cormorant	3.69	Acs	52.9	C	0.44	Acc	21.05	Acc	2.80	Acs	100.0	C
Red-legged cormorant					0.18	Acc	10.53	Acc				
Guanay cormorant					2.06	Acc	26.32	Acs				
Great egret	0.19	Acc	82.35	C	0.44	Acc	42.11	Acs	0.23	Acc	92.0	C
Snowy egret	1.05	Acc	82.35	C					0.22	Acc	83.0	C
Little blue heron	0.06	Acc	47.06	Acs	0.04	Acc	5.26	Acc				
Cattle egret	0.53	Acc	35.29	Acs	0.09	Acc	10.53	Acc	0.01	Acc	8.0	Acc
Cocoi heron									0.14	Acc	83.0	C

(continued)

Table 5.14 (continued)

Name	Lluta[a]				Loa[b]				Huasco[c]			
Indicators	n%	D	f%	c	n%	D	f%	c	n%	D	f%	C
Black-crowned night	0.54	Acc	58.8	C	0.57	Acc	36.84	Acs	0.65	Acc	83.0	C
Stripe-backed bittern									0.03	Acc	25.0	Acc
Puna Ibis		Acc	23.53	Acc								
Black-faced ibis									0.02	Acc	42.0	Acs
Turkey vulture	3.71	Acs	70.59	C	11.41	D			1.18	Acc	100.0	C
Black vulture									0.02	Acc	17.0	Acc
Osprey					0.09	Acc	10.53	Acc				
Cinereous harrier									0.02	Acc	17.0	Acc
White-tailed kite									0.02	Acc	17.0	Acc
Variable hawk					0.04	Acc	5.26	Acc	0.01	Acc	8.0	Acc
American kestrel	0.03	Acc	11.76	Acc	0.26	Acc	31.58	Acs	0.06	Acc	58.0	C
Peregrine falcon					0.22	Acc	26.32	Acs	0.02	Acc	17.0	Acc
Aplomado falcon					0.04	Acc	5.26	Acc				
Southern crested caracara					0.53	Acc	42.11	Acs	0.01	Acc	8.0	Acc
Chimango caracara									0.08	Acc	42.0	Acs
Common gallinule	0.40	Acc	58.8	C	1.05	Acc	47.37	Acs	0.18	Acc		
Spot-flanked gallinule									0.05	Acc	67.0	C
Andean coot					0.13	Acc	15.79	Acc			33.0	Acs
White-winged coot									0.18	Acc	50.0	C
Red-gartered coot									24.27	D	100.0	C
Red-fronted coot									1.93	Acc	83.0	C
Plumbeous rail									0.01	Acc	8.0	Acc
Black rail									0.04	Acc	8.0	Acc
Andean lapwing	0.01	Acc	5.9	Acc								
Southern lapwing									0.87	Acc	92.0	C
Semipalmated plover	0.47	Acc	41.18	Acs								

Name	Lluta[a]				Loa[b]				Huasco[c]			
Indicators	n%	D	f%	c	n%	D	f%	c	n%	D	f%	C
Kentish plover	0.06	Acc	5.9	Acc	0.26	Acc	10.53	Acc	2.81	Acs	92.0	C
Two-banded plover	0.04	Acc	17.65	Acc								
Collared plover	0.31	Acc	29.41	Acs					0.02	Acc	8.0	Acc
Killdeer	0.19	Acc	41.18	Acs								
Rufous-chested plover									0.21	Acc	42.0	Acs
Grey plover	0.42	Acc	23.53	Acc								
American golden plover	0.15	Acc	17.65	Acc								
American oystercatcher	1.04	Acc	64.71		3.34	Acs	94.74	C	3.63	Acs	25.0	Acs
Blackish oystercatcher	0.01	Acc	5.9	Acc	0.18	Acc	15.79	Acc	0.03	Acc	100.0	C
Black-necked stilt									1.27	Acc	100.0	C
Lesser yellowlegs	0.27	Acc	17.65	Acc					0.23	Acc	33.0	Acs
Greater yellowlegs					0.31	Acc	21.05	Acc	0.13	Acc	33.0	Acs
Willet									0.03	Acc	25.0	Acs
Spotted sandpiper	0.04	Acc	17.65	Acc								
Ruddy turnstone	0.07	Acc	11.76	Acc	1.58	Acc	36.84	Acs	0.25	Acc	33.0	Acs
Surfbird									0.05	Acc	8.0	Acc
Least sandpiper	0.31	Acc	29.41									
Semipalmated sandpiper	0.31	Acc	23.53	Acc	0.09	Acc	5.26	Acc				
Baird's sandpiper	0.31	Acc	64.71		1.05	Acc	21.05	Acc	0.06	Acc	17.0	Acc
Sanderling	2.18	Acc	23.53	Acc	0.18	Acc	5.26	Acc	2.79	Acs	58.0	C
Pectoral sandpiper	0.02	Acc	11.76	Acc	0.09	Acc	5.26	Acc				
Stilt sandpiper	0.41	Acc	23.53	Acc								
Whimbrel	0.22	Acc	23.53	Acc	0.92	Acc	47.37	Acs	1.62	Acc	100.0	C
Wilson's phalarope	0.01	Acc	5.9	Acc	0.18	Acc	10.53	Acc				
Andean Gull	0.60	Acc	47.06									
Brown-hooded Gull									0.02	Acc	8.0	Acc

(continued)

Table 5.14 (continued)

Name	Lluta[a]				Loa[b]				Huasco[c]			
Indicators	n%	D	f%	c	n%	D	f%	c	n%	D	f%	C
Grey gull	30.52	D	76.47		14.98	D	89.47	C	2.48	Acs	67.0	C
Laughing gull												
Franklin's gull	26.29	D	47.06		1.01	Acc	10.53	Acc	0.35	Acc	42.0	Acs
Belcher's gull	4.50	Acs	88.24		5.62	D			1.18	Acc	58.0	C
Kelp gull	10.55	D	70.59		13.08	D	78.95	C	28.71	D	100.0	C
Peruvian tern					0.22	Acc	5.26	Acc				
Arctic tern	0.08	Acc	5.9	Acc					0.02	Acc	17.0	Acc
South American tern	0.01	Acc	5.9	Acc								
Elegant tern	2.68	Acs	11.76	Acc								
Rhynchops niger	0.65	Acc	23.53	Acc					0.03	Acc	25.0	Acs
Croaking ground dove	0.01	Acc	5.9	Acc	0.18	Acc	10.53	Acc				
West Peruvian dove	0.18	Acc	47.06						0.58	Acc	83.0	C
Eared dove	0.05	Acc	17.65	Acc					0.51	Acc	83.0	C
Andean swift	0.08	Acc	5.9	Acc								
Giant hummingbird									0.02	Acc	17.0	Acc
Green-backed firecrown									0.02	Acc	17.0	Acc
Buff-winged cinclodes					0.26	Acc	26.32	Acs	0.18	Acc	58.0	C
Dark-bellied cinclodes									0.36	Acc	42.0	Acs
Greyish miner					1.49	Acc	63.16	C				
Plain-mantled tit-spinetail					1.67	Acc	47.37	Acs				
Wren-like rushbird					1.67	Acc	68.42	C	0.02	Acc	17.0	Acc
Dark-faced ground tyrant	0.01	Acc	11.76	Acc	9.00	D	73.68	C				
Spot-billed ground tyrant	0.02	Acc	5.9	Acc	1.76	Acc	31.58	Acs				
Vermilion flycatcher	0.02	Acc	17.65	Acc								
Many-coloured rush tyrant									0.08	Acc	33.0	Acs
Myiophobus fasciatus	0.01	Acc	5.9	Acc								

Name	Lluta[a]				Loa[b]				Huasco[c]			
Indicators	n%	D	f%	c	n%	D	f%	c	n%	D	f%	C
Austral negrito									1.16	Acc	100.0	C
Andean negrito					0.04	Acc	5.26	Acc				
Spectacled tyrant									0.12	Acc	50.0	C
Blue-and-white swallow	0.17	Acc	29.41	Acs	3.56	Acs	36.84	Acs				
Barn swallow	0.05	Acc	5.9	Acc					0.23	Acc	8.0	Acc
Chilean swallow									0.58	Acc	83.0	C
Yellowish pipit	0.01	Acc	11.76	Acc								
Correndera pipit					0.35	Acc	26.32	Acs				
House wren	0.01	Acc	5.9	Acc	8.25	D	100.00	C	0.20	Acc	50.0	C
Sedge wren									0.06	Acc	17.0	Acc
Rufous-tailed plantcutter									0.07	Acc	33.0	Acs
Cinereous conebill	0.01	Acc	5.9	Acc								
Rufous-collared sparrow	0.35	Acc	76.47	Acs	3.47	Acs	73.68	C	0.98	Acc	100.0	C
Slender-billed finch	0.10	Acc	11.76	Acc	0.09	Acc	5.26	Acc				
Mourning sierra finch									0.02	Acc	8.0	Acc
Peruvian meadowlark	0.07	Acc	29.41	Acs								
Long-tailed meadowlark									0.05	Acc	17.0	Acc
Austral blackbird									0.02	Acc	17.0	Acc
Yellow-winged blackbird									0.14	Acc	58.0	C
Black-chinned siskin									0.06	Acc	42.0	Acs
Passer domesticus	0.03	Acc	11.76	Acc	0.57	Acc	15.79					

D Dominance in abundance, *C* constancy, *n%* Percentage of abundance, *F%* frequency in percentage
D dominant, *C* constant, *Acs* accesorial, *Acc* accidental
[a]Miranda (2003)
[b]Malinarich (2005)
[c]Luna and Cortez (2007)
[d]Olivares (2006)

Table 5.15 Diversity indexes for the bird species occurring in each wetland

index	Lluta	Loa	Copiapo	Huasco
Shannon	2.3	2.9	2.0	2.6
Simpson	0.8	0.9	0.7	0.9

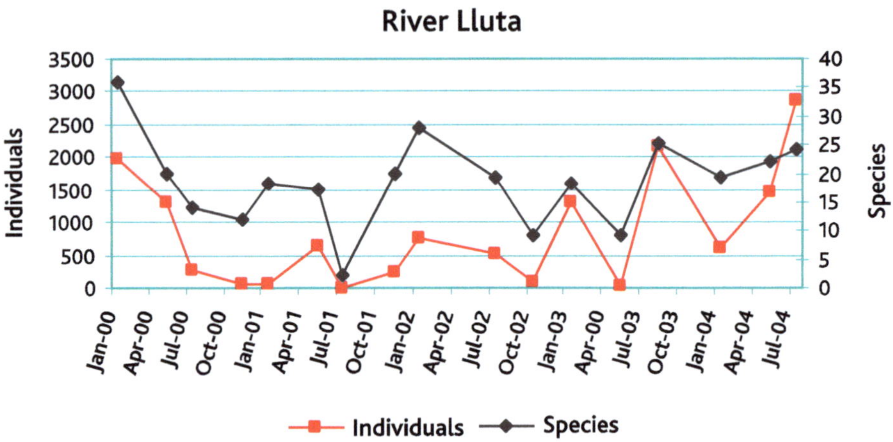

Fig. 5.4 Seasonal trends of the species richness and abundance on the Lluta River between 2000 and 2004

The Size of the Bird Populations that Congregate in the Wetlands

Bird nesting in the wetlands studied here is generally variable, depending on a number of factors, such as disturbances, climatic conditions, human presence, among others. Apart from this, it is possible to recognize seasonal behaviors in the size of the bird endowments, which apparently would be related to the sizes of the mirrors of water derived in turn from the rise of the respective rivers. Figures 5.4, 5.5, and 5.6 show the situation found in the wetlands of Lluta, Loa and Huasco.

In the Lluta river wetland, 2871 individuals (July 2004, post-flood period of the river) and 36 species were found, at most, during the 2000/2004 study period; however, timely records and sampling outside of the period reported, account for significant abundance of 8000 up until 20,000 birds.

In the Loa river wetland, there were 161 individuals (July 2005) and 14 species, and in the wetland of the Huayco River, there were 1149 individuals (September 2004) of 45 species. Similarly, Luna et al. (2007) reported, for the Copiapo River wetland, a total of 1115 individuals from 33 species associated with the lagoon and contiguous sandy beach.

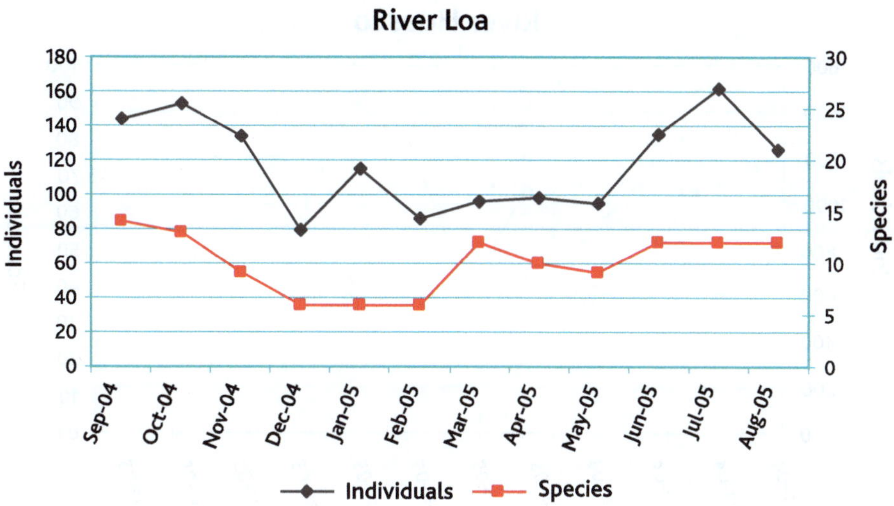

Fig. 5.5 Seasonal trends of the species richness and abundance on the Loa River between 2004 and 2005

General Conclusions About Flora and Avifauna of the Northern Chilean Wetlands

The northern areas of Chile (Regions I-III and XV) comprise 194 species of birds associated with coastal wetlands, corresponding to 41 families and 16 orders, equivalent to 39.8% of the 487 species of birds currently known for national territory, and Without doubt transforming it into one of the most diverse areas of wetland birds.

The OAU list of birds for Chile and the South American Classification Committee, version 2010, considers 487 species (residents: The northern zone of Chile (Regions I-III and XV), comprises 194 species of birds associated with coastal wetlands, corresponding to 41 families and 16 orders, equivalent to 39.8% of the 487 bird species currently known on the national territory; without a doubt, transforming it into one of the most diverse areas of wetland birds.

It should be known that the OAU list of birds for Chile and the South American Classification Committee, version 2010, considers 487 species (residents: 315; visitors: 71; accidental: 71; introduced: 4; extinct: 1; hypothetic: 14; and endemic: 11).

The analysis of the structure of the bird assemblages associated with the wetlands studied, allows to identify two faunal groups segregated from each other, which correspond to the gorges of the Loa River to the north and the set of coastal wetlands of the rivers with Mediterranean influence from Chañaral and further south. The main differences are found in the species of the families Rallidae, Ardeidae, Anatidae, Laridae and the terrestrial Passeriformes species, commonly associated with wetlands.

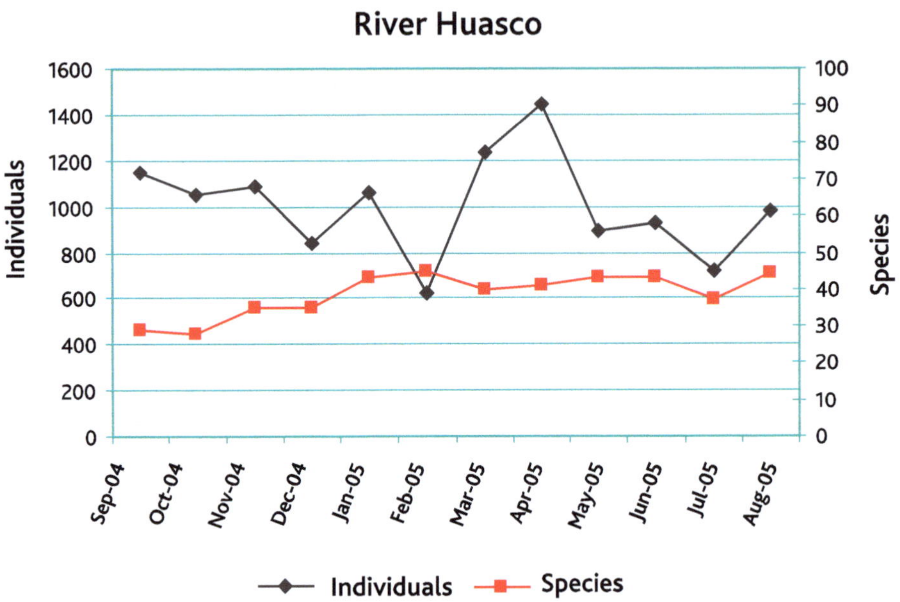

Fig. 5.6 Seasonal trends of the species richness and abundance in the Huasco river between 2004 and 2005

When considering the number of species, number of individuals as well as diversity, the numbers indicate that the wetland associated with the Lluta river is the most important in the northern part of Chile, gathering at least 25 visiting species (mainly chorlos and beachbirds) and 25 species of rare or accidental species, many of whom are only registered in that sector in national territory. Wetlands associated with the Vitor, Camarones and Tana rivers have little relevance as avian concentration sites, but are probably part of biological corridors necessary to maintain the presence of birds in the most important sectors.

The Loa river wetland, currently proposed as a priority area in the Conservation Strategy of the Antofagasta Region II, does not represent an outstanding site for wetland birds, but it does, however, correspond to an important biological corridor between the highland/altiplanic plateau and the seacoast, and as well as a resting place for migrating birds along the seaboard. This aspect fully justifies a deeper analysis of the contamination of this river and eventual measures leading to its recovery.

In the wetlands associated with the rivers of the Region lll, in particular the Salado River, which was violently contaminated in the past with mining tailings, and at the present seems practically irrecoverable; but this same river is probably a resting place for birds in latitudinal migration. This aspect should also be evaluated in the future as part of the protection plans for the other environmentally healthy wetlands in the region.

As far as the wetlands of the Copiapo and Huasco rivers, the first one presents important signs of desiccation due to extraction of water for mining and agricultural use, despite being a protected area; and the second, suffers from mainly urban pollution. Nevertheless, both of them still support diverse fauna groups, abundant and representative of the Chilean Mediterranean tendencies, whose protection measures must be strengthened and monitored more firmly.

Regarding contamination, the northern gorges of the Loa River generally have medium to poor bacterial contamination and water quality problems; the Loa river has high bacterial and chemical contamination, the Copiapo River suffers medium bacteriologic contamination, and the Huasco River has chemical and bacteriologic contamination; all of which constitutes a potential danger for the ecological sustainability of the wetlands of the northern Chile.

The effect of the economic development of the area (mining, agriculture), urbanization, modification of river systems and reduction of flows for various purposes, adds to the effect of bacteriological and chemical contamination, becoming the main reasons that should be taken into account in regional bird conservation strategies and biodiversity in general.

References

Aguirre, J., and H. Torres. 2005. Aves de los Humedales Altoandinos del Norte de Chile. 180 pp.

Andrade, C. 2007. Área Marina Costera Protegida Isla Grande, Caldera III Región Limnología del Humedal del Río Copiapó. Limnología de la desembocadura del río Copiapó. Proyecto GEF Conservación de la biodiversidad de importancia mundial a lo largo de la costa chilena.

Aramayo, O., F. De La Barrera, P. Aranda, M. Asmüssen, D. Moreira, F. Vargas, F. Zorondo, A. Hoffmann, L. Faúndezy, and J. Rodríguez. 2005. Levantamiento de información, estudio de línea base e instrumento legal para la conservación de la biodiversidad en los sitios priorizados en la ERB: Aguas calientes IV y sector costero de Paposo, II Región de Antofagasta. Documento Técnico. Vicerrectoría de Investigación y Desarrollo, Universidad de Chile. 125 pp.

Araya, B., G. Miliie, and M. Bernal. 1995. Guía de campo de las aves de Chile. Séptima Edición. Santiago: Editorial Universitaria. 406 pp.

Avendaño, V., and F. Sáiz. 1977. Taxocenosis de rotíferos limnéticos de la laguna El Plateado. *Ans. Mus. Hist. Nat. Valparaíso* 10: 121–133.

Cabrera, A.L. 1957. La Vegetación de la Puna Argentina. *Rev. Invest. Agrícola* 11 (4): 317–512.

———. 1971. Fitogeografíadela República Argentina. *Bol. Soc. Arg. Bota.* 14 (1–2): 1–42.

Cabrera, A. L., and A. Willink. 1980. Biogeografía de América Latina. Secretaría General de los Estados Americanos. Programa Regional de Desarrollo Científico y Tecnológico. Washington D.C. Segunda Edición. 122 pp.

Cade-Idepe Consultores. 2004. Diagnóstico y clasificación de los cursos y cuerpos de agua según objetivos de calidad: cuenca Río Loa. GORE, Ministerio de Obras Públicas, Dirección General de Aguas. Informe Técnico. 155 pp.

Campos, G. 2005. Localización de un Embalse en el río San José y su Impacto Económico. Informe Ingeniería Civil Industrial. Universidad de Tarapacá.

Campos, H., Díaz, G. Y., Campos, C. 2007. Aportes Sedimentarios de los Ríos Lluta y San José en la Zona Costera de la Rada de Arica, Chile. Idesia, Vol. 25, N° 2, pp. 37-48.

Cea. 2006a. Conceptos y Criterios para la Evaluación Ambiental de Humedales. CONTRATO SAG.

Clarke K.R., and R.N. Gorley. 2006. Primer V6: User manual/tutorial, 192 pp. Plymouth: PRIMER-E.

———. 2007. Revisiónde Sitios Prioritariosparala Conservacióndela Biodiversidad, Región de Tarapacá.

Cortez, M. 2010. Capítulo 1.8 ANDES COPPER COMPANY y codelcodestruyenelrío Salado y la Bahía de Chañaral. In *Conflictos por el agua en Chile: Entre los derechos humanos y las reglas del mercado*, ed. S. Larraín & D. Poo, 128–139. Gráfica Andes, Santiago, 360 pp.

Dirección general de aguas (DGA). 2004a. Diagnóstico y clasificación de los cursos y cuerpos de agua según objetivos de calidad. Cuenca del río Copiapó. Documento Técnico, Diciembre 2004, 21 pp.

———. 2004b. Diagnóstico y clasificación de los cursos y cuerpos de agua según objetivos de calidad. Cuenca del río Loa. Documento Técnico, Diciembre 2004, 149 pp.

Di Castri, F.. (1969) Esquisse ecologique du Chili. *Biologie de la Amerique Australe* 4: 7–52. Editions du Centre National de la Recherche Scientifique 15, Quay Anatole- France – Paris VII.

Fittkau, E.J. 1969. The fauna of South America. In Biogeography and ecology in South America, ed. E.J. Fittkau et al., vol. 2, 624–658. New York: Hillary.

Fuller, R.J., and D.R. Lanslow. 1984. Estimating number of birds by point counts: how long should counts last? *Bird Study* 31: 195–202.

Gajardo, R. 1995. *La Vegetación Natural de Chile, Clasificación y Distribución Geográfica*. Santiago: Editorial Universitaria. 165 pp.

Godoy, J. 2010. Copiapó: seco por indiscriminado otorgamiento de derechos de agua. In *Conflictos por el agua en Chile: Entre los derechos humanos y las reglas del mercado*, ed. S. Larraín & D. Poo, 159–170. Santiago: Gráfica Andes. 360 pp.

Jarvinen, Olli. 1978. Species-specific census efficiency in line transects. *Ornis Scandinavica* 9: 164–167.

Jeannel, R. 1942. La Genèse des Faunes terrestres. Élements de Biogéographie. Presses Univ. Francaises, 514 pp.

———. 1967. Biogéographie del'amérique australe. *Biologique de l'Amérique australe. Centre Nacional de la Recherche Scientifique, Paris* 3: 401–460.

Kuschel, G. 1964. Problems concerning an austral region, pp. 443–449 en: Gressit J. L., C. H. Lindroth, F. R. Forsberg, A. Fleming, and E.G. Turbott (eds.), Pacific basin biogeography, Bishop Museum Press, Honolulu.

Lloyd, M., J.Y. Zar, and J. Karr. 1968. On the calculation of information theoretical measures of diversity. *Amer. Mid. Nat.* 79 (2): 257–272.

Luebert, F., and P. Pliscoff. 2006. *Sinopsis bioclimática y vegetacional de Chile*. Santiago: Editorial Universitaria. 318 pp.

Luna, G., and M. Cortés. 2007. Preinforme Final Consultoría Estudio del ensamble de aves y mamíferos marinos al interior del Área Marina y Costera Protegida de Múltiples Usos Isla Grande de Atacama. Proyecto Conservación de la Biodiversidad de Importancia Mundial a lo largo de la Costa Chilena. Informe Técnico, 34 pp.

Malinarich, V. 2005 Ensamble aviar, sector desembocadura del río Loa. II Región, Chile. Período 2004–2005. Tesis. Universidad Arturo Prat, Chile.

Mann, G. 1960. Regiones Biogeográficas de Chile. *Invest. Zool. Chilenas* 6: 15–49.

Miranda, L. 2003. Dinámica Espacio Temporal del Ensamble Avial Presente en la Desembocadura del Río Lluta (18° 23′ S - 69° 45′ W) durante 1997 a 2001. Seminario de Investigación para optar al Grado Académico de Licenciado en Ciencias del Mar. Universidad Arturo Prat.142 pp.

Monrós, F. 1958. Consideraciones sobre la fauna del sur de Chile y revision de la tribu Stenomelini (Col. Chrysomelidae) bacta. *Zool Lilloana* 15: 143–153.

Morrone, J.J. 1996. The biogeographical Andean subregion: A proposal exemplified by arthropod taxa (Arachnida, Crustacea and Hexapoda). *Neotropica* 42: 103–114.

Müller, P. 1973. *The dispersal centres of terrestrial vertebrates in the Neotropical realm: A study in the evolution of the Neotropical biota and its native landscapes*. La Haya: W. Junk Publishers.

Myers, N. 1994. Global biodiversity II: Iosess. In Principles of conservation biology, ed. G.K. Meffe & C.R. Caroll 110–142. Sunderland: Sinauer Associates, Inc.
Olivares, F. 2006. Dinámica espacio temporal del ensamble alvial del humedal de la desembocadura del río Huasco 3ª Región (28°27′S - 71°12'O) durante el período 2004–2005. Informe Final de Tesis para optar al Título de Biólogo Marino. Universidad Arturo Prat. 165 pp.
Orians, G. 1994. Global biodiversity I: patterns and processes. In *Principles of conservation biology*, ed. G.K. Meffe and C.R. Caroll, 78–109. Sunderland: Sinauer Associates, Inc.
Peña, l. 1966. Ensayo preliminar para dividir Chile en Regiones entomofaunísticas, basadas especialmente en la familia Tenebrionidae (Coleoptera). *Revista Universitaria (Universidad Católica de Chile)* 50-51 (2): 109–220.
Peña, H. Recursos hídricos. Serie documentos técnicos. DGA-1993.
Peredo, R., and N. Amado. 1995a. Segundo registro de *Egretta tricolor* para Chile. *Boletín Chileno de Ornitología* 2: 27–28.
———. 1995b. Primer registro de *Larus cirrocephalus* para Chile. *Boletín Chileno de Ornitología* 2: 36.
Peredo, R., and L. Miranda. 2000a. Importancia del estuario del Lluta (I Región, Chile) para la avifauna marina. Libro Resúmenes XX Congreso Ciencias del Mar, Concepción.
———. 2000b. Importancia del Estuario del Lluta (I Región, Chile) para la avifauna marina. VI Congreso Nacional de Profesores de Ciencias y II Congreso Internacional.
———. 2001. Nuevos registros para la avifauna del estuario del río Lluta (Arica, Región de Tarapacá). Boletín Chileno de Ornitología N° 8.
Pisano, E. 1956. Esquema de clasificación de las comunidades vegetales de Chile. *Agronomía* 2 (1): 30–33.
Raunkiaer, C. 1934. *The life forms of plants and statistical plant geography*. London: Oxford University Press.
Remsen, J.V., Jr., C.D. Cadena, A. Jaramillo, M. Nores, J.F. Pacheco, M.B. Robbins, T.S. Schulenberg, F.G. Stiles, D.F. Stotz, and K.J. Zimmer. 2010. A classification of the bird species of South America. American Ornithologists' Union.
Roig Juñent, S. 1994. Historia Biogeográficade Américadel Sur. Austral. *Multequina* 3: 167–203.
Reynolds, R.T., J.M. Scott, and R.A. Nussbaum. 1980. A variable circular-plot method for estimating bird numbers. *Condor* 82: 309–313.
Salazar, C. 2003. Situación de los recursos hídricos en Chile. The Nippon Foundation, Informe Técnico, 109 pp.
Schwabe, G.H. 1969. Toward an ecological characterization of the South American continent. *Biogeography and Ecology in South America* 1: 230–250.
Silob C. 2001. Estudio de Línea Base Ambiental para el Proyecto: "Entierro de Línea Ossa 2 en el Río Lluta". TRANSREDES S. A. 122 pp.
Squeo, F.A., G. Arancio, A. Cortés, D. Hiriart, and D. López. 2006. Estudio de Línea de Base de Recursos Bióticos Terrestres del AMCP Isla Grande de Atacama (Punta Morro –Desembocadura del Río Copiapó). Informe Final. 41 pp.
World conservation And Monitoring Center (WCMC). 1992. *Global biodiversity: State of earth's living resources*. London: Chapman & Hall.

Chapter 6
Natural History and Environmental Patterns in the El Yali Coastal Wetland, Central Chile

José Miguel Fariña, Mark D. Bertness, Brian Silliman, Nuria Aragoneses, and Eugenia Gayo

Abstract The coastal wetland of El Yali (33.82–33.72 ° S) is a priority site for the conservation of the flora and fauna of central Chile. In fact, it has been declared the Ramsar wetland and National Reserve by the Chilean authorities. In this chapter, a synopsis on the natural history and ecological patterns of the wetland is presented. Therefore, this effort aims to provide tools and implements that cover all the elements necessary to build up a conservation vision for the wetlands in Chile. Our background compilation demonstrates that the El Yali flora and fauna exhibit extraordinary biogeographic singularities and conservation states. However, our knowledge and understanding of the mechanisms that determine the functioning and structure of their biological communities is limited, especially for fauna. The scarcity of information is critical considering that since pre-hispanic times, the anthropic pressures on the El Yali wetland ecosystems have been particularly strong. For this reason, a series of recommendations and future projections are proposed, all of which seek to improve and suggest new lines of action when it comes to the preservation and sustainable development of the coastal wetland.

J.M. Fariña (✉)
Center of Applied Ecology and Sustainability-CAPES, Pontificia Universidad Católica de Chile, Santiago, Chile

Department of Ecology and Evolutionary Biology, Brown University, Providence, RI, USA
e-mail: jfarina@bio.puc.cl

M.D. Bertness
Department of Ecology and Evolutionary Biology, Brown University, Providence, RI, USA

B. Silliman
Department of Zoology, University of Florida, Gainesville, FL, USA

N. Aragoneses
Center for Advanced Studies in Ecology and Biodiversity (CASEB), Pontificia Universidad Católica de Chile, Santiago, Chile

E. Gayo
Instituto de Ecología y Biodiversidad (IEB), Santiago, Chile

Center for Advanced Studies in Ecology and Biodiversity (CASEB), Pontificia Universidad Católica de Chile, Santiago, Chile

J.M. Fariña, A. Camaño (eds.), *The Ecology and Natural History of Chilean Saltmarshes*, DOI 10.1007/978-3-319-63877-5_6

Keywords Wetland • Central Chile • National Reserve • Biodiversity conservation

Introduction

The Yali wetland is located in the commune of Santo Domingo (San Antonio Province, Valparaiso Region, Chile). This basin, situated in the coastal plain between 33.82 ° S – 33.72 ° S and 71.77 ° W –71.37 ° W, has its border to the north marked with the basin of the river Maipo and to the south with that of the river Rapel. With its 11,500 hectares of land, it involves about 14 different bodies of water which together form part of the "Wetlands Complex of the Central Coast". Within this complex, there is an important diversity of wetlands, represented by estuaries (El Yali, Santa Rosa, El Peuco, Tricao, Maitenlahue), vegas (El Convento, Vega de Talca), salt marshes (El Convento and Bucalemu), the Molles reservoir, sand dune areas, and about 8 lagoons (Fig. 6.1, Valverde 1996; CONAMA-CONAF 2005). Among the latter ones are the freshwater lagoons Cabildo, Seca, Matanzas, Colejuda, Guaraivo, El Rey, Maura and the brackish lagoon or Albufera, which runs parallel to the coastline and is separated from the sea by a sand barrier (Galdames 2007; Sun 2008; Dussaillant et al. 2009).

On May 23, 1996, by the *Supreme Decree No. 41* of the Ministry of Agriculture, about 520 hectares of the wetland belonging to the El Convento community became part of the El Yali National Reserve, which on December 2nd of that same year was integrated as the only Mediterranean neotropical wetland to the list of Wetlands of International Importance of the Ramsar Convention (site No. 878). Both conservation measures are based on the following characteristics: (i) El Yali wetland has an excellent representation of the national avifauna, as well as the freshwater fauna of central Chile; (ii) this area includes the most important lake complex in the central zone of Chile: Matanzas and Colejuda lagoons, in addition to Costera Lagoon or Albufera; (iii) within this area, about 13 endangered species are distributed; (iv) the Albufera is a rare ecological and hydrological system in Chile; (v) it is one of the 20 conservation priority sites of Chile's biological diversity; (vi) this area has a high educational, recreational and scientific potential.

Following the accession of Chile to the Ramsar Convention, our country is committed to protect and conserve the El Yali and 8 other wetlands. Subsequently, one of the first wetland conservation acts was to cover a total area of 34,000 hectares, including the Protected Zone with a ban on hunting and fishing amphibians, fish, reptiles, birds and mammals within the El Yali estuary and its adjacent lagoons (Ministry of Agriculture Decree No. 143/1998). Further initiatives include the declaration of El Yali as a priority site of regional conservation through Exempt Resolution No. 739/2007 "Sites of the Regional Strategy for Biodiversity, Valparaiso Region" and the environmental action plan of the San Antonio Province (PAPSA 2009). In addition to the training in 2003 of the Local Advisory Council for the

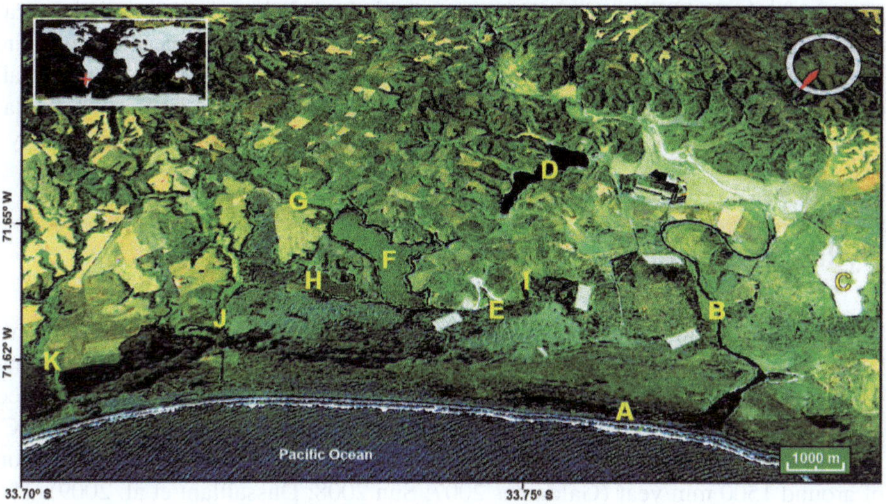

Fig. 6.1 Landast image (NASA/JPL) of El Yali wetland. The principal water bodies are shown. (*A*) Albufera; (*B*) El Yali marsh; (*C*) El Rey lagoon; (*D*) El Molle reservoir; (*E*) Colejuda lagoon; (*F*) Matanzas lagoon; (*G*) Santa Rosa marsh; (*H*) Cabildo lagoon; (*I*) Guaraivo lagoon; (*J*) El Peuco marsh and (*K*) Tricao marsh

conservation and development of the El Yali National Reserve, which involves the participation of social players from the public, community and private sectors. This last sector, represented by the company Agrosuper S. A, has allocated 750,000 m3 of water for conservation of the Los Molles dam, and also developed basis for protection of flora and fauna, in addition to the Reserve. However, due to the environmental conflict between the community and the company that started in 2006, these conservation measures have fallen into the background. The community claims that the collection of organic material derived from the breeding sites would be contaminating the Maitenlahue and Yali estuaries, both wetland tributaries and both crossing Agrosuper SA properties. At the national level, conservation strategies for priority wetlands such as El Yali are addressed in the "National Strategy for conservation and wiser use of wetlands in Chile"(CONAMA-CONAF 2005), whose second objective is to fundamentally improve and increase basic and applied knowledge about wetlands. Specifically, this action alines itself with the objective to encourage permanent scientific research and to develop and disseminate technical information on the factors that determine the structure and functioning of Chilean wetlands (CONAMA-CONAF 2005). Given the relevance of the El Yali wetland, one could expect important advances in the knowledge of the ecological characteristics of the system. However, studies on the mechanisms that affect the biodiversity and functioning of El Yali are scarce, generally limited to flora and/or fauna listings. On the other hand, the information generated by different existing studies is biased, depending on a source, and in many cases difficult to access. This chapter provides a synthesis that contextualizes all the priors derived from the research done in the last

15 years in the El Yali wetland. This summary is intended to provide political, academic and social actors with all the necessary information to successfully plan, develop and maintain the conservation strategies for the ecosystems of the El Yali coastal wetland, thus contributing to the theoretical and practical framework fundamental for the development of a vision of conservation of wetlands in our country.

Hydroclimatic Priors

The El Yali wetland is located in the Mediterranean region of central Chile, presenting a rainy season in the winter and a dry period during the Patagonian summer (Fernández 2004). Specifically, the climate in the wetland area has been classified as subhumid-temperate Mediterranean, with annual mean temperature of 12 °C, with an average annual rainfall of 481 mm/year, and an average annual evaporation of around 1500 mm/year (Galdames 2007; Sun 2008; Dussaillant et al. 2009). The marked seasonality of rainfall in this region is determined by the annual cycle of the South Pacific Subtropical Anti-cyclone (ASPS, Aceituno 1990). Therefore, between September and April, the anti-cyclone activity is strengthened and the ASPS moves its area of influence to the south, generating a dry season. In contrast, during the winter months, the ASPS presents its maximum penetration towards the north, allowing rainfall from the west winds of the temperate latitudes in the central Chile region (Aceituno 1990).

Interannual variables, variations in precipitation course are determined by the "El Niño Southern Oscillation" (ENSO). Thus, during the positive ENSO phase (El Niño events) there is a weakening of the ASPS, which increases the frequency and magnitude of rainfall over central Chile (Aceituno 1990), while during the years of the influence of La Niña, the ASPS is invigorated, predominantly cold with dry winter conditions (Aceituno 1990).

The Yali basin is pluvial and is hydrologically isolated from the basins of the Maipo, Alhue and Rapel rivers by the geological faults of Melipilla, Puangue and Maipo (Galdames 2007). The persistence of the wetland complex is related to (i) the low slope of the relief (Fernández 2004); (ii) reduced flow velocity presented by the tributary estuaries of the lagoons, vegas, dunes (MOP 2005); (iii) upwelling (MOP 2005; Galdames 2007); (iv) development of a salt wedge that would enter the aquifer from the sea (MOP 2005).

Due to the pluvial regime of the different systems that integrate into the wetland, the hydrological patterns are determined by the intra and interannual variations in the precipitations. Indeed, it has been observed that in anomalously rainy winters, the floodwaters of the Santa Rosa estuary allow the Matanzas lagoon to drain into the sea (Biblioredes 2010). The same is true for the Laguna Albufera (Galdames 2007; Sun 2008; Dussaillant et al. 2009). Due to the increased flow of the El Yali estuary in the winter, the sandy bar separating the Albufera from the sea is eroded, establishing a phase of connection with strong maritime influence. The opposite occurs during the dry season and/or anomalously dry years (details in Chap. 7 of this book).

Historical Background

The development of the "Central Coastal Wetlands Complex" is due to the interaction of several factors that have determined the geomorphological evolution of the coastal landscape of central Chile during the last 18,000 years (Gana et al. 1996). In the late Pleistocene, this complex is formed by deposition of continental sands and gravels on the strata of the Natal Formation, a Miocene-Pliocene sequence that constitutes the main landfill between the Maipo and Rapel rivers (Gana et al. 1996). Marine transgressions occurring after the last ice age (~14,000 years ago) and during the mid-Holocene (~6000 years ago) would have allowed the accumulation of marine and/or fluviomarine sediments through the basin (Paskoff 1970; Hervé et al. (2003). The neotectonic activity of the coseismic type that characterizes the last 3000 years (lower Holocene), has produced the regional uprising of the coast and caused the emergence of a coastal terrace on which the wetland lies. As a result of coastal uprising, a marine regression occurs that initiates progradation of the coastline through sedimentary eolic and fluvial processes (Paskoff et al. 2000; Hervé et al. 2003). In fact, the existence of transgressive and parabolic dunes in the area indicates that the progradation occurred as a result of contribution of the Rapel and Maipo rivers sedimentary loads through coastal drifts and wind transport (Paskoff et al. 2000; Piñones et al. 2005).

The palaeoclimatic studies of the lake surroundings and contributors, or "historical witnesses", in the Matanzas (Jenny et al. 2000; Villa-Martínez 2002) and Colejuda lagoons (Valero-Garcés et al. 2010), suggests that the evolution of the wetlands complex during the last 3000 years has been modeled mainly by changes in eolic sedimentation and precipitation variations by ENSO. This is how pollinic, sedimentological, geochemical and subfossil analyzes of Chironomids indicate that between 3000 and 1300 calendar years before the current era (YBP – years before presengt), the Matanzas lagoon was a littoral lagoon with moderate to low salinity (Jenny et al. 2000; Villa-Martínez 2002). Throughout this period, rainfall shows significant fluctuations, with two rainy intervals at 2900–2800 and 2500–1300 YBP separated by a drier interval between 2800 and 2500 YBP (Villa-Martínez 2002). Variations in concentrations of Pediastrum boryanum detected at 2500–1300 YBP suggested fluctuations in salinity, although the peak reached by this microalga during this interval would indicate predominance of conditions of low salinity in the Matanzas coastal.

Different paleoenvironmental indicators indicate that from 1300 and up to 600 YBP would have occurred a significant increase in salinity, probably representing a change in the system from the coastal lagoon to the hypersaline lake (Jenny et al. 2000; Villa-Martínez 2002). This change would be related to the closure of the basin of the Matanzas lagoon (~ 600 YBP) due to the predominance of eolic sedimentary processes that led to the formation of the dune cord that currently dam the system (Jenny et al. 2000; Villa-Martínez 2002). The sedimentological analysis suggests that during this process, the fluvial sedimentation caused by floods of the Santa Rosa estuary would have also been important. The occurrence of Chironomid species intolerant to salinity and the low concentrations of chlorine detected in the sediments suggest that from the 400 YBP the Matanzas lagoon would have maintained moderate to low salinity conditions (Jenny et al. 2000). During this time, sediments show at least 20 different flood periods, indicating marked fluctuations in winter rains (Villa-Martínez 2002).

For the last 150 years, Villa-Martínez (2002) recognizes a deterioration in forest vegetation around the Matanzas lagoon, which attributes to anthropic activity. However, the Colejuda lagoon surroundings and "historical witnesses" (Valero-Garcés et al. 2010) indicate that from the late nineteenth century onwards rainfall shows a clear tendency towards decline, providing an alternative explanation, although not exclusive, to the pattern recognized by Villa-Martínez (2002).

The available archaeological record for the Yali wetland suggests a continuous occupation for the last 6000 years (Brito 2004). Nonetheless, analysis of depletion of soil particles in the Matanzas lagoon suggests that these settlement patterns may have varied in frequency and magnitude over the last 3000 years (Villa-Martínez 2002). Therefore, the fact of the low concentrations of microcarbons starting at 600 YBP, concomitant with the closure of the lagoon, has been interpreted as an abandonment of the area forced by the formation of sand dunes (Villa-Martínez 2002).

In the El Yali area, traces of hunter gatherers from the Archaic period have been found (~ 8000 years old), especially in the vicinity of El Convento (Falabella and Planella 1980). In addition to permanent camps of the later agrofood traditions such as Llo-Lleo and Aconcagua (Falabella and Planella 1980; Brito 2004), the archaeological remains of the Aconcagua culture recovered in El Rey lagoon represent the southern coastal limit of this culture (Brito 2004).

Flora and Fauna

In this section we provide an inventory of the animal and plant taxa recorded in the El Yali wetland (Valverde 1996; Vilina and López-Calleja 1996; Brito 1999; Vilina and Cofré 2000; Vilina et al. 2002; Villa-Martínez 2002; Fariña et al. 2005, 2006, 2007, 2009; Cofré et al. 2009; Figueroa et al. 2009; Biblioredes 2010). Based on a revision of the range distribution for each taxa, biogeographic origins were established (Donoso-Barros 1970; Marticorena and Quezada 1985; Figueroa et al. 1999; Díaz-Páez and Ortiz 2003; Victoriano et al. 2006; Sernapesca 2008; Zunino et al. 2009; Avesdechile 2010; Fuentealba et al. 2010; Lib 2010). We have omitted from this analysis all those taxa recorded at the gender level. This taxonomic unit generally has wider distribution ranges than the species it contains, limiting the recognition of endemic or native taxa in studies at regional or local scales (Brown and Lomolino 1998). Therefore, the inclusion of gender in the present analysis does not contribute much to the knowledge about the biogeographic singularities of the wetland. Therefore, the species recorded in the area were put into six biogeographical element groups:

1. *Endemic:* range of natural distribution restricted to our continental territory.
2. *Native:* range of natural distribution that includes Chile and other regions of South America.
3. *Broad*: range of natural distribution that includes the entire Southern Hemisphere.
4. *Cosmopolitan:* range of natural distribution spanning vast regions of the globe.
5. *Exotic:* all those species that have been introduced in Chile accidentally or intentionally and that currently maintain natural populations within our territory.
6. *Visitor:* species of birds that migrate from the Northern Hemisphere to El Yali.

For each taxa, the conservation status is also given according to the conservation categories established in the "Wild Species Classification Regulations" (D. S. No. 75 of 2005, Presidential General Ministry, Chile). Due to the fact that conservation states are specific for each species, we have excluded from this analysis those records at the generic level. Those species not yet classified under any category are designated as "Undefined". The classification "Low threat" indicates that the specialists, after the evaluation, have concluded that the species is not in risk or that the ethreat is low. Information on the conservation status of fauna was obtained from bibliographic sources (Campos et al. 1988; Glade 1993; Díaz-Páez and Ortiz 2003; Victoriano et al. 2006; Tala et al. 2009; Zunino et al. 2009) and from the government sites such as "Servicio Nacional de Pesca" (SERNAPESCA 2008) and "Servicio Agricola Ganadero" (SAG 2004). In general, the conservation states considered are of those described for the central region of Chile, including regions IV, V, Metropolitan, VI and VII (SAG 2004). For the flora, conservation status was obtained from the Benoit (1989) red books and those available for the regions III and IV (Squeo et al. 2001, 2008a, b). This is because the Region of Valparaiso does not yet have a list of endangered flora (CONAF regional in: http://otros.conaf.cl/cd_sitio_web_flora_regional/index.htm).

The wetland flora is composed of 32 spermatophytes, including Gymnosperms and Angiosperms, both Dico and Monocotyledonous (Table 6.1). About 53% of these taxa are found in the areas bordering the wetlands, forming a part of the coastal sclerophyllous scrub or of plantations near the Matanzas and Colejuda lagoons (Villa-Martínez 2002). The remaining percentage (~ 46%) corresponds to taxa hygrophytes and/or halophytes directly associated with the existence of the wetland. In the Matanzas and Cabildo lagoons, a filamentous microalgae of the genus Anabaena (Cyanophyceae), which gives a green color to these bodies of water, is recorded (Fig. 6.1, Figueroa et al. 2009).

Within the vascular plants, only one Monocotyledon is classified at the generic level (Table 6.1), so the total species for the flora is 31. 42%, and 39% of the flora of the area is native or endemic, respectively. The 19.3% (n = 6 spp) corresponds to introduced species. Conservation status for most species (>41%) has been defined under the category "Out of risk". There is one specie for which there are no significant risk. For 35% of the flora, the status has not been evaluated or there is no

Table 6.1 Plant species registered in el Yali wetland. (*) Palustrine vegetation (Valverde 1996; Villa-Martínez 2002; Fariña et al. 2005, 2006, 2007)

Specie	Family	Element	Conservation Status
Acacia caven	Mimosoideae	Native	Out of risk
Aristeguietia salvia	Asteraceae	Endemic	Out of risk
Aristotelia chilensis	Elaeocarpaceae	Native	Out of risk
Azara celastrina	Flacourtiaceae	Endemic	Endangered
Chusquea cumingii	Poaceae	Endemic	Vulnerable
Colliguaja odorífera	Euphorbiaceae	Endemic	Out of risk
*Cotula coronopifolia**	Asteraceae	Exotic	Undefined
Cryptocarya alba	Lauraceae	Endemic	Vulnerable
*Distichlis spicata**	Poaceae	Nativo	Out of risk
Escallonia pulverulenta	Escalloniaceae	Endemic	Out of risk
Eucalypus globulus	Myrtaceae	Exotic	Undefined
*Frankenia salina**	Frankeniaceae	Native	Out of risk
*Galega officinalis**	Fabaceae	Exotic	Undefined
*Juncus balticus**	Juncaceae	Native	Undefined
*Juncus cf acutus**	Juncaceae	Native	Insufficiently known
Kageneckia oblonga	Rosaceae	Endemic	Vulnerable
*Lolium temulentun**	Poaceae	Exotic	Undefined
Luma chequen	Myrtaceae	Endemic	Out of risk
Nolana crassulifolia	Nolanaceae	Endemic	Out of risk
Nolana sedifolia	Nolanaceae	Native	Out of risk
Peumus boldus	Monimiaceae	Endemic	Vulnerable
Pinus radiata	Pinaceae	Exotic	Undefined
*Sarcocornia fruticosa**	Chenopodiaceae	Native	Out of risk
Schinus latifolius	Anacardiaceae	Endemic	Out of risk
*Scirpus californicus**	Cyperaceae	Native	Insufficiently known
*Scirpus cernuus**	Cyperaceae	Native	Low threat
Scirpus sp *	Cyperaceae		
*Selliera radicans**	Goodeniaceae	Native	Insufficiently known
Sophora macrocarpa	Fabaceae	Endemic	Endangered
*Spartina densiflora**	Poaceae	Native	Undefined
*Spergularia media**	Caryophyllaceae	Exotic	Undefined
*Typha angustifolia**	Typhaceae	Native	Out of risk

adequate information for classification. The missing 20% of species present some degree of risk. Therefore, 13% has been classified as "Vulnerable" and 6.5% is under the category "Endangered".

The Conifers are represented by the Pine Insigne, exotic species of the Pinaceae family. Within the Magnoliophytes, the important ones are Dicotyledons, with 22 taxa in 17 families. The distinguished ones are the Asteraceae, Fabaceae, Myrtaceae and Nolanaceae, all of them represented by two species. 81% of the Dicotyledonous species are native to the southern forests of South America, being 52.4% and 28.5% endemic and native, respectively. 19% of the species in the group are exotic.

Regarding the conservation status, 52% of the species are out of risk. This percentage includes almost all of the woody species. Approximately 14% of Magnoliophytes are "Vulnerable", 9.5% "Endangered" and 5% do not have enough data for classification. The missing 19% comprise species that have not been categorized so far.

Monocotyledons include 10 taxa (9 species). Of the 4 families present in the area, the Poaceae (n = 4 taxa) and Ciperaceae (n = 3 taxa) are important. Virtually all species are autochtonous, 7 native and 1 endemic. Only one exotic species is recorded. The conservation status for 33.3% of the species has not been evaluated, while for almost 22% of the species there is insufficient data. About 33% of Monocotyledons are out of risk (22.2%) or do not present significant risk (S. cernuus). Only the endemic species C. cumingii is listed as endangered, and classified as "Vulnerable".

Tables 6.2, 6.3, 6.4 and 6.5 show the fauna richness of the El Yali wetland. In this area there are 209 invertebrates and vertebrates taxa recorded. Out of these, 199 taxa have been reported at the species level. The taxonomic identification for the remaining 10 taxa has only reached the generic level. The native element dominates in the fauna assembly, and it includes 142 species. The following elements are "Visitor" (23 sp.) and "Endemic" (18 sp.). About 9% of species have been introduced by anthropic activities. Only 3.5% of the fauna has wide distribution (2 sp) or cosmopolitan range (5 sp). For 72% of the species, their degree of endangerment is not yet defined, while for 5.5% the information is sufficient for categorization. 12.6%, 5% and 3.5% of the fauna have been classified as "Vulnerable", "Endangered" and "Rare", respectively. The "No threat" and "Out of risk" categories represent 1%.

Freshwater invertebrates (n = 12 taxa, Table 6.2) include insects (n = 6 taxa), annelids (n = 1 taxa), mollusks (n = 4 taxa) and crustaceans (n = 1 taxa). Unfortunately, the records of invertebrates for the wetland have low taxonomic resolution. This limits the analysis of the endemism patterns and threat to local invertebrate fauna. Within this group it is possible to recognize only one endemic species in the Los Molles and El Rey lagoon *(P. chilensis,* Figueroa et al. 2009), and a native species in the El Convento (A. franciscana, De los Ríos 2000, Crespo 2001). However, neither specie is under any conservation status (Table 6.2).

About 95% of the remaining fauna is composed of terrestrial and aquatic vertebrates. This group includes fish, amphibians, reptiles, birds and mammals. Among fish (n = 11 spp), the most representative families are Sciaenidae (n = 3 spp), Atherinidae (n = 2 spp) and Poeciliidae (n = 2 spp.; Table 6.2). Virtually all species are freshwater, with the exception of the members of the families Sciaenidae and Clupeidae, which correspond to marine fish observed in the Laguna Albufera. About 45% of ichthyic fauna is native (n = 5 spp), 27% endemic (n = 3 spp), 18% exotic (n = 2 spp), and 9% cosmopolitan (n = 1 spp). 45% of these species have been classified as "Vulnerable". For the remaining percentage of taxa, the conservation status is unknown.

Amphibians are the group underrepresented in the area, registering only 5 species in 3 families (Table 6.3). 80% of the taxa are endemic or native to southern South America and have been classified under the categories of "Vulnerable" or "Endangered" (*C. caudiverbera*). The exotic species *X. laevis* has been considered

Table 6.2 Aquatic invertebrates and fish species registered in el Yali wetland. (*) indicates marine fish registered in Albufera lagoon (Brito 1999)

Specie	Family	Element	Conservation status
Invertebrados			
Artemia cf franciscana	Artemiidae	Native	Undefined
Berosus sp	Hydrophilidae		
Biomphalaria sp	Planorbidae		
Chironomus sp	Chironomidae		
Enochrus sp	Hydrophilidae		
Gundlachia sp	Ancyliidae		
Helobdella sp	Glossiphoniidae		
Littoridina sp	Amnicolidae		
Nothiphila sp	Ephydridae		
Physa Chilensis	Physidae	Endemic	Undefined
Thermonectus sp	Dysticidae		
Tropisternus sp	Hydrophilidae		
Peces			
Basilichthys australis	Atherinidae	Endemic	Vulnerable
Cheirodon pisciculus	Characidae	Endemic	Vulnerable
Cnesterodon decemmaculatus	Poeciliidae	Exotic	Undefined
Eleginops maclovinus	Eleginopsidae	Native	Vulnerable
Gambusia holbrooki	Poeciliidae	Exotic	Undefined
*Menticirrhus ophicephalus**	Sciaenidae	Native	Undefined
Mugil cephalus	Mugilidae	Cosmopolitan	Vulnerable
Odontesthes mauleanum	Atherinidae	Endémico	Vulnerable
*Sciaena deliciosa**	Sciaenidae	Nativo	Undefined
*Stellifer minor**	Sciaenidae	Nativo	Undefined
*Strangomera bentincki**	Clupeidae	Nativo	Undefined

unhealthy or harmful (SAG 2004). Therefore, according to the Hunting Law Regulations (DS No. 53 of 2003, General Ministry of Chile), they can be hunted or captured at any time of the year, throughout the national territory and without any limitations.

In the Yali wetland, 8 reptile species are reported in four families, with *Tropidurids* being the most important (Table 6.3). The *Cheloniidae and Dermochelyidae* families include species of sea turtles with cosmopolitan distribution that are currently under threat of extinction worldwide (SERNAPESCA 2008). Therefore the local herpetic fauna is dominated by terrestrial taxa, incorporating endemic (n = 4 spp) or native (n = 2 spp) elements under the "Vulnerable" conservation status, except *L. chiliensis*, which is categorized as "Insufficiently known".

Within vertebrates, birds are definitely the group with the highest specific richness in the area, with 148 species in 42 families (Table 6.4). The records of Valverde (1996) and Brito (1999) indicate that about 62% of the species depend directly on the aquatic habitats of the wetland. While 38% of the taxa are associated with

Table 6.3 Amphibians and reptiles registered in el Yali wetland. (*) occasional records in the wetland coast (Brito 1999)

Specie	Family	Element	Conservation status
Anfibios			
Batrachyla taeniata	Leptodactylidae	Native	Vulnerable
Bufo chilensis	Bufonidae	Endemic	Vulnerable
Caudiverbera caudiverbera	Leptodactylidae	Endemic	Endangered
Pleurodema thaul	Leptodactylidae	Native	Vulnerable
Xenopus laevis	Pipidae	Exotic	Undefined
Reptiles			
Dermochelys coriaceae *	Dermochelyidae	Cosmopolitan	Endangered
*Lepidochelys olivaceae**	Cheloniidae	Cosmopolitan	Endangered
Liolaemus chiliensis	Tropiduridae	Native	Insufficiently known
Liolaemus lemniscatus	Tropiduridae	Native	Vulnerable
Liolaemus nitidus	Tropiduridae	Endemic	Vulnerable
Liolaemus tenuis	Tropiduridae	Endemic	Vulnerable
Philodryas chamissonis	Colubridae	Endemic	Vulnerable
Tachymenis chilensis	Colubridae	Endemic	Vulnerable

coastal sclerophyllous scrub that develops around the wetland. Representative families are Anatidae (13 spp), Scolopacidae (12 spp), Furnariidae (10 spp), Tyrannidae (10 spp), Charadriidae (9 spp) and Laridae (8 spp). 78% of the birds are native to southern South America, 15% are visitors, 3.3% endemic, 3% are introduced and 0.7% are cosmopolitan. About 83% of taxa present "undefined" conservation status, a percentage that includes all endemic species. The Magellanic Penguin presents a low risk of threat (SERNAPESCA 2008), although in the rest of South America it is considered almost threatened since 1994 (IUCN 2010). The remaining species are classified as "Inadequately known" (2.7%), "Endangered" (2.7%), "Rare" (3.4%) and "Vulnerable" %). It is important to note that the last category includes the parrot Tricahue (C. patagonus), a species that in the 1980s was extinguished locally from the wetland (Brito 1999).

Mammals comprise 25 species, in 16 families, including a family of marsupials (*Didelphidae*, Table 6.5). In general, families are represented by one or two species, except for the Dolphins (n = 3 sp). The native element dominates the biogeographic spectrum (n = 17 sp., 68%). 32% of the species are endemic (n = 3 sp), exotic (n = 2 sp), cosmopolitan (n = 1 sp), and widely distributed (n = 2 sp). Exotic species (hare and rabbit) are considered bad or harmful (SAG2004). For 36% and 24% of the species their conservation status has not been defined or there are insufficient data for their classification, respectively. 16% of the taxa have been classified under the categories "Vulnerable" and "Rare" (8% each). Within the mammalian assembly, the felids and the endemic specie *S. Cygnus* are in risk of extinction, while conservation status for the common wolf is "Out of risk".

Table 6.4 Birds registered in el Yali wetland. (*) indicates those species with strict association to the existence of the wetland and water bodies (Valverde 1996; Brito 1999)

Specie	Family	Element	Conservation status
*Agelaius thilius**	Emberizidae	Native	Undefined
Agriomis livida	Tyrannidae	Native	Undefined
Anairetes parulus	Tyrannidae	Native	Undefined
*Anas bahamensis**	Anatidae	Visitor	Rare
*Anas cyanoptera**	Anatidae	Native	Undefined
*Anas flavirostris**	Anatidae	Native	Undefined
*Anas georgica**	Anatidae	Native	Undefined
*Anas platalea**	Anatidae	Native	Insufficiently known
*Anas sibilatrix**	Anatidae	Visitor	Undefined
*Anas versicolor**	Anatidae	Native	Undefined
Anthus correndera	Motacillidae	Native	Undefined
*Ardea cocoi**	Ardeidae	Native	Rare
*Arenaria interpres**	Scolopacidae	Visitor	Undefined
*Asio flammeus**	Strigidae	Native	Insufficiently known
Asthenes humicola	Furnariidae	Endemic	Undefined
Athene cunicularia	Strigidae	Native	Undefined
Bubo virginianus	Strigidae	Exotic	Undefined
*Bubulcus ibis**	Ardeidae	Exotic	Undefined
Buteo polisoma	Accipitridae	Native	Undefined
*Calidris alba**	Scolopacidae	Visitor	Undefined
*Calidris bairdii**	Scolopacidae	Visitor	Undefined
*Calidris canutus**	Scolopacidae	Visitor	Undefined
*Calidris melanotos**	Scolopacidae	Visitor	Undefined
*Calidris pusilla**	Scolopacidae	Visitor	Undefined
Callipepla californica	Phasianidae	Exotic	Undefined
Caprimulgus longirostris	Caprimulgidae	Native	Undefined
Carduelis barbata	Fringillidae	Native	Undefined
*Casmerodius albus**	Ardeidae	Native	Undefined
*Catharacta chilensis**	Stercoraciidae	Native	Undefined
Cathartes aura	Cathartidae	Native	Undefined
*Catoptrophorus semipalmatus**	Scolopacidae	Visitante	Undefined
*Charadrius alexandrinus**	Charadriidae	Native	Undefined
*Charadrius collaris**	Charadriidae	Native	Undefined
*Charadrius falklandicus**	Charadriidae	Native	Undefined
*Charadrius modestus**	Charadriidae	Native	Undefined
*Charadrius semipalmatus**	Charadriidae	Visitor	Undefined
*Chloephaga melanoptera**	Anserinae	Native	Rare
*Chloephaga poliocephala**	Anserinae	Native	Undefined

(continued)

Table 6.4 (continued)

Specie	Family	Element	Conservation status
*Cinclodes nigrofumosus**	Furnariidae	Endemic	Undefined
Cinclodes oustaleti	Furnariidae	Native	Undefined
*Cinclodes patagonicus**	Furnariidae	Native	Undefined
*Circus cinereus**	Accipitridae	Native	Undefined
*Cistothorus platensis**	Troglodytidae	Native	Undefined
Colaptes pitius	Picidae	Native	Undefined
Colorhamphus parvirostris	Hirundinidae	Native	Undefined
*Columba araucana**	Columbidae	Native	Endangered
Columbina picui	Columbidae	Native	Undefined
Coragyps atratus	Cathartidae	Native	Undefined
*Coscoroba coscoroba**	Anatidae	Native	Endangered
Curaeus curaeus	Emberizidae	Native	Undefined
Cyanoliseus patagonus	Psittacidae	Native	Vulnerable
*Cygnus melancoryphus**	Anatidae	Native	Vulnerable
Diuca diuca	Fringillidae	Native	Undefined
*Egretta thula**	Ardeidae	Native	Undefined
*Elaenia albicep**	Tyrannidae	Native	Undefined
Elanus leucurus	Accipitridae	Native	Undefined
Falco femoralis	Falconidae	Native	Undefined
Falco peregrinus	Falconidae	Native	Vulnerable
Falco sparverius	Falconidae	Native	Undefined
*Fulica armillata**	Rallidae	Native	Undefined
*Fulica leucoptera**	Rallidae	Native	Undefined
*Fulica rufifrons**	Rallidae	Native	Undefined
*Gallinago paraguaiae**	Scolopacinae	Native	Vulnerable
*Gallinula melanops**	Rallidae	Native	Undefined
Geositta cunicularia	Furnariidae	Native	Undefined
*Geositta maritima**	Furnariidae	Native	Undefined
Geranoaetus melanoleucus	Accipitrinae	Native	Undefined
Glaucidium nanum	Strigidae	Native	Undefined
*Haematopus palliatus**	Haematopodidae	Native	Undefined
*Heteronetta atricapilla**	Anatidae	Native	Rare
*Himantopus melanurus**	Recurvirostridae	Native	Undefined
*Hirundo rustica**	Hirundinidae	Visitor	Undefined
Hymenops perspicillatus	Tyrannidae	Visitor	Undefined
*Ixobrychus involucris**	Ardeidae	Native	Rare
*Larosterna inca**	Strigidae	Native	Vulnerable
*Larus dominicanus**	Laridae	Native	Undefined
*Larus maculipennis**	Laridae	Native	Undefined
*Larus modestus**	Laridae	Native	Vulnerable
*Larus pipixcan**	Laridae	Visitor	Undefined

(continued)

Table 6.4 (continued)

Specie	Family	Element	Conservation status
Leptasthenura aegithaloides	Furnariidae	Native	Undefined
*Lessonia rufa**	Tyrannidae	Native	Undefined
*Limosa haemastica**	Scolopacidae	Visitor	Undefined
Milvago chimango	Falconidae	Native	Undefined
Mimus thenca	Mimidae	Endemic	Undefined
Molothrus bonariensis	Emberizidae	Native	Undefined
Muscisaxicola macloviana	Tyrannidae	Native	Undefined
*Netta peposaca**	Anatidae	Native	Undefined
Nothoprocta perdicaria	Tinamidae	Endemic	Undefined
*Numenius phaeopus**	Scolopacidae	Visitor	Undefined
*Nycticorax nycticorax**	Ardeidae	Native	Undefined
*Oxyura jamaicensis**	Anatidae	Native	Undefined
*Oxyura vittata**	Anatidae	Native	Undefined
Parabuteo unicinctus	Accipitridae	Native	Undefined
*Pardirallus sanguinolentus**	Rallidae	Native	Undefined
Patagona gigas	Trochilidae	Native	Undefined
*Pelecanus thagus**	Pelecanidae	Native	Undefined
*Phalacrocorax bougainvillii**	Phalacrocoracidae	Native	Vulnerable
*Phalacrocorax brasilianus**	Phalacrocoracidae	Native	Undefined
*Phalacrocorax gaimardi**	Phalacrocoracidae	Native	Insufficiently known
*Phalacrocorax olivaceus**	Phalacrocoracidae	Endemic	Undefined
*Phalaropus tricolor**	Phalaropodinae	Visitor	Undefined
Phleocryptes melanops	Furnariidae	Native	Undefined
*Phleocryptes melanops**	Furnariidae	Native	Undefined
*Phoenicopterus chilensis**	Phoenicopteridae	Native	Vulnerable
Phrygilus alaudinus	Fringillidae	Native	Undefined
Phrygilus fruticeti	Fringillidae	Native	Undefined
Phrygilus gayi	Fringillidae	Native	Undefined
Phytotoma rara	Phytotomidae	Native	Undefined
*Plegadis chihi**	Threskiornithidae	Native	Endangered
*Pluvialis dominica**	Charadriidae	Visitor	Undefined
*Pluvialis squatarola**	Charadriidae	Visitor	Undefined
*Podiceps major**	Podicipedidae	Native	Undefined
*Podiceps occipitalis**	Podicipedidae	Native	Undefined
*Podilymbus podiceps**	Podicipedidae	Native	Undefined
*Porphyriops melanops**	Rallidae	Exotic	Undefined
*Pygochelidon cyanoleuca**	Hirundinidae	Native	Undefined

(continued)

Table 6.4 (continued)

Specie	Family	Element	Conservation status
*Rollandia rolland**	Podicipedidae	Native	Undefined
*Rostratula semicollaris**	Rostratulidae	Native	Endangered
*Rynchops niger**	Laridae	Visitor	Undefined
Scelorchilus albicollis	Rhinocryptidae	Native	Undefined
Scytalopus magellanicus	Rhinocryptidae	Native	Undefined
*Sephanoides galeritus**	Trochilidae	Native	Undefined
Sicalis luteola	Emberizidae	Native	Undefined
Spheniscus humboldti	Spheniscidae	Native	Vulnerable
Spheniscus magellanicus	Spheniscidae	Native	Low threat
*Sterna hirundinacea**	Laridae	Native	Undefined
*Sterna paradisaea**	Laridae	Visitor	Undefined
*Sterna trudeaui**	Laridae	Native	Undefined
*Stumella Ioyca**	Emberizidae	Native	Undefined
*Sula variegata**	Sulidae	Native	Insufficiently known
Sylviorthorhynchus desmursii	Furnariidae	Native	Undefined
*Tachuris rubrigastra**	Tyrannidae	Native	Undefined
*Tachycineta meyini**	Hirundinidae	Native	Undefined
*Theristicus melanopis**	Threskiornithidae	Native	Vulnerable
Thinocorus rumicivorus	Thinocoridae	Native	Undefined
*Tringa flavipes**	Scolopacidae	Visitor	Undefined
*Tringa melanoleuca**	Scolopacidae	Visitor	Undefined
Troglodytes aedon	Troglodytidae	Native	Undefined
Turdus falcklandii	Muscicapidae	Native	Undefined
Tyrannus tyrannus	Tyrannidae	Visitor	Undefined
Tyto alba	Tytonidae	Cosmopolitan	Undefined
*Vanellus chilensis**	Charadriidae	Native	Undefined
*Veniliornis Iignarius**	Picidae	Native	Undefined
Xolmis pyrope	Tyrannidae	Native	Undefined
Zenaida auriculata	Columbidae	Native	Undefined
*Zonibyx modestus**	Charadriidae	Native	Undefined
Zonotrichia capensis	Emberizidae	Native	Undefined

Environmental Patterns

In this section we present a synthesis of the works that have evaluated the factors and mechanisms that determine the structure and dynamics of the birds and vegetation assemblies present in the El Yali wetland (Vilina and Cofré 2000; Vilina et al. 2002; Fariña et al. 2005, 2006, 2007, 2009).

When it comes to birds, the studies focus on the effects of the exogenous ENSO variable on the dynamics of the populations of four species of grebes (*R. rolland, P.*

Table 6.5 Mammals registered in el Yali wetland. (*) occasional records in the wetland coast (Brito 1999)

Specie	Family	Element	Conservation state
Akodon olivaceus	Muridae	Native	Not defined
*Arctocephalus australis**	Otariidae	Ample	Out of risk
Cephalorhynchus eutropia	Delphinidae	Endemic	Insufficiently known
Conepatus chinga	Mephitidae	Native	Rare
*Eubalenea australis**	Balaenidae	Ample	Not defined
Galictis cuja	Mustelidae	Native	Vulnerable
*Grampus griseus**	Delphinidae	Cosmopolitan	Insufficiently known
Hydrurga leptonyx *	Phocidae	Ample	Insufficiently known
Lasiurus borealis	Vespertilionidae	Native	Not defined
Lasiurus cinereus	Vespertilionidae	Native	Not defined
Lepus capensis	Leporidae	Exotic	Not defined
Lynchailurus colocolo	Felidae	Native	In risk of extinction
Myocastor coypus	Myocastoridae	Native	Vulnerable
Octodon degus	Octodontidae	Endemic	Not defined
Oligoryzomys longicaudatus	Cricetidae	Native	Not defined
Oncifelis guigna	Felidae	Native	In risk of extinction
Oryctolagus cuniculus	Leporidae	Exotic	Not defined
Otaria flavescens	Otariidae	Native	Insufficiently known
Phocoena spinipinnis	Phocoenidae	Native	Insufficiently known
Phyllotis darwini	Muridae	Native	Not defined
Pseudalopex culpaeus	Canidae	Native	Insufficiently known
Pseudalopex griseus	Canidae	Native	Not defined
Spalacopus cyanus	Octodontidae	Endemic	In risk of extinction
Tadarida brasiliensis	Molossidae	Native	Not defined
Thylamys elegans	Didelphidae	Native	Rare

occipitalis, P. major and P. Podiceps) and the Black Neck Swan (*C. melancoryphus*). The changes in the patterns of abundance, spatial-temporal distribution and reproduction associated with positive precipitation anomalies during the warm ENSO phase have been evaluated (Vilina and Cofré 2000, Vilina et al. 2002). In general, these studies show that the birds population variables evaluated in the El Yali wetland, correlate positively with the precipitation anomalies that occurred during the El Niño 1992–1993 and 1997–1998 events. However, for grebes, there are no changes in habitat preference during the El Niño years, remaining in Los Molles reservoir and El Rey lagoon (Vilina and Cofré 2000). Apparently, these systems are able to maintain the spaces with open water and at the same time an abundant vegetation cover; both of these variables are considered the most relevant for the continuity of this type of birds (Vilina and Cofré 2000).

The censuses carried out at Los Molles, El Yali estuary, Albufera and El Rey lagoons, show that grebes increase their numbers in the spring of the year immediately after the ENSO event, and then descend in density (Vilina and Cofré 2000).

This is particularly significant for *P. occipitales* ($R2 = 0.8$) and *P. podiceps* ($R2 = 0.9$). The positive correlation between rainfall in the ENSO years and the abundance of grebes observed during the post-ENSO years imply that there is a one-year gap in the population response (Vilina and Cofré 2000). This in turn suggests that the abundance of grebes in the spring of 1993 and 1998 was related to the greater availability of prey (Vilina and Cofré 2000), which in turn is related to the high productivity reached by the semi-arid systems of central Chile during the ENSO years (Jaksic 2001). It is probable that the higher densities of *P. occipitales* observed in the post-ENSO spring are also favored by the appearance of new breeding areas after the increase in the extended flooded areas of the wetland (Vilina and Cofré 2000). Indeed, the Silvery Grebe, unlike the other species of grebes present in the wetland, reproduce in this sector, exclusively during the summers of the ENSO years and the year immediately afterwards (Vilina and Cofré 2000).

The monthly censuses carried out between 1993 and 1999 indicate that the populations of *C.melancoryphus* in El Yali present an important intra- and inter-annual fluctuations related to variations in rainfall and their effects on local hydrology (Vilina et al. 2002). At the seasonal scale, it has been observed that the abundance of Black Neck Swans is higher during the winter, when the different bodies of water in the system maintain high water levels. In contrast, the marked decline in these levels during the dry season (summer) determines that the local abundance of these swans decreases as a result of the migration to the wetlands of south-central Chile.

At interannual scale, the Black Neck Swan abundance patterns in the wetland are related to rainfall anomalies occurring during the cold and warm ENSO phases. Thus, extreme droughts during La Niña events lead to the disappearance of the local population or a substantial reduction in their numbers (Vilina et al. 2002). On the other hand, the positive anomalies during ENSO events maintain an abundant colony of *C. melancoryphus* throughout the year, but also in the year following the event. Nevertheless, the analysis carried out by Vilina et al. (2002) indicate that only the abundance observed in the post-ENSO years show a significant correlation with the positive precipitation anomalies ($R2 = 0.65$). This correlation implies a connection between the abundant presence of the Black Neck Swans population and the rain accumulated in the previous year, with a one year gap. This response would be mediated by increases in water levels and availability of resources. That is why Vilina et al. (2002) found a positive and significant correlation between rainfall accumulated in the previous year and the number of months in which *C. melancoryphus* is reproduced in the wetland ($R2 = 0.74$; Vilina et al. 2002). It appears that the greatest extent and coverage of *S. californicus* facilitates and favors in situ reproduction in El Rey lagoon, a system that supports the most important breeding colony in the south central part of the country.

The studies on the mechanisms that determine the specific diversity and structuring of the vegetation communities of the Albufera lagoon in the El Yali, have been systematically carried out between 2003 and 2006 at the request of CONAF (Fariña et al. 2005, 2006, 2007, 2009). Therefore, from the edge from the edge of the Albufera to the dune areas (~ 180 m from the edge of the lagoon), three vegetation zones along a flood gradient have been identified (Fig. 6.2). These vegetation strips result from spatial variations in physicochemical variables and nutrients (phosphate and nitrate) through the flood gradient. Also, the intensity and relative importance of biological interactions as competition and herbivory (Fariña et al. 2005, 2006, 2007, 2009).

The first zone extends from the edge of the lagoon (0 m) to 40 m, in the sector that regularly suffers floods and consistently presents higher salinity (20–40 ppm), soil moisture (70–80%), percolation (1 l/h), and low redox potential (−240 to −80 Eh). Here nitrate and phosphate concentrations are particularly low (<0.035 mg/ g dry soil). Within this range, *S. fruticosa* (*Sarcocornia* area) dominates widely, a species that reaches 100% coverage through the first 20 m (Fig. 6.2). The gradual development of *S. densiflora* between 30 and 40 m determines the decrease in *S. fruticosa* cover, which is finally displaced by competition at 80 m (Fig. 6.2). However, herbivore exclusion experiments in the area, where both species coexist, indicate that the interaction between *S. densiflora* and *S. Fruticosa* is further modulated by herbivory (Fariña et al. 2005, 2006, 2007). In fact, it has been observed that in the absence of horses and cows, the biomass of *S. fruticosa* is increased despite the reduction of *S. densiflora*.

The strong dominance of *S. densiflora* between 40 and 80 m determines the second vegetational strip *(Spartina* zone). In this sector of the Albufera, floods occasionally occur after rainfall episodes, and physical-chemical variables have intermediate values: salinity (25–10 ppm), soil moisture (60–70%), percolation (2 l/h), and redox potential (−5 to 160 Eh). In comparison with the previous strip, the nitrate concentration is higher (0.06–0.08 mg/g dry soil), while in the case of phosphate, there are no significant differences. The highest coverage of *S.densiflora* is reached at 60 m (73.3%). From this distance, their coverage falls to values close to 40% (Fig. 6.2). This decrease is concomitant with the occurrence of *Scyrpus californicus* and *Selliera radicans*. The latter species tends to be abundant (19.2%) at 80 m, distance to which *S. densiflora* and *S. californicus* coexist and facilitate each other's growth (Fariña et al. 2009). As in the *Sarcocornia* area, herbivory is an important factor within the second vegetational fringe. The artificial exclusion of herbivores reduces the coverage of *S. densiflora*, determining in turn the occurrence of different weeds and the increase in abundance of competitively subordinated species (Fariña et al. 2005, 2006, 2007).

Between 80 and 180 m, in the area where experimentation is carried out only during heavy storms, the third vegetation zone is established. This sector is characterized by high redox potential (150–320 Eh), and higher concentrations of nitrate (0.12–0.15 mg/g dry soil) and phosphate (0.12–0.15 mg g dry soil), with measurements of salinity (10–0 ppm), percolation (15 L / h) and humidity (20–30%). Within this range dominates *S. californicus* (Scyrpus zone), and is particularly abundant between 100 and 120 m (60% and 68.3%, respectively). Although *S.californicus* has been classified as competitively superior (Fariña et al. 2005, 2006, 2007, 2009), *S. densiflora* and six other species can be found within this strip, although all these represent less than 15% of Coverage (Fig. 6.2). However, herbivore exclusion experiments demonstrate that these taxa may undergo a significant increase, in particular *S. densiflora, D. spicata and C. cronopifolia*. From the 140 m, the reduction

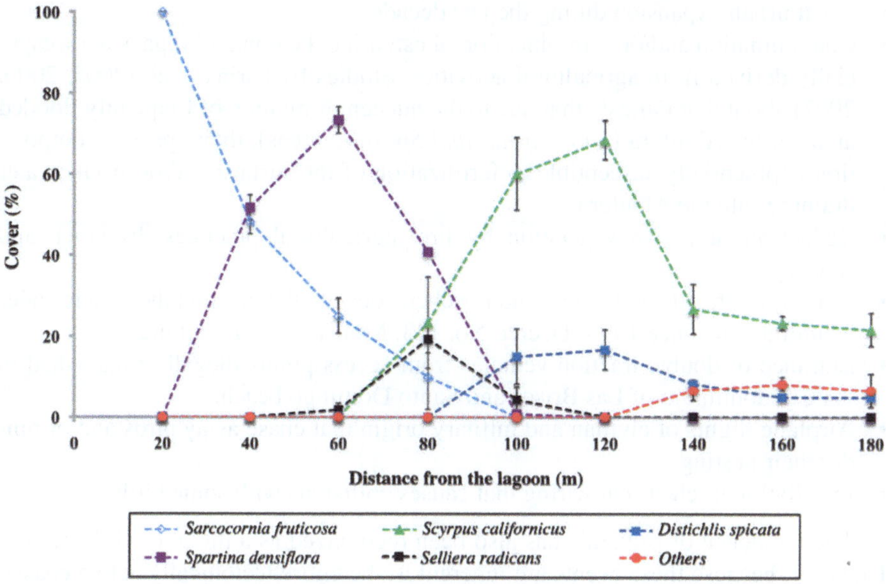

Fig. 6.2 Plant zonation pattern throughout the environmental gradient that develops from the edge of Albufera lagoon to the dune zone (Modified from Fariña et al. 2009). The species under the category of "Others" include *F. salina*, *C. cronopifolia*, *S. media*, *L. temulentum* and *J. Balticus*

of moisture content in the sediments determines a marked decrease in the total coverage of vegetation (<35%) and the predominance of the bare soil, giving rise to the development of sandy dunes (Fariña et al. 2005, 2006, 2007, 2009).

Threats

With the arrival of sedentary humans to the Yali wetland about 6000 YBP, begins the pressure on the biotic and abiotic resources of the system (Falabella and Planella 1980). In fact, the Llo-Lleo and Aconcagua cultures, that show clear evidence of settlement in the area, apparently have modified and intervened in vegetation, soils systems as well as bodies of water, for agricultural purposes that supported their economy and society (Falabella and Planella 1980, Brito 2004). Currently, despite the existence of specific conservation and management plans for the El Yali wetland, there are still some factors that threaten its biodiversity, functioning and structure. The INTOSAI report (2007) and the "Green Network Innovation Agenda for the San Antonio - Seacoast of the Poets" (CODESSER 2009) recognize the following anthropic pressures on the wetland:

- Extraction of groundwater and change of course for surface water, for agricultural and industrial use.

- Great urban expansion during the last decades.
- Contamination and/or eutrophication of estuarine, lagoon and vega waters, especially derived from agricultural activities. Studies by Fariña et al. (2005, 2006, 2007) about that indicate that due to the nutrient limitation of frequently flooded areas in the Albufera (*Sarcocornia and Spartina* areas), their specific composition is potentially susceptible to fertilization of the surface and/or groundwater draining into the Albufera.
- Reduction of native vegetation by fire, agricultural practices, livestock and forestry.
- Actions by the illegal hunters and poachers, despite the fact that the area is under a hunting ban since 1998 (Decree No. 143, Ministry of Agriculture).
- Entrance of double traction vehicles from access points illegally established in the Condominium of Las Brisas and Santo Domingo beach.
- Airplane flights of civilian and military origin that chase away birds and/or hinder their nesting.
- Installation of electrical wiring that causes collisions with some birds.

The incidence of tsunamis has also been recognized as a threat (PAPSA 2009). However, because these events are inherent to the entire tectonically active coastal edge of Chile, they are part of the natural coastline variability and must be treated as exogenous disturbances. In fact, for the coast of San Antonio we know of several tidal waves: May 13th, 1647; July 8th, 1730; November 19th, 1822; August 16th, 1906; March 3rd, 1985; and February 27th, 2010.

For this last tsunami, there is a record of the violent impact the flood had on the ecosystems of the El Yali wetland, specifically on the vegetation bands recognized in the Albufera (Fariña et al. 2010). Before February 27th, 2010, in the Albufera area there was a sandy beach, some dunes and an area of wetland vegetation. The violent entry of tsunami sea water has destroyed part of the dune and beach area, spreading a large amount of sand over the lagoon. This led to an advance of 70 m of the inland coastline, the widening of the lagoon (~30 m) and the creation of several points of connection with the sea. As a consequence, vegetation bands in the Albufera were practically buried, except in those areas where Fariña et al. (2005, 2006, 2007) had conducted their experiments to exclude furtive herbivores. The permanent connection between the lagoon and the sea produced important changes in the salinity of the lagoon, causing the migration of some birds that are commonly sighted in the area.

Fariña et al. (2010) point out that 3 months after the tsunami, there were clear signs of recovery in this ecosystem. In May 2010, almost all the connecting roads between the Albufera and the sea had been lost, while the bands of vegetation in the wetland seem to be regenerating from areas where livestock had been excluded. Based on these findings, the authors indicate that the high resilience observed of the system, and through its adaptive mechanisms, the wetland species have learned to deal with this type of natural disturbance. On the other hand, this also demonstrates the need to keep these systems in pristine conditions. Their observations suggest that the effects of activities of anthropic origin (e.g. livestock) have the potential to alter the system's strength to recover from exogenous shocks.

Conclusions

The success of wetland conservation in other countries and nations has been determined by the joint effort and commitment of social, business, academic and political players (NAWCC 2001). In the academic arena, science has played a key role in providing tools, knowledge and experience to address and propose conservation actions relevant to each wetland (NAWCC 2001).

In this chapter, we present a synthesis that integrates the existing information on the natural history and ecological patterns of the wetland, and we intend to offer a tool that includes all the necessary elements to develop, in Chile, a conservation vision concerning wetlands. Although the conservation initiatives developed by the Government of Chile in the case of the El Yali wetland have been celebrated by international organizations (e.g. INTOSAI 2007), our level of knowledge about the operation of this system is rather limited. In fact, the bases on which conservation measures for the coastal wetland have been established are nothing more than isolated cadaster of some components of the ecosystems, with special and exaggerated emphasis on avifauna. On the other hand, we know the structuring and functioning of the vegetation communities that are developed around the Albufera, including their dynamics and response to exogenous perturbations and possibly to anthropic threats.

In light of the background presented in this synthesis, we believe that studies focused on ecosystem processes at different spatial and temporal scales are urgently needed and that they also evaluate the interaction between the different components of ecosystems (e.g. the relationship between the balance of nutrients in water and soil, and the composition of flora and fauna species). This way, an adequate characterization of the dynamics and natural diversity of the biological communities of the wetland can be obtained. Therfore, we propose the following recommendations:

- It is necessary to establish the ecological patterns (species composition, diversity) in the fauna component, evaluating the mechanisms that determine them. This approach is particularly urgent in invertebrates, as this group is commonly used as a bio-indicator (e.g. Lindegaard 1995), providing information on the "health" and quality of water bodies. In the specific case of avifauna, it is necessary to know the direction and magnitude of intra and inter-specific biological interactions, as well as their effects on the community (e.g., abundance, distribution).
- It is important to know and evaluate the ENSO effects, inter-annual variations in rainfall and hydrological regimes on the short- and long-term dynamics in wetland communities.

- At the ecosystem level, it is urgent to establish and understand the interaction between native fauna and flora components, considering also the factors and mechanisms that influence this relationship.
- It is of priority to encourage studies that characterize the ecological patterns of marsh vegetation in other bodies of water in the system (e.g. Matanzas and Colejuda lagoons), as well as the coastal and dune matorral and vegetation systems.
- Given the magnitude of anthropogenic activities in the wetland and surrounding areas, it is urgent to establish and to maintain new monitoring networks to evaluate the effect of pollution of water resources in the distribution and abundance of flora and fauna. In particular, those relating to the eutrophication of water.
- Continuing the long-term monitoring of the wetland. They have provided valuable information about the system's resilience to exogenous shocks and how it can be modified through the anthropic activities. These have also demonstrated the extraordinary adaptive capacity of the Albufera plant species.
- Encourage interdisciplinary studies on the mechanisms that have determined the historical evolution of El Yali. This type of study is a necessary tool to establish long-term trends and to predict the possible effects of global change on the coastal wetland.
- Generate stronger conscience and awareness, as well as citizen participation in the conservation of the El Yali through implementation of educational and recreational activities whose focus is on the value of the flora and fauna present in the wetland according to their conservation status, and just as wel highlighting its historical richness represented by the presence of unique archaeological sites and biogeographic singularities of all its biota.

Acknowledgments E. M.G is grateful for the CONICYT funding (Scholarship # 24080156) and IEB ("ICM Project Doctorate Scholarship, code P05-002").

References

Aceituno, P. 1990. Anomalías climáticas en la Región Sudamericana durante los extremos de la Oscilación Austral. *Revista de Geofísica* 32: 65–78.

Avesdechile. 2010. *Guía web de las aves de Chile.*. http://www.avesdechile.cl.

Benoit, I. L. (1989). Red book on Chilean terrestrial flora (Part one). CONAF. pp. 151. Santiago.

Biblioredes. (2010). Antecedentes básicos del área denominada Reserva Nacional El Yali. Http://www4.biblioredes.cl/bibliored/Nosotros+en+Internet/Humedal+El+yali/reser.

Brito, J.L. 1999. Vertebrados del humedal La Reserva Nacional El Yali y su costa, Santo Domingo, Chile central. *Anales del Museo de Historia Natural de Valparaíso* 24: 121–126.

Brito, J. L. (2004). Prehistoria de San Antonio. Antípoda, Grupo de Acción Ecológica- Cultural y Amigos del Museo de San Antonio. San Antonio.

Brown, J.H., and M.V. Lomolino. 1998. *Biogeography*. 2nd ed. Sunderland: Sinauer Associates Inc.

Campos, H., G. Dazarola, B. Dyer, L. Fuentes, J.F. Gavilán, L. Huaquín, G. Martínez, R. Meléndez, G. Pequeño, F. Ponce, V.H. Ruiz, W. Siefeld, D. Soto, R. Vega, and I. Vila. 1988. Categorías de

Conservación de peces nativos de aguas continentales de Chile. *Boletín del Museo Nacional de Historia Natural, Santiago de Chile* 47: 101–122.
Codesser. (2009). Agenda de Innovación para la Red Verde San Antonio-Litoral de los Poetas. Corporación de Desarrollo Social del Sector Rural & CORFO. pp. 372. San Antonio.
Cofré, H.L., Y.A. Vilina, J. Aguirre, W. Egli, M.D. García, H. Kocksch, J. Rottmann, H. Seeger, and C. Tala. 2009. Local distribution, abundance and conservation of Black-Headed Duck in the Mediterranean wetlands of Chile. *Waterbirds* 30: 412–416.
Conama-conaf. (2005). Estrategia nacional para la conservación y uso racional de los Humedales en Chile. Comisión Nacional de Medio Ambiente (CONAMA) y Corporacion Nacional Forestal de Chile (CONAF). pp. 30. Santiago.
Crespo, J.E. 2001. Occurrence of Artemia (Branchiopoda, Anostraca) in the "Salinas El Convento", Central Chile. *Crustaceana* 74: 1393–1395.
De Los Ríos, P. 2000. Descripción morfométrica de la población de Artemia (Leach 1819) de salina el Convento (Valparaíso, Chile). *Gayana* 64: 59–62.
Díaz-Páez, H., and J.C. Ortiz. 2003. Evaluación del estado de conservación de los anfibios en Chile. *Revista Chilena de Historia Natural* 76: 509–525.
Donoso-Barros, R. 1970. Catálogo Herpetológico chileno. *Boletín Museo Nacional de Historia Natural* 31: 49–124.
Dussaillant, A., P. Galdames, and C.-L. Sun. 2009. Water level fluctuations in a coastal lagoon: El Yali Ramsar wetland, Chile. *Desalination* 246: 202–214.
Falabella, F., and M.T. Planella. 1980. Secuencia Cronológica Cultural para el Sector de la Desembocadura del río Maipo. *Revista Chilena de Antropología* 3: 87–107.
Fariña, J. M., Bertness, M. D., and Silliman, B. R. (2005). Efectos de la competencia, estrés ambiental y herbivoría en los patrones de zonación del ensamble de plantas presentes en el Humedal costero de El Yali, V Región, Chile. Informe final. CONAF. pp. 21.
———. (2006). Efectos de la competencia, estrés ambiental y herbivoría en los patrones de zonación del ensamble de plantas presentes en el Humedal costero de El Yali, V Región, Chile. Segunda parte. CONAF. pp. 21.
———. (2007). Efectos de la competencia, estrés ambiental y herbivoría en los patrones de zonación del ensamble de plantas Presentes enel Humedal costero de El Yali, V Región, Chile. Segunda parte (informe 2). CONAF. pp. 16.
Fariña, J.M., B.R. Silliman, and M.D. Bertness. 2009. Can conservation biologists rely on established community structure rules to manage novel systems? ... Not in salt marshes. *Ecological Applications* 19: 413–422.
Fariña, J. M., Aragoneses, N., Armesto, S., Almar, R., Bertness, M. D., and Silliman, B. R. (2010). Cambios en la estructura de hábitat de la Laguna Albufera, Humedal El Yali (V Región), luego del Tsunami del 27 de Febrero del 2010. Boletín Red Universitaria Cruz del Sur: Ciencias del Mar 1.
Fernández, J. V. (2004). Proposición de un plan de manejo integral de la cuenca hidrográfica del Estero Yali, V y Región Metropolitana. Tesis Ingeniero Forestal, Universidad Mayor.
Figueroa, R., Araya, E., Parra, O., and Valdovinos, C. (1999). Macroinvertebrados Bentónicos como indicadores de calidad de agua. VI Jornadas del Conaphi-Chile.
Figueroa, R., M.L. Suárez, A. Andreu, V.H. Ruiz, and M.R. Vidal-Abarca. 2009. Caracterización ecológica de humedales de la zona semiárida en Chile central. *Gayana* 73: 76–94.
Fuentealba, C., R. Figueroa, and J.J. Morrone. 2010. Análisis de endemismo de moluscos dulceacuícolas de Chile. *Revista Chilena de Historia Natural* 83: 289–298.
Galdames, P. (2007). Caracterización hidrológica de laguna albufera en humedal El Yali con énfasis en flujos subterráneos. Magíster en Ciencias de la Ingeniería, Pontificia Universidad Católica de Chile. pp 188.
Gana, P., Gutiérrez, A., and Wall, R. (1996). Mapa geológico del área de San Antonio – Melipilla. Mapa geológico n° 2. Regiones de Valparaíso, Metropolitana y Del Libertador G. B. O'Higgins. Escala 1:100000. Sernageomin.

Glade, A. 1993. *Libro rojo de los vertebrados terrestres de Chile*. Santiago: Corporación Nacional Forestal. pp. 68.

Hervé, F., Encinas, A., Villa-Martínez, R., Nielsen, S., Finger, K., and Peterson, D. (2003). Registro sedimentológico de la transgresiónmarina del holoceno medio en el área de Algarrobo (33° 22′ sur), Chile Central. Implicaciones tectónicas y Paleoecológicas. 10° Congreso Geológico Chileno. Concepción.

INTOSAI. (2007). Evolution and trends in environmental auditing. International Organization of Supreme Audit Institutions (INTOSAI) – Working Group on Environmental Auditing (WGEA). pp. 88.

Iucn. (2010). The IUCN red list of threatened species Version 2010.3. http://www.Iucnredlist.org.

Jaksic, F.M. 2001. Ecological effects of El Niño in terrestrial ecosystems of western South America. *Ecography* 24: 241–250.

Jenny, B., Massaferro, J., Valero-Garcés, B., Geyh, M., and Rizzo, A. (2000). Evidence of salinity changes due to marine influence during the last 3000 years in Laguna Matanzas (Central Chile) from sedimentological, geochemical and chironomid analyses. Paleolimnology Congress. Kingston.

Lib. (2010). Generaciónydesarrollodebasededatosdeespeciesexóticasen Chile. Labora- torio de Invasiones Biológicas. Universidadde Concepción. www.lib.udec.cl.

Lindegaard, C. (1995). The Chironomidae: Biology and ecology of nonbiting midges. In *Classification of water-bodies and pollution*, 385–404. Chapman & Hall.

Marticorena, C., and Quezada, M. (1985). Catálogo de la flora vascular de Chile. *Gayana (Botánica)* 1–152.

Mop. (2005). Informe técnico N°421: Area de Restricción Sector Hidrogeológico de Aprovechamiento Común de Yali Bajo El Prado. Ministerio de Obras Públicas. pp. 25. Santiago.

Nawcc. (2001). A wetland conservation vision for Canada. North American Wetlands Conservation Council (Canada). pp. 9. Ottawa.

Papsa. (2009). *Plan de Acción Provincial San Antonio*. San Antonio: Corporación Nacional Forestal (CONAF). pp. 84

Paskoff R. (1970). Geomorfología de Chile semiárido (Traducción de E. Novoa). Universidad de la Serena. Pp. 321.

Paskoff, R., H. Manríquez, L. Cuitiño, and R. Petiot. 2000. Características, origen y cronología de los cordones dunares de la playa de Santo Domingo, Región de Valparaíso, Chile. *Revista Geológica de Chile* 27: 121–131.

Piñones, A., A. Valle-Levinson, D.A. Narváez, C.A. Vargas, S.A. Navarrete, G. Yuras, and J.C. Castilla. 2005. Wind-induced diurnal variability in river plume motion. *Estuarine, Coastal and Shelf Science* 65: 513–525.

Sag. (2004). Vida Silvestre. Servicio Agrícola Ganadero. Gobierno de Chile. Http://www.sag.cl.

Sernapesca. (2008). Especies Hidrobiológicas en Estado de Conservación en Chile. Servicio Nacional de Pesca. Gobierno de Chile. www.sernapesca.cl.

Squeo, F.A., G. Arancio, and J.R. Gutiérrez. 2001. *Libro Rojo de la Flora Nativa y de los Sitios Prioritarios para su Conservación: Región de Coquimbo*. La Serena: Ediciones Universidad de La Serena. pp. 371.

Squeo, F., G. Arancio, J.R. Gutiérrez, L. Letelier, M.T.K. Arroyo, P. León-Lobos, and L. Rentería-Arrieta. 2008a. *Flora Amenazada de la Región de Atacama y Estrategias para su Conservación*. La Serena: Ediciones Universidad de La Serena. pp. 72.

Squeo, F.A., G. Arancio, and J.R. Gutiérrez. 2008b. *Libro Rojo de la Flora Nativa y de los Sitios Prioritarios para su Conservación: Región de Coquimbo*. La Serena: Ediciones de la Universidad de La Serena. pp. 456.

Sun, C.-L. (2008). Caracterización hidrológica y fisicoquímica de la laguna albufera del humedal El Yali con énfasis en flujos superficiales. Magíster en Ciencias de la Ingeniería, Pontificia Universidad Católica de Chile. pp. 171.

Tala, C., Guerrero, S., Avilés, R., and Stutzin, M. (2009). Especies amenazadas de Chile: Protejámoslas y evitemos su extinción, pp. 122. CONAMA.

Valero-Garcés, B., Latorre, C., Morellón, M., Corella, P., Maldonado, A., Frugone, M., and Moreno, A. (2010). Recent climate variability and human impact from lacustrine cores in central Chile. II International Symposium PAGES. Valdivia.

Valverde, V. (1996). Ficha informativa sobre humedales RAMSAR: Humedal el Yali. CONAF. pp. 20.

Victoriano, P.F., A.L. González, and R. Schlatter. 2006. Estado de conocimiento de las aves de aguas continentales de Chile. *Gayana* 70: 140–162.

Vilina, Y.A., and H.L. Cofré. 2000. "El Niño" Effects on the abundance and habitat association patterns of four Grebes species in ChileAn wetlAnds. *Waterbirds* 23: 95–101.

Vilina, Y., and V. López-Calleja. 1996. The Neotropical plovers of estero El Yali in central Chile. Shorebird Ecology and Conservation in the Western Hemisphere. *International Wader Studies* 8: 85–92.

Vilina, Y.A., H.L. Cofré, C. Silva-García, M.D. García, and C. Pérez-Friedenthal. 2002. Effects of El Niño on abundance and breeding of Black-Necked Swans on El Yali wetland in Chile. *Waterbirds* 25: 123–127.

Villa-Martínez, R. P. (2002). Historia del clima y la vegetaciónde chile central durante el holoceno: una reconstrucción basada en el análisis de polen, sedimentos, microalgas y carbón. Tesis de Doctorado, Universidad de Chile. pp 350.

Zunino, S., C. Aliaga, and P. Da Venezia. 2009. Comunidades de peces en desembocaduras de ríos y esteros de la Región de Valparaíso, Chile central. *Revista de Biología Marina y Oceanografía* 44: 123–130.

Chapter 7
Hydrology of the Central Chile Coastal Wetlands: The Case of the Yali Reserve

Alejandro Dussaillant

Abstract The El Yali lagoon is a wetland of great importance in the central zone of Chile, especially due to its relation with the native and foreign avifauna that arrives there periodically, and being a habitat for about 25% of the birds' biodiversity in Chile. This study aims to identify and quantify the main variables of the hydrological cycle of the lagoon and to characterize the main physicochemical variables of surface and groundwater, in order to estimate a hydrological balance of the wetland that will allow, in the future, to evaluate the vulnerability of this body of water. The data still do not allow to identify the hydrological cycle of the lagoon, but even though, the records collected will be presented, including the "behavior" on the coastal and interior sides of the lagoon. A significant concentration of phosphates from the aquifer has been detected, whose magnitude (above 2 mg (PO) 43-L-1) triples in the body of water (about 1 mg (PO) 43-L-1).

Keywords Wetland • Eutrophication • Coastal lagoon • Water balance • Nutrients

Introduction

In a number of countries, research on coastal wetlands has developed because of the importance they place locally for both, the ecosystem contributions to the society and the fact that they are shelters for a large number of flora and fauna (Mitsch and Gosselink 2000). For example, studies on natural wetlands in eastern Spain (López-Flores et al. 2003; Pérez-ruzafa et al. 2005), Portugal (Newton and Mudge 2005), several places in the interior of the United States (Hussein and Rabenhorst 2001; Winter et al. 2001) Canada (Crowe et al. 2004), on the coasts of Brazil (Suzuki et al. 1998), Italy (Bianchi et al. 2004), Greece (Sylaios and Theocharis 2002; Kormas et al. 2001), South Africa (WRC 2007), Australia (Marimuthu et al. 2005),

A. Dussaillant (✉)
Water Resources Associates LP, King's College of London, London, UK

Patagonia Research Center, Patagonia, Chile
e-mail: ale.dussillant@gmail.com

J.M. Fariña, A. Camaño (eds.), *The Ecology and Natural History of Chilean Saltmarshes*, DOI 10.1007/978-3-319-63877-5_7

Uruguay (Bonilla et al. 2005; Mao et al. 2006; Aubriot et al. 2005; Sommaruga and Pintos 1991), and Mexico (Contreras-Espinoza and Zabalegui-Medina 1991).

The main water supply to the systems such as these are rainfall, rivers and other water beds, and often the groundwater that may be discharging into the sector; the importance of each one of them will depend on the specific conditions of the place of study (Winter et al. 2001). Knowing and quantifying flooding processes is vital, in order to understand the functioning of a wetland and to predict certain responses to natural changes and anthropic effects, such as changes in land use, pressures on water resources, climate change and management of basins (Krasnostein and Oldham 2004; Ursino et al. 2004).

Usually the methodology used to characterize the hydrological cycle in a system of this type consists of identifying the water inlet and outlet routes, and completing a mass balance by trying to measure most of the flows directly (precipitation, surface flows), except for subterranean flows and evapotranspiration, which are estimated indirectly or are considered as a difference between the balance of all other flows (Winter et al. 2001). Some studies, such as Sylaios and Theocharis (2002), perform a modeling of the hydrological cycle neglecting the underground contribution, proposing that it should be at least one order of magnitude smaller than the other flows; the authors justify this decision, alluding to the lack of relevant information in the particular case analyzed, which is debatable.

Common techniques for measuring and sampling underground flows are the installation of a network of piezometers that penetrate a little deeper than the location of the lower level of the surface that is groundwater free, to record the position of the layer in time, in addition to measuring the relevant hydrogeological parameters in order to quantify the subsurface runoff applying Darcy's law (Vidon and Hill 2004; Galdames 2007). In some cases, due to environmental, hydrological and/or chemical conditions, these measurements may be insufficient to correctly estimate the actual flow, so they are complemented by additional chemical parameter measurements, including isotopic measurements (Krasnostein and Oldham 2004). However, the latter have not yet been sufficiently studied in terms of knowing the transient interactions between groundwater and surface waters that occur in coastal waters (Krasnostein and Oldham 2004; Marimuthu et al. 2005).

Field studies and numerical models have shown that the relationships established between groundwater and a wetland are highly dynamic and complex, and depend on both the regional and local scale, and it could be said that the local conditions of a wetland can reverse the flow direction (Krasnostein and Oldham 2004; Marimuthu et al. 2005; Mao et al. 2006). Groundwater flows depend to a greater or lesser extent on a number of factors, such as hydrogeological formations, infiltration of rainwater into the entire basin, nearby surface flows that may be recharging or discharging the lagoon, extractions or anthropogenic injections (Tyler et al. 1997) and tides (e.g. Sun 1997, Sylaios and Theocharis 2002), so that these variables must be taken into account when estimating the subsurface flow.

It is important to note that in the case of water that is submerged in the soil, it is continually recharged, and the surface waterlogging is relatively uniform because of a damping effect produced by the underground flow, which produces a more continuous water intake over time rather than through precipitation. Just as well, those wetlands disconnected from subsurface recharge can show great variation in their water levels and their surface hydrology is strongly influenced by the temporal variations of precipitation, basin and runoff (Marimuthu et al. 2005; Galdames et al. 2006).

In the particular case of the coastal lagoons, these form very delicate and dynamic ecosystems, exposed to frequent fluctuations and alterations presenting peculiar characteristics due to their location between the sea and the mainland (Suzuki et al. 1998). Depending on the prevailing hydrological and climatological conditions, these bodies of water can harbor water with dissolved salt contents that can vary from fresh water to hyper saline (even in cases where the water is brackish or salty, the body of water is still classified as lagoon). In addition, these same conditions, together with the geomorphology of the place, imply that the predominant hydrodynamic processes include effects due to meteorology, tides, winds and spatial or temporal variations of salinity and temperature (Ursino et al. 2004; Kormas et al. 2001).

General functioning of the coastal lagoons is conditioned by the periodic connection with the ocean, through a channel that opens through the sandy bar (Sommaruga and Pintos 1991; Galdames et al. 2006; WRC 2007). This periodic phenomenon conditions three states in the hydrological dynamics of the coastal lagoons: a first state, when the opening of the bar occurs as a consequence of the increase of the water level in the lagoon (frequently by rainfall events, in this phase the system is homogenized spatially and discharges the water into the ocean); a second state, which occurs after the partial emptying of the system and consists of the entry and advance of oceanic water (marine intrusion), and finally a third state, which occurs when the salt influence affects the most remote areas of the lagoon, which commonly present a lower level of salinity (Conde et al. 2003; Bonilla et al. 2005; WRC 2007).

The factors mentioned above make it difficult to estimate the hydrological cycle of any coastal lagoon, in different measure according to the characteristics of each site, so it is important to evaluate each site as isolated as possible – each of the effects that occur to predict, to some extent, the behavior of the entire system before a particular event.

The rise and fall of the sea water level, due to the tides, makes the zones closer to the coast variable in height as a result of this oscillation. Because during the high tides (also known as flood tides) water covers areas of the unsaturated soil producing vertical infiltration, that raises the level of the layer under the flooded zone; and due to the energy gradient established by this route, a wave surge of the napa that propagates into the ground is generated (Mao et al. 2006; Sun 1997; Galdames 2007), which can trigger an increase of the submerged soil portion in the wetland. Subsequently, as the tide decreases, the inverse effect occurs at the border with the sea, as it now generates a strong energy gradient that favors the flow from the raised napa to the sea, whose level has dropped. The interaction between groundwater, the sea that oscillates according to the tides, rainwater and evaporation forces saline movements by advective and diffusive effects in the flood-plain and in the surrounding areas, in a dynamic equilibrium that favors the establishment of different habitats and biodiversity of the wetland (Greenblatt and Sobey 1999).

In the central zone of Chile, several coastal lagoons are identified, as it follows: Conchali lagoon (Estero Pupio), in Region IV; El Yali lagoon (Maitencillo), Tunquen (Casablanca estuary), El Peral (El Tabo) and De los Cisnes (estero Cartagena), in Region V; Topocalma (Topocalma estuary), Petrel (Pichilemu estuary) and Bucalemu, in Region VI. Each lagoon has its own characteristics that make them individually relevant to the ecosystem of the low basin where they are located, but they also have common characteristics according to the area in which they are located, so any study for any particular lagoon, serves as much for that particular as for the group as a whole, and may even be useful for other similar systems worldwide in geography, ecology, anthropic impact level or other aspects. I will use the El Yali lagoon as an example, and later mention observations in some similar systems, as those mentioned above.

El Yali Lagoon (Region V)

El Yali lagoon corresponds to part of a coastal wetland of national importance (National Reserve), as well as international (Ramsar site). Despite the importance of this ecosystem, there is little research on balance and water quality of this location. In the area, there are only studies carried out by Galdames (2007), who investigated the interaction between the surface lagoon and groundwater, and have characterized the main physicochemical variables of groundwater for a year (2005–2006), and the research of Sun (2008), on the surface water entering the lagoon, and its quality.

In the past, has been a concern about the possibility of accelerated eutrophication of the lagoon (Natural History Museum of San Antonio, personal communication, 2004). Even though, there has been no clear evidence in respect to it The possible causes of it are :

(i) contamination by discharges to the Yali estuary from agricultural activities and subsequent surface entry to the lagoon; (ii) contamination of groundwater by agricultural and livestock activities, and following discharge to the lagoon and (iii) minor surface and groundwater inflow due to natural fluctuations (ENSO, in particular La Niña); anthropogenic ones (extraction of groundwater for drinking and consumption, by new urban developments and increased demand, for agro-industrial use, and the extraction of water from the Yali estuary for agricultural activities).

The Yali is a complex of more than a dozen bodies of water and waterways, and is the only Mediterranean wetland of the Neotropics adhering to the Ramsar Convention (Möder et al. 2002). It is an established area of concentration and rest for several migratory species at altitudinal, latitudinal and inter-hemisphere levels.

This complex has a high seasonal (winter-summer) and cyclical dynamics, the latter associated with the ENSO phenomenon (Möder et al. 2002). Part of the El Yali wetlands complex is a protected wilderness area managed by the National Forestry Corporation (CONAF), the Chilean Forestry Service, and has an area of 520 ha, called "El Yali National Reserve", located in the Santo Domingo, V Region, Chile, 33° 44′ to 33° 47′S and 71° 39′ to 71° 46′N (see Chapter 6). It highlights the long and narrow form of the coastal lagoon or wetland that extends parallel to the coastline.

The climate of the area is Mediterranean, attenuated by the marine influence, presenting a marked seasonality in relation to the precipitations, with autumn and rainy winters and dry summers. As a typical coastal basin of the area, the wetland has a purely pluvial regimen. Both El Yali and its tributaries remain dry for long periods of time, presenting surface runoff only after intense precipitation (see Chapter 6).

In general, the sector has been of particular interest (and therefore protected), basically due to the abundance and diversity of native and migratory avifauna, about 115 species, representing approximately 25% of Chilean avifauna native diversity (Leiva et al. 1995). The sector is the source of an ongoing ecological study on the response of vegetation communities to gradients of physicochemical variables and in particular of nutrients (see Chapter 6).

The hydrological dynamics that are due to the characteristics of the basin, and added to the effects of the climate, winds, tides, etc., produce frequent level changes in the wetland that is of particular importance, as it consequently affects the distribution and abundance of biotic (zonations of plant and animal species), as well as abiotic (temperature and dissolved salts) elements. This same dynamic can be potentially affected by natural disturbances such as those caused by climate change (precipitation, coastal storms, ENSO phenomenon), and anthropic pressures, such as contamination of incoming water to the lagoon, extraction of groundwater by growth of rural urbanizations, and tourism (Bianchi et al. 2004; Fernández 2004; López-Flores et al. 2003) (Fig. 7.1).

Hydrological Background

The climate of the area is Mediterranean, attenuated by the marine influence, showing a marked seasonality in relation to precipitation, with autumn and winter periods during which 90% of the rainfall is accumulated in 4 months, and long periods of dry spring and summer that reach 7 months without precipitation (Fernández 2004). This climate behavior influences greatly the El Yali estuary and its tributaries to remain dry for long periods, presenting surface runoff only after intense rainfall. As a typical coastal basin in the area, the wetland basin has a pure rainfall regimen (APR Ingeniería 2000; Geofun 2001). Taking into account the collection of hydrometeorological antecedents developed by APR Ingenieria S.A. and AC Ingenieros Consultores Ltda. In 2000, and Geometra in 1995, based on tray evaporation that

Fig. 7.1 Satellite image of El Yali (CONAF/CONAMA, March, 1999)

reaches 1000 mm/year and characteristics of the area, it concludes with some information on the annual real evapotranspiration values in the lagoons, which is estimated to be greater than 300 mm/year (Fig. 7.2).

Inside the El Yali basin, there is a meteorological station of the General Water Direction "Fundo Las Dos Puertas", at the UTM coordinates: North 6,258,612 m and East 253,070 m, which collects daily information of precipitations since the June of 1990, annual records that are indicated in Table 7.1, and which show an average annual precipitation of 481 mm. Because this station is the closest to the study area, it is chosen to obtain representative rainfall data. The station is in the general environment that can be seen in Fig. 7.3.

Close to the study area, which can also be seen in Fig. 7.3, there is the Santo Domingo Aerodrome (UTM coordinates: North 6.279.028 m, East: 255.639 m, elevation 71msnm), where the Meteorological Directive of Chile registers the hourly wind data values, among other things. This information complements the pluviometric measurements of the above mentioned DGA station.

From the background compilation by APR Ingenieria SA and AC Ingenieros Consultores Ltda (2001) and the study by Geometra Ltda. (1995), comes the information about the underground basin of El Yali, showing it isolated from the neighboring basins of the Rivers Maipo, Alhue and Rapel due to a series of geological faults (Melipilla, Puangue and Maipo), and which houses an aquifer of importance that favors the agriculture of the area, through the exploitation of groundwater to the

Fig. 7.2 Detail of the wetland lagoon El Yali (CONAF/CONAMA, March, 1999)

Table 7.1 Annual precipitation at Las Dos Puertas (DGA, 10/12/2006)

Year	Precipitation (mm)
1990	207.0
1991	398.9
1992	664.0
1993	412.7
1994	317.0
1995	269.0
1996	315.0
1997	852.0[a]
1998	151.0
1999	411.5
2000	630.0
2001	682.0
2002	789.4[a]
2003	407.0
2004	514.0
2005	678.5

[a]Years of El Niño

Fig. 7.3 Location of the weather station Las Dos Puertas (pluviometer) and Aerodrome of Santo Domingo (wind speed and direction)

middle zone of the basin, where the aquifer decreases its average width from at least 1000 m to about 300 m as it crosses the coast range, which implies smaller extraction wells. Following the longitudinal direction of the valley along the estuary, there is an identified decrease in the gravel and pebbles stratum as it descends towards the sea, reaching its final conformation almost exclusively as sand, with some portions of silt and clay. This configuration is complemented by gravel and coarse sand from the basin tributaries. According to probing conducted in sectors near the lagoon (about 5 km inland of the basin), there is a phreatic aquifer composed of a mixture of medium-coarse sand, gravel and silt, and bellow it there are semi confined aquifers according to sediments of thick granulometry interleaved with abundant, fine deposits. It is estimated that the permeability of the water table is 10^{-5} to 10^{-4} [m/s], while the storage coefficient is close to 10–15%, and the transmissiveness could vary between 100 and 250 [m^2/day], reaching higher values towards the interior of the valley (APR Ingeniería 2000; Geometra Ltda. 1995).

According to a study carried out by the Ministry of Public Works at the end of 2005, in the lower part of the basin there are terraces of tertiary marine deposits on which are laid out recent marine deposits of less than 40 m in thickness, eolic deposits in the coastal zone, and present fluvial deposits. Therefore, in the lower section of the basin there is a free aquifer whose transmissiveness covers the range of 1000–2000 m^2/day, with a storage coefficient of 0.1 (MOP 2005a). In the same study, it is indicated that the impermeable bottom of the aquifer in the low zone of the basin presents thicknesses close to 300 m, which decreases in the mouth to values close to 100 m, reason why the underground water flow is diverted to the surface of the territory, thus supporting the formation of the lagoon system in the area. This effect is reinforced by the effect of saline intrusion from the ocean.

It should be emphasized that the lagoon, due to its location in the final section of the basin described, is vulnerable to the consequences of what happens in terms of management and development, concerning current extraction of waters of the El Yali estuary, extraction of groundwater, and urban and real estate expansion within the area.

Based on field visits, background and data collection, as well as analysis, it is estimated that the hydrological balance of the lagoon is a function of the following elements: rainfall (mostly winter), estuary basin El Yali (ditto), drainage of rainwater from ditches above, sporadic surface connection to the ocean (tidal floods and winter storms), subsurface water intrusion from (and into) the ocean, fresh groundwater flow from the slopes above, and evapotranspiration.

The hydrological dynamics of the lagoon is of singular importance, as it affects at least the following elements: water height in the lagoon and its effect on the distribution and abundance of plants; abiotic elements: temperature, pH, redox potential, solutes (salts, nutrients) – also with zoning effect – and abundance of vegetal (and animal) species. Hydrological dynamics can potentially be affected by the following natural and anthropogenic disturbances: extraction of groundwater due to the growth of tourism and crops (avocado) in the water balance of the lagoon; climate change and El Niño (precipitation, coastal storms); erosion and deposits of sediment; contamination of incoming water to the lagoon.

Since the water balance corresponding to the El Yali Reserve inflows and outflows of the lagoon was unknown, two studies were carried out with the objective to identify and quantify the main water flows of the lagoon, establishing the quality of the incoming and outgoing waters of the system, to perform a mass balance of physicochemical variables (dissolved oxygen, pH, nitrogen, phosphorus, salinity) and to determine the seasonal variables in water quantity and quality, as well as the key processes that affect these (Galdames 2006; Sun 2008).

The lagoon has an elongated shape, parallel to the beach, connected at one end with the El Yali estuary, and eventually connected superficially with the sea. A satellite image of the lagoon is shown in Fig. 7.1. Given its configuration, two zones with different characteristics stand out: slopes above the lagoon, where the flow is related to the aquifer of the basin that, in theory, would be discharging towards this area, and the sector between the lagoon and the sea, where there is an influence of the latter in both the magnitude of flows and the quality of these.

Due to the strong dependence on the lagoon level, and presumably that of the adjacent area to the tide level, the monitoring was carried out monthly, during the full moon, because during those periods the lagoon is livelier, registering major amplitudes in daily oscillations, which is why the hypothesis is that in these occasions a greater movement of the water should be produced and the substances present in it would be transported in greater volume. A conceptual model of the water balance of the lagoon is shown in Fig. 7.4, where QSM corresponds to the surface

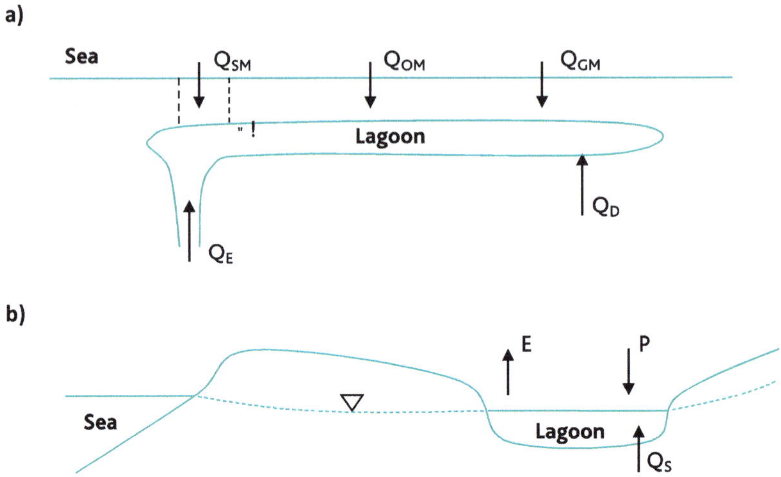

Fig. 7.4 Scheme of the hypothetical water cycle in the wetland lagoon (**a**) plan view, (**b**) transversal section view

water flow that enters from the mouth of the estuary to the lagoon or vice versa. This flow can be temporarily blocked due to the formation of sand barriers, which prevent the connection between the lagoon and the sea. QGM corresponds to the underground flow between the sea and the lagoon, whose direction varies depending on the levels of both. QOM corresponds to the occasional flow that occurs when the tidal waves surpass the sand dune and arrive all the way to the lagoon.

Figure 7.4a presents schematically the flow coming from the El Yali (QE) estuary during the rainy months and some surface drainage (QD) that also flow into the lagoon (some throughout the year and others only sporadically), and the surface routes through the mouth that opens in winter (QGM) and occasional inflows due to tidal waves that surpass the dune (QSM). Figure 7.4b includes contributions from mostly winter precipitation (P) and hypothetically permanent subsurface (QS) outcrops, together with losses due to evaporation over the year (E).

Based on this hydrological cycle scheme we propose Eq. 7.1 that summarizes the water balance to which the amount of water present to the interior of the lagoon is due. The development of this research identifies the behavior of this balance during the year of study and will provide quantitative information of the magnitude, occurrence and frequency of the flows identified, in addition to the characteristics of the flow oscillation in those presented with a "±" sign.

$$\Delta S = P + Q_E + Q_D + Q_{OM} \pm Q_{GM} \pm Q_{SM} \pm Q_S - E \qquad (7.1)$$

A network of piezometers has been installed on both sides of the lagoon: towards the above slopes to get to know the connection with the basin aquifer, and towards the sea to estimate flows to and from the sea. The piezometers are located a few centimeters below the level phreatic level, to obtain information on the position of the free surface at all times, and to access information on the dependence of these levels with the fluctuations of the tides.

Water Quality Measurement

In addition to determining the amount of water entering or exiting, we are monitoring the concentration gradient of salts and nutrients present (we are specifically analyzing concentrations of dissolved nitrates and phosphates), to obtain information about dependence with the sea and to have some indications of the vulnerability of this wetland to the possibility of undergoing natural or anthropogenic eutrophication. In addition, measured values of dissolved oxygen and acidity (pH) were recorded to obtain more information on the existing environmental conditions.

Currently, the available data do not allow definite conclusions, but they do suggest that there is a good drainage in the entire sector adjacent to the lagoon, since it is a thick soil containing gravel and fine sand whose saturated hydraulic conductivity varies between 2.5 and 4.5 m d^{-1} on the east side of the lagoon for the tests performed, while on the west side (Fig. 7.1), the range is between 0.5 and 0.6 m d^{-1}. Based on hydrogeological information of the area where the aquifer reaches the sea, it has an average depth of 100 m (MOP b), and with levels recorded in observation wells during the summer, it is estimated that during this period the sea recharges the aquifer with flows between 1.8 and 2.2 m^3d^{-1} per meter width of the flow zone, depending on the phase of the Moon. In winter, there is a direct connection of the surface water between the sea and the lagoon, inducing a daily oscillation of the level of the lagoon, about a magnitude greater than that registered in summer (Fig. 7.5).

Sea Connected Lagoon Position

The connection of the lagoon with the sea is achieved when the El Yali estuary leads the response of the basin to intense precipitation, transporting a flow that can erode the sand bar located at the mouth of the same, which brings as a consequence the free entry of the sea to the lagoon. This flow occurs only when the level of the tide exceeds a certain threshold, corresponding to the level of the mouth. This situation was already enunciated in some of the studies done in the area that referred to the

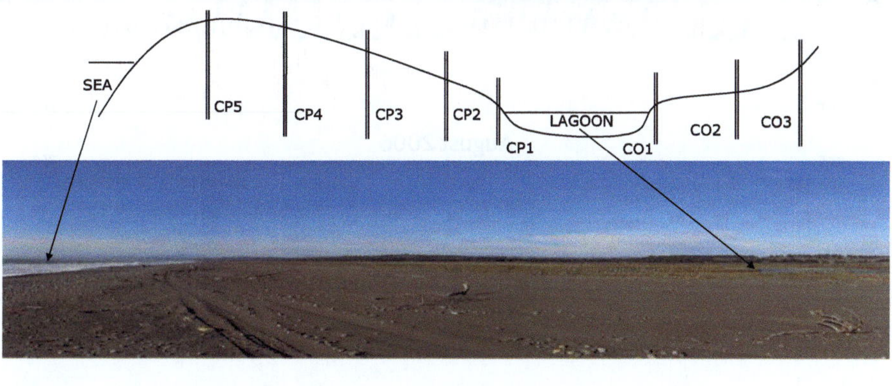

Fig. 7.5 Location of piezometers in the central transect of the lagoon El Yali

fact that only during the rainy season the El Yali estuary empties into the sea (APR Ingeniería 2000; MOP 2005a).

In 2006, the opening of the lagoon to the sea occurred on June 7th and 8th (See Fig. 7.6), due to the effect of a storm (whose data are analyzed in detail in the next sections), which added 77 mm of water during those 2 days, with winds that reached up to 14 knots (26 km/h).

The main characteristic of this period is the great daily oscillation of the water level that suffers this body, which varies about 80 cm in the water mirror height in a 4 h time lapse (for example, the event recorded on August 11, 2006, see Fig. 7.7), in periods when the effects caused by high tides and rains are added. The elevations of the water level have great repercussion in the flooded area, given the relatively flat topography features and the low depth of the lagoon, thus strongly affecting the surrounding ecosystem. The consequences of the tides, precipitation and winds in the lagoon, when connected to the sea, are reviewed separately on the basis of analysis,

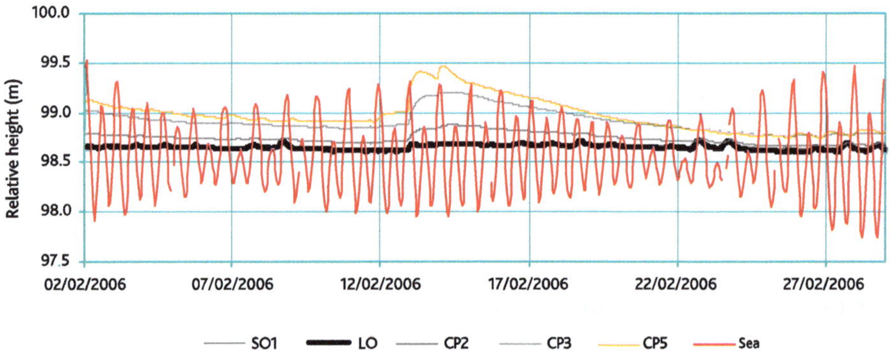

Fig. 7.6 Water level in the lagoon (*red line*) compared with tidal oscillations (*gray line*) during a period of disconnection from the sea (Austral summer 2006; February)

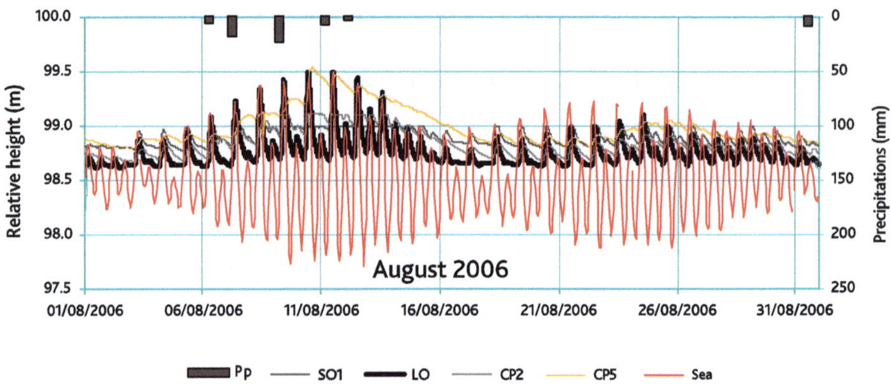

Fig. 7.7 Water level in the lagoon (*red line*) compared with tidal oscillations (*gray line*) during a period of disconnection and connection with the sea (Austral winter; August)

in which, each of the factors can be appropriately isolated to identify their individual repercussions on the system.

Tidal Effect

Observing broader data, recorded in June, it can be seen that since the lagoon is connected to the sea, the daily variation of the tide has a direct impact on the level of the lagoon, thus increasing its volume of water, and since there is a part of the dune located at the mouth of the estuary that has been eroded, whenever the tide exceeds the present land threshold, the sea water will flow freely into the lagoon if the latter one is at a lower level.

Even when the tide begins to decrease from the high tide, the lagoon will continue to be fed by the sea until the levels of both bodies of water are equal, after which the direction of the flow will be inverted and the lagoon will be delivered by a certain flow towards the sea. At the time of descent, it is perceived that the lagoon initially delivers a flow at a higher rate, while the closer it is to the maximum reached. This situation bears much resemblance to what could be associated with a discharge curve from the mouth of the lagoon, delivering a proportionally greater water flow while there is a greater level of difference exists between the free surface and the bottom. In the upper sections of the curves the average rate of descent is around 5.0–6.0 [cm/h], while the final sections of the same curves only reach 1.8–2.0 [cm/h]. In this last section, the slopes tend to be more and more parallel to those recorded in the downward curves of wells SO1 and CP2, which could be indicating that the flow from the lagoon only or mainly occurs in this last phase towards the area through the soil, considerably decreasing the flow to the sea.

Rainfall Effect

In reviewing the data of the transducers for the days in which there were precipitations and the days following them, during the period of surface connection, it is confirmed that the underground layer has a rise in direct relation to the falling rain and then a slow descent by the lateral flow that occurs towards the borders (see Fig. 7.6). However, the lagoon does not exhibit a similar behavior, presumably due to the fact that when the El Yali estuary conducts the runoff of the basin under intense precipitation circumstances, so the flow can further erode the mouth of the estuary, and this way expanding the section of connection to the sea, making the lagoon even more dependent on sea level to establish its own level, especially at the time of the tide rise. When the tide begins to descend, once it falls below the level reached by the lagoon, the latter is not able to sustain the same rate of influence established for the high tide, thus giving indications of the link of the wave surge during high tide at the opening of the sand bar and the establishment of the surface flow. This showing of a very high tidal connection will decay as the link is partially closed once the flow decreases in the days following the precipitations, producing a partial sedimentation in the zone of the opening of the sand bar.

As for the effects of rain on the lagoon, the situation can be recognized circumstance during the time when the lagoon is not connected to the sea, since during the period of connection the influence of the tide is such that Magnitude in the establishment of normal conditions, which limits clarity for the individual analysis of this effect.

Wind Effect

The influence of the wind during the period of the lagoon connection, as well as the implication of the precipitations, is not an easy subject to evaluate, because during these ties the lagoon is strongly dependent on tides, as already explained above.

In order to understand what the wind effect in this situation was, there was an hourly information of the direction and speed of the wind at Santo Domingo aerodrome (Meteorological Direction of Chile), with the limitation factor of measuring conducted only between 8:00 and 7:00 h, which prevents any further analysis or involving conditions that include nighttime measurements. Data were requested for those dates which have shown, according to transducer information, an increase of levels that are not consistent with recorded tidal variations (days before and after June 7th and 17th, July 8th, 12th, 19th and 26th, August 11th, and September 7th, 11th, 23rd). The detailed direction and strength of the measured winds can be reviewed in the study by Galdames (2007). Contrasting the wind data with the periods in which rainfall information is available (only up to August, as already indicated), in order to effectively isolate events only from strong winds from those that include storms, it was confirmed that only the events of June 16th and 17th correspond to the influence of winds without precipitation or syzygy tide.

For the days studied there are no precipitations and the tide is descending progressively from the Full Moon on day 12; but nevertheless, the product of the relatively strong winds that begin to blow at noon (above 6 knots), the sea level presents an increase of its maximum level compared to the previous day as well as the tendency to descent. In quantitative terms, the sea on days 15 and 16 registers its maximum, noting that on the 16th the winds make the high tide slightly higher (than expected) and longer in the permanence in this range, inducing a higher income of water to the lagoon and adjacent area.

From the above, the following hypothesis can be considered: during the time of lagoon-sea surface connection, the wind would produce an effect on the water level inside the lagoon that implies a bank at the sea level (producing tidal waves) which promotes the entry of seawater through the mouth of the estuary by raising the ocean level above what would be expected, due to daily fluctuations of the tides, provided that this condition is maintained during the flood-tide that produces marine flows to the lagoon.

In addition to the abovementioned effect, the direction of the wind blowing at a bank level of the lagoon was analyzed too. When the wind blows mainly from the west, there is a certain uprising of the level of the lagoon; however, this effect is certainly dependent on the tidal fluctuations, which dominate the system at this period.

Based on the results obtained from ascents and descents that continually register in the lagoon during this period, a pattern of behavioral parallels of fluctuations was identified. This pattern was taken into account in the period (from June 10 until September 15, specifically). The minimum values recorded by the lagoon from day to day were identified, and from those, the filling and emptying cycle was established. These values were averaged and the mean minimum start value of the cycle (98.674 m) was obtained, then the same procedure was repeated for the consecutive values in time, which gave rise to the pattern of behavior observed in Fig. 7.8.

When making a spectral analysis (Fig. 7.9) of the data entered, three frequencies of great importance are recognized in the cyclical phenomenon, as well as two of less significance, all linked to the most representative periodicities detected. The first period to be detected is of 16.44 days, which is attributed to the lunar cycle, whose characteristics are similar in each cycle of quadrature tides and the next syzygy. The periods of 24.16 h and 12.46 h, again attributable to the semidiurnal characteristics of the local tide, stand out as the greater of the previous ones, which would mean a greater importance of the processes that occur twice a day, in the water levels of the lagoon.

Sea-Disconnected Lagoon Position

During the period when there is no heavy rainfall, the El Yali estuary does not produce a large enough flow (when there is runoff in the channel) to effectively erode the dune at the mouth and create a superficial connection between the sea and the lagoon, and this shows a much more attenuated behavior than in the time of connection, produced mainly by the independence that reaches the fluctuations of the tide (Fig. 7.10).

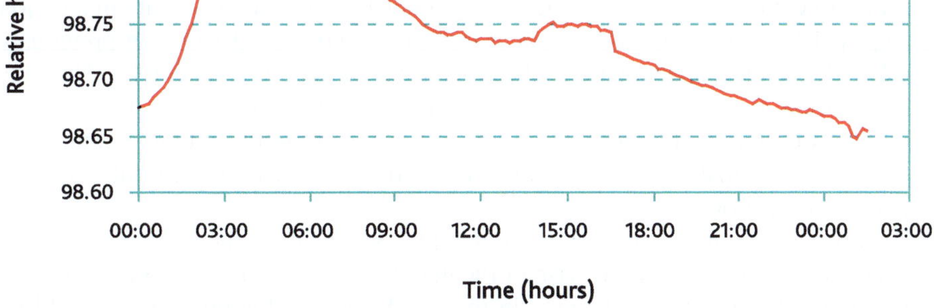

Fig. 7.8 Lagoon dynamic during connection with the sea

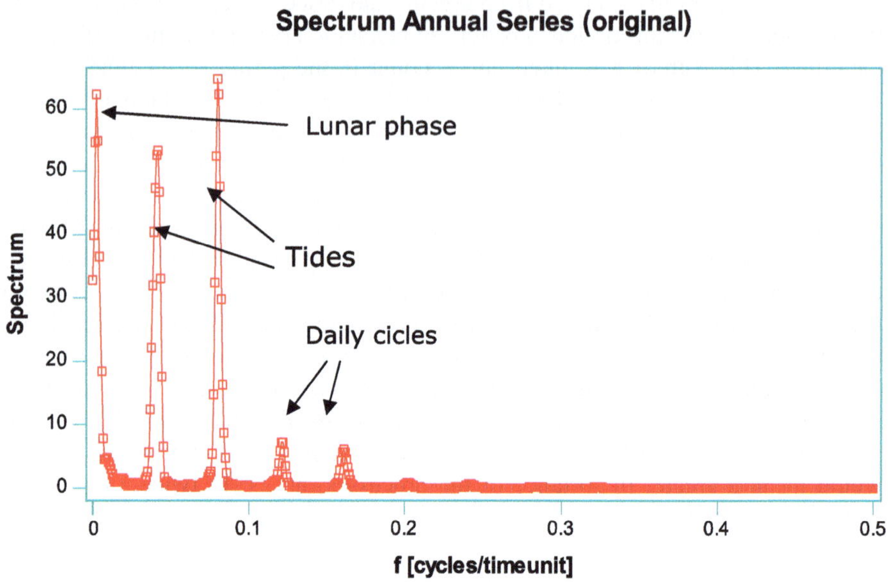

Fig. 7.9 Spectral analysis of the lagoon data

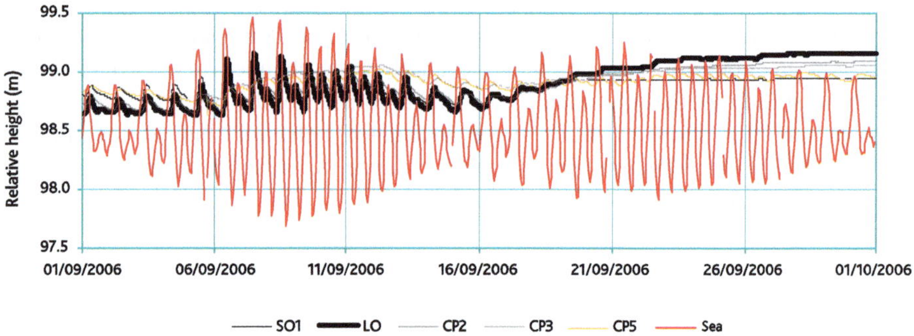

Fig. 7.10 Water level in the lagoon (*red line*) compared with tidal oscillations (*gray line*) during a period of connection to disconnection with the sea (September 2006)

The ascent of the lagoon, besides being smaller in magnitude, occurs in a period of up to 10 h, unlike the 4–6 h that it took in the connected situation previously analyzed. Another aspect that must be emphasized is that these lagoon level ascents do not occur simultaneously with the greater amplitude of the tide, which gives reason to discuss the influence of the tide in periods of disconnection in contrast to other processes that may be affecting the lagoon, such as winds, tidal waves or human interventions in the basin.

Unlike the previous situation, when the tidal waves occur, the levels of the wells located in the dune register the rise of the maximum sea level that would induce an infiltration of sea water to the area in the beach sector. This infiltration would be greater than the volume that can slip sideways when the tide goes down, which would raise the levels of the lagoon cumulatively in that border, producing an ascent wave of the subterranean level that would propagate the dune dampening as it advances through the floor. This phenomenon has a similar behavior to those reported by Ursino et al. (2004) in their work, in which they model computationally a problem of similar characteristics. When comparing this elevation wave of the nape due to the tide, with the fluctuation of the lagoon, the behavior of the lagoon that is identified during the sea disconnection event is practically independent of the occurring tides.

Effect of Rainfall

The precipitations which take place during the disconnection period, allow for a better analysis than those during the connection period, because it is easier to isolate this effect. Although during the summer there are no precipitations in Fundo Las Dos Puertas, towards the end of May (day 25) a relatively important precipitation phenomenon occurs, which will still be used in this section of the investigation, even though it does not generate, in the estuary, a flow sufficiently powerful to open the dune and to connect the lagoon to the sea.

The rainfall occurring on May 25 (with tidal waves) corresponds to 29 mm, but the rise of the lagoon is only 8 mm throughout the day, which is practically negligible when compared to the fluctuations recorded in winter, reaching 100 times the value, due to the combined effects of precipitation and tidal waves (see Fig. 7.6). The difference is attributed to being the first rain of considerable magnitude since last year, (Except for an event, on a smaller scale that reaches 16 mm during the month of April), most of which are probably intercepted by vegetation and infiltrated on the surface of the basin, whose soils are relatively dry, and would be absorbed very quickly.

It is possible to say that in this period, the level of the lagoon is little sensitive to precipitation occurring in the basin, unlike what happens in the neighboring wells located in the dunes, which perceive the ascent of the area favored by influx of the tides. In addition, with respect to the eastern shore of the lagoon, it can be seen that the sandy soil allows an easy displacement of the rainfall down to the lagoon, increasing its level quickly according to the precipitations perceived, and a lateral flow is likely to be established soon discharging towards the lacustrine body, if its ascent has been smaller than the one of the area.

Wind Effect

During the disconnection period it is easier to isolate the effect produced by the wind on the surface mirror level of the lagoon, because there are no factors that are preponderant over the others, as are the tides in the winter season. It was observed that for some windy days, the surface level of the lagoon increases in the area of study, this behavior being attributed to the fact that the water is disturbed by the winds from the sector of the mouth to the instrumentalized sector. From the above, it can be concluded that the effect of the wind on the lagoon level will depend on the direction from which it comes with great intensity, since it is these strong winds that induce the free surface banks heaving in one direction or another.

ENSO Effect (Fig. 7.11)

Water Quality

In terms of water quality, it can be appreciated that those from the upper aquifer are clearly freshwater, with average salinity less than 3‰, whereas the lagoon shows mean values of 35‰ in the east and 45‰ in the west zone, close to the sea, with a subsequent a salinity gradient inside the body of water (Fig. 7.12). In addition, in the subterranean zone between the lagoon and the sea, it was detected that the salinity of the water decreases as it moves away from both bodies of water reaching a minimum average of about 10‰, approximately 27 m from the western shore of the lagoon, so it would be in the presence of a double salt wedge produced by the presence of the brackish bodies of water on both sides of the area, in this sector of study, as can be seen in Fig. 7.12, of a transect representative of the area.

During the summer season the lagoon is disconnected from the sea, and is exposed to the strong sun and wind, which would favor the evaporation of water present inside it, concentrating more and more on the dissolved salts in the water and with it becoming more and more salty. On the other hand, during the months of precipitation, the fresh water intake that the lagoon receives along this route, added to the flow from the estuary, some surface drainage and the discharge of the aquifer, tend to generate an inverse behavior to the previous one, diluting the ions dissolved in a greater amount of water.

As for the acidity of the waters, all values are recorded in the area in neutral levels, reaching, in the wells studied, a minimum average of 6.53 and maximum of 7.76, while in the lagoon itself the averages are of 8.60 and 8.62 for the east and west banks, respectively, while the sea stands at 8.06.

Dissolved oxygen does not appear to follow a defined spatial pattern. The values in the lagoon are between 2 and 4 mg L^{-1}, while the surface waters of the lagoon (measured maximum of 12 mg L^{-1}) and the sea are in the order of 6–9 mg L^{-1}. Out of the nutrients evaluated in the samples (Fig. 7.13), nitrate is present in a relatively constant form in both the piezometers and on surface of the bodies of water, taking values below 1.0 mg N-NO3-L-1 in most of the evaluated samples. On the other hand, the phosphate shows a very different behavior to that of nitrate, since the averages of the recorded concentrations seem to manifest a clear tendency to go down as the water approaches the sea. The most eastern points of measurement are practically all about 2.0 mg (PO) 43-L-1 (except for the eastern margin well, a situation

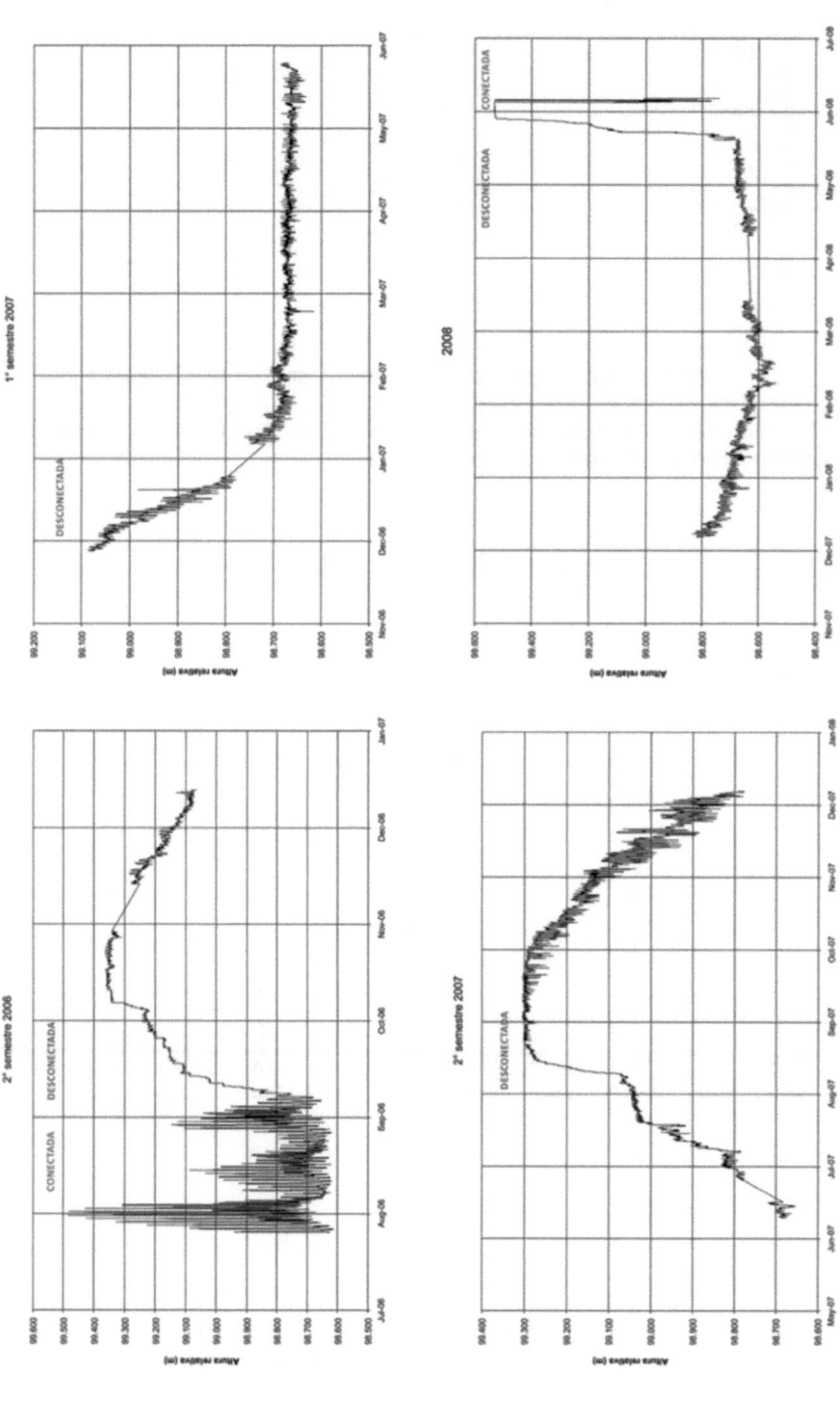

Fig. 7.11 Total disconnection year 2007 (La Niña) and reconnection in 2008

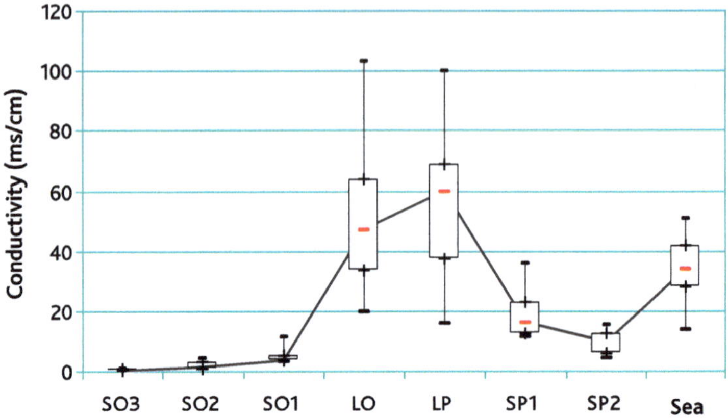

Fig. 7.12 Conductivity profile in the hillside-lagoon-bank-ocean transect, with evidence of a less salty-water lens between the lagoon and the sea

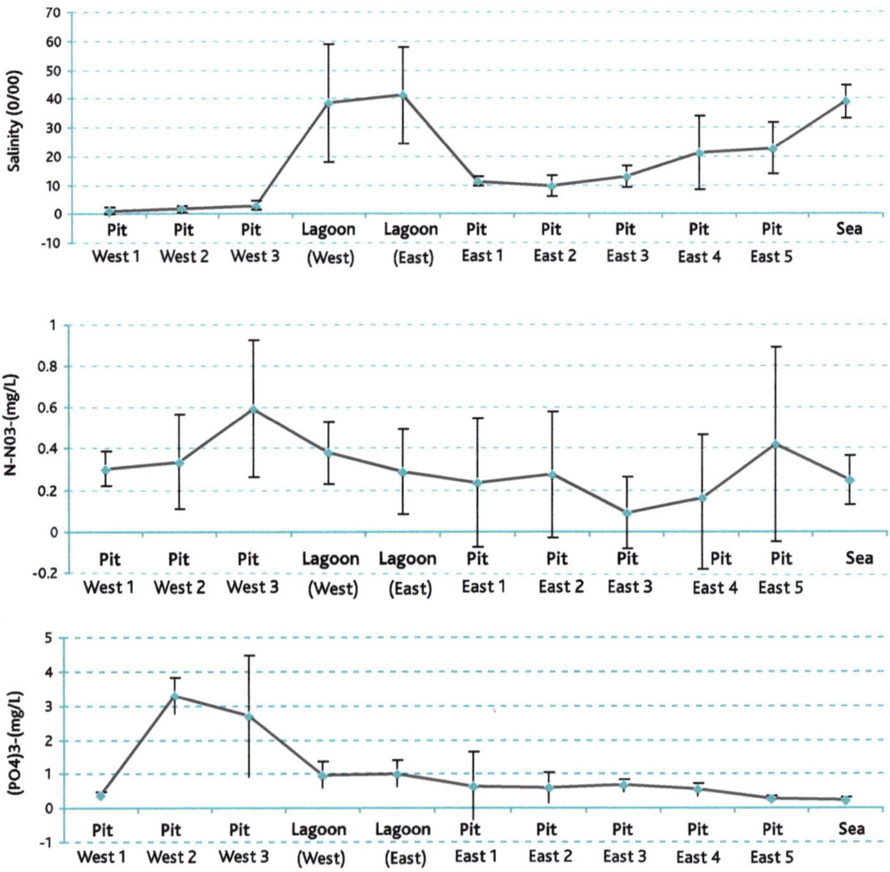

Fig. 7.13 Quality of underground water in a representative transect with piezometers: salinity (*top*), nitrate (*middle*), phosphate (*bottom*). Data shown are means ±1 standard deviation

that is still under study) showing a point with a maximum of 25.6 mg (PO) L-1, with an average around 0.97 mg (PO) 43-L-1 in the lagoon and values of the western area that decrease until reaching the 0.21 mg (PO) 43- L-1 in the sea.

In the case of waters flowing through the ditches, the samples have a higher concentration of nitrate compared to the groundwater and the lagoon, and varies depending on the sampling point. One possible explanation for this variation is the presence, occasionally during the day, of cattle in the wetland that uses the ditches as sources of water. The results of the analysis at the sample points, from east to west, yielded mean values of 1.09, 1.97 and 2.00 mg N-NO3-L-1, with a maximum of 4.6 mg N-NO3-L-1, measured at the central ditch. Phosphate also exhibits spatial variation and with a realy high concentrations, except at a sampling point, as measured in groundwater. The mean concentrations are 2.73, 4.19 and 2.03 mg (PO) 43-L-1, with a maximum of 9.28, recorded in the southern sampling dyke.

Petrel Lagoon (Pichilemu, Region VI)

The Petrel Lagoon (Figs. 7.14, 7.15, and 7.16) flows into the sea in winter, with the floodwaters of the micro-basins that are tributaries of the San Antonio de Petrel estuary and in turn, of the lagoon too. Concordant with other similar systems of the central coast, as for example El Yali (Galdames 2006), it is very probable that there is a great amount of sediments, product of the erosive process and desertification of these micro basins starting in XVIII century, that along with the Mediterranean hydrology characteristic of this area, determines that during the dry season

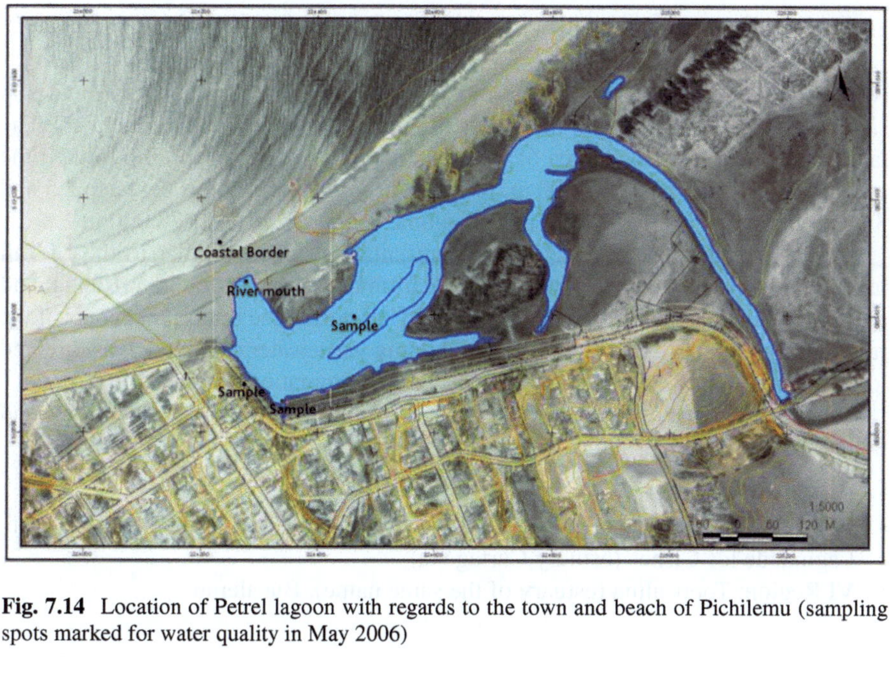

Fig. 7.14 Location of Petrel lagoon with regards to the town and beach of Pichilemu (sampling spots marked for water quality in May 2006)

Fig. 7.15 Panoramic view from the south of Petrel lagoon (June 10, 2006)

Fig. 7.16 View from the northeast of Petrel lagoon (June 10, 2006)

(October-May) a bar is formed that blocks the mouth, creating a lagoon, and that same mouth opens again with the floods of the rainy season (June-September). The lagoon has an approximate area of 21 hectares (which varies according to the time of year) and its perimeter is 7000 m. Its depth in some areas can be between 2.5 and 3 m; assuming an average of 1.8 m, its volume is approximately 378,000 m^3.

It is observed that the Petrel lagoon presents a situation similar to the El Yali lagoon, being fed by an estuary that drains from the Coastal Mountain Range, and therefore only manages to break with the sand bar during the important events of the rainy season. Because for a long time Pichilemu wastewater discharge was directed to the lagoon, it has suffered from significant contamination in the stable periods.

Other systems whose general hydrology should have similarities to El Yali and Petrel, given the author's field observations, include:

- IV Region: Laguna de Conchali (Pupio estuary)
- V Region: La Laguna (Maitencillo), Tunquen (estuary Casablanca), El Tabo, Laguna de los Cisnes (estuary Cartagena)
- VI Region: Topocalma (estuary of the same name), Bucalemu

- VII Region: Boyeruca
- VIII Region: Buchupureo

However, it should be noted that each wetland can present particularities that cause its hydrology to be diverse, and therefore a case-by-case study is required.

Conclusions and Recommendations

In the central zone of Chile there are coastal wetlands that form similar systems where coastal marshes are setting off coastal estuaries with longitudinal transport.

These show periods of steady disconnection, interrupted by important events at the beginning of the rainy season, although in dry years (La Niña), the breaking of the bar may not occur, as it has occurred in the El Yali during 2007–2008. In light of the background and analysis shown in this paper, it is suggested:

- The need for water studies at the basin level is imperative. It is essential to consider pressures on water resources, such as increasing water demand for urban development, agricultural and industrial activities, and possible threats to wetlands if these pressures are not controlled in time.
- Studies that combine hydrology, ecology and basin management are required, i.e. using a local (trans) disciplinary approach but connected to the context of the basin, where many of the pressures occur.
- Finally, it is critical and fundamental to have territorial planning and basin management plans, adaptive, and informed by local studies on the hydrology and ecology of coastal wetlands.

Acknowledgements To the National Forestry Corporation (CONAF-Chile) V Region, for permission and support in the field, as well as the collaboration of the ecosystems ecology group of Dr. Jose Miguel Fariña of CASEB PUC.

References

APR Ingeniería S.A. 2000. Evaluación de recursos hídricos subterráneos cuenca del Estero Yali.

APR Ingeniería S.A. & AC Consultores Ltda. 2001. Pozos Agrícola Super Ltda. Sector Maitenlahue y su relación con Humedales de El Yali.

Aubriot, L., D. Conde, S. Bonilla, V. Hein, and A. Britos. 2005. Vulnerabilidad de una laguna costera en una Reserva de Biosfera: indicios recientes de eutrofización. En: Vila, I. Y J. Pizarro (eds.) Taller Internacional de Eutrofización de Lagos y Embalses. CYTED XVIIB. Patagonia Impresores Chile. P. 65–85.

Bianchi, F., E. Ravagnan, F. Acri, F. Bernardi-Aubry, A. Boldrin, E. Camatti, D. Cassin, and M. Turchetto. 2004. Variability and fluxes of hydrology, nutrients and particulate matter between the Venice Lagoon and the Adriatic Sea. Preliminary results (years 2001–2002). *Journal of Marine Systems* 51 (1–4): 49–64.

Bonilla, S., D. Conde, L. Aubriot, and M.C. Pérez. 2005. Influence of Hydrology on Phytoplankton Species Composition and Life Strategies in a Subtropical Coastal Lagoon Periodically Connected with the Atlantic Ocean. *Estuaries* 28 (6): 884–895.
Conde, D., L. Rodríguez-Gallego, and L. Rodríguez-Graña. 2003. Análisis Conceptual de las Interacciones Abióticas y Biológicas entre el Océano y las Lagunas de la Costa Atlántica de Uruguay: Informe Final. Facultad de Ciencias, de la Universidad de la República, Sección Limnología. Montevideo. 76 pp.
Contreras-Espinoza, F., and L.M. Zabalegui-Medina. 1991. Hidrología, nutrientes y productividad primaria en la laguna La Joya-Buenavista, Chiapas, México. Anales del Instituto de Ciencias del Mar y Limnología.
Crowe, A.S., S.G. Shikaze, and C.J. Ptacek. 2004. Numerical modelling of groundwater flow and contaminant transport to Point Pelee Marsh, Ontario, Canada. *Hydrological Processes* 18 (2): 293–314.
Fernández, J.V. 2004. Proposición de un plan de manejo integral de la cuenca hidrográficadel Estero Yali, V y Región Metropolitana. Tesispara título de Ingeniero Forestal, Escuela de Ingeniería Forestal, Facultad de Ciencias Silvo agropecuarias, Universidad Mayor, Chile.
Galdames, P., C. Sun, and A. Dussaillant. 2006. Hidrología de laguna albufera Reserva El Yali, humedal RAMSAR, Chile. In *Climate Variability and Change-Hydrological Impacts (Proceedings of the Fifth FRIEND World Conference held at Havana, Cuba, November 2006)*, ed. IAHS press, 1st ed., 563–568. Wallingford: IAHS Publication. 308.
Galdames, P.F. 2007. Caracterización Hidrológica de Laguna Albufera en Humedal Yali con Énfasis en Flujos Subterráneos. Tesis de Magíster en Ciencias de la Ingeniería, Pontificia Universidad Católica de Chile, Santiago, Chile, 188 pp.
Geofun Ltda. 2001. Estudio integral de optimizacióndel regadío de la 3° seccióndel río Maipo y valles de Yali y Alhué. Comisión Nacional de Riego. Gobierno de Chile.
Geometra Ltda. 1995. Evaluación hidrológica e hidrogeológica en la cuenca del estero El Yali.
Greenblatt, M.S., and R.J. Sobey. 1999. Subsurface flow and transportintidal wetlands: Marsh plain equations. *Journal of Engineering Mechanics* 125: 971–974.
Hussein, A.H., and M.C. Rabenhorst. 2001. Tidal inundation of transgressive coastal areas: Pedogenesis of salinization and alkalinization. *Soil Science Society of America Journal* 65 (2): 536–544.
Kormas, K.A., A. Nicolaidou, and S. Reizopoulou. 2001. Temporal variations of nutrients, chlorophyll a and particulate matter in three coastal lagoons of Amvrakikos Gulf (Ionian sea, Greece). *Marine Ecology-Pubblicazioni Della Stazione Zoologica Di Napoli I* 22 (3): 201–213.
Krasnostein, A.L., and C.E. Oldham. 2004. Predicting wetland water storage. *Water Resources Research* 40 (10): W10203.
Leiva, I., J. Meza, and L. Möder. 1995. Boletín Técnico N° 59: Fundamentos para la creación de la Reserva Nacional Yali. CONAF V Región.
López-Flores, R., X.D. Quintana, V. Salvadó, M. Hidalgo, Ll Sala, and R. Moreno- Amich. 2003. Comparison of nutrient and contaminant fluxes in two areas with different hydrological regimes (Empordà Wetlands, NE Spain). *Water Research* 37: 3034–3046.
Mao, X., P. Enot, D.A. Barry, L. Li, A. Binley, and D.S. Jeng. 2006. Tidal influence on behaviour of a coastal aquifer adjacent to a low-relief estuary. *Journal of Hydrology* 327 (1–2): 110–127.
Marimuthu, S., D.A. Reynolds, and C.L. La Salle. 2005. A field study of hydraulic, geochemical and stable isotope relationships in a coastal wetlands system. *Journal of Hydrology* 315 (1–4): 93–116.
Mitsch, W.J., and J.G. Gosselink. 2000. *Wetlands*. New York: John Wiley & Sons, Inc..
Möder, L., H. Gallardo, and Y. Vilina. 2002. Ficha Informativa de los Humedales de Ramsar: Humedal Yali.
MOP. 2005a. Determinación de la disponibilidad de derechos de aprovechamiento de aguas subterráneas en la cuenca del estero Yali hasta sector El Prado Región Metropolitana. Gobierno de Chile.
———. 2005b. Informe técnico N° 421: Área de restricción sector hidrogeológico de aprovechamiento común de Yali bajo El Prado. Gobierno de Chile.

Newton, A., and S.M. Mudge. 2005. Lagoon-sea exchanges, nutrient dynamics and water quality management of the Ria Formosa (Portugal). *Estuarine Coastal and Shelf Science* 62 (3): 405–414.

Pérez-ruzafa, A., A.I. Fernández, C. Marcos, J. Gilabert, J.I. Quispe, and J.A. García-Charton. 2005. Spatial and temporal variations of hydrological conditions, nutrients and chlorophyll a in a Mediterranean coastal lagoon (Mar Menor, Spain). *Hydrobiologia* 550: 11–27.

Sommaruga, R., and W. Pintos. 1991. Laguna de Rocha. En *Databook of world lake environments A survey of the state of world lakes*, ed. ILEC, 7 pp. ILEC/ UNEP/SAM-9 Otsu.

Sun, C.-l. 2008. Caracterización hidrológica y físicoquímica de la laguna albufera del humedal El Yali con énfasis en flujos superficiales. Magister en Ciencias de la Ingeniería, Pontificia Universidad Católica de Chile. 171 pp.

Suzuki, M.S., A.R.C. Ovalle, and E.A. Pereira. 1998. Effects of sand bar openings on some limnological variables in a hypertrophic tropical coastal lagoon of Brazil. *Hydrobiologia* 369: 111–122.

Sylaios, G., and V. Theocharis. 2002. Hydrology and nutrientenrichmentattwo coastal lagoon system in northern Greece. *Water Resources Management* 16: 171–196.

Tyler, S.W., S. Kranz, M.B. Parlange, J. Albertson, G.G. Katul, G.F. Cochran, B.A. Lyles, and G. Holder. 1997. Estimation of groundwater evaporation and salt flux from Owens lake, California, USA. *Journal of Hydrology* 200 (1–4): 110–135.

Ursino, N., S. Silvestri, and M. Marani. 2004. Subsurface flow and vegetation patterns in tidal environments. *Water Resources Research* 40 (5): W05115.

Vidon, P.G.F., and A.R. Hill. 2004. Landscape controls on the hydrology of stream riparian zones. *Journal of Hydrology* 292 (1–4): 210–228.

Winter, T.C., D.O. Rosenberry, D.C. Buso, and D.A. Merk. 2001. Water source to four US wetlands: Implications for wetland management. *Wetlands* 21 (4): 462–473.

WRC. 2007. A review of information on temporarily open/closed Estuaries in the warm and cool temperate biogeographic regions of South Africa, with particular emphasis on the influence of river flow on these systems. A. Whitfield & G. Bate [Ed.] Interim report to the Water Research Commission on the Project "The freshwater requirements of intermittently open Cape estuaries". WRC Report No 1581/1/07.

Chapter 8
Characterization of the Coastal Wetland Mataquito, Curico Province, Maule Region

Patricio Torres-Ramírez, Jose Ortiz, Sergio Figueroa, Rodolfo Jerez, and Dagoberto Arcos

Abstract Based on a hydrological, physicochemical and biotic characterization, this chapter presents the factors that affect the short-term dynamics of the Mataquito wetland. The results indicate that the wetland corresponds to an estuary of the "salt wedge" type. The decrease in the influence of oceanic waters determines this wetland's semi-stratified characteristics towards the sectors located upstream. Water and sediment quality analysis, as well as toxicological tests, indicate that this system generally has a good environmental quality, according to the CONAMA classification. The salinity stratification and water quality are largely responsible for the intra- and inter-annual fluctuations observed through different variables analyzed between 2006 and 2009. Hence the main drivers of the wetland dynamics are climatic factors, physiographic, type/state of vegetation/soil, besides the anthropic activities that have developed in the basin. Our results indicate that in response to changes in dynamics controllers, the ecosystem associated with the wetland is able to respond quickly and with a high degree of resilience. These antecedents would suggest that the coastal wetland of Mataquito river is highly adjusted to the action of its diverse controllers, especially those that modify the balance and mixture of the continental and oceanic waters.

Keywords Coastal wetland • Central Chile • Estuary • Resilience • Mataquito river

Introduction

In the Maule Region (VII Region of Southern Chile), 183 natural and artificial wetlands have been identified, which occupy an area of more than 27,000 hectares (CONAMA 2000). These wetlands have been grouped into six "complexes"

P. Torres-Ramírez (✉) • J. Ortiz • S. Figueroa • R. Jerez • D. Arcos
Regional Center for Environmental Studies (CREA), Universidad Católica de la Santísima Concepción (Catholic University of the Sacred Conception), Talcahuano, Chile
e-mail: ptorres@ucsc.cl

J.M. Fariña, A. Camaño (eds.), *The Ecology and Natural History of Chilean Saltmarshes*, DOI 10.1007/978-3-319-63877-5_8

(Vichuquen, Mataquito-Maule, Reloca lagoon and river estuary, Cienaga del Name, Colbun-Machicura and Maule lagoon), identifying a distinction according to their location in coastal zone wetlands, in the interior, and Andes highlands. Based on the attributes and functions identified in this cadaster and in several studies developed by the agencies such as UNORCH,[1] CONAF, SAG2[2] and CODEFF,[3] the Maule Region has been presented as the "wetland network under official protection". The network consists of placing under one category of official protection 12 wetlands identified as having the greatest ecological relevance and which are representative of the six complexes present in the region (Olivares 2007).

Among the wetlands of the Maule Region with ecological relevance, the Mataquito coastal wetland (Curico Province) stands out, having developed at the river mouth bearing the same name (Fig. 8.1). This wetland has been defined as a unit of high biogeographical value and biological conservation. For this reason, the system has been decreed a hunting ban area through the instrument D. E 169/00. However, at the present, both the wetland and the Mataquito River continue to be affected by the anthropogenic actions. In fact, the Mataquito River, which drains a 6190 km^2 hydrographic basin and receives through its route the discharge of domestic wastewater from an estimated total urban population of 151,000 inhabitants, which are mainly associated with the Curico, Lontue, Teno, Romeral, Hualañe and Licanten locations (Fig. 8.1a, DGA & Cade-Idepe Consultores 2004). It also receives the contribution from various diffused sources of public agricultural activity (e.g., pesticides, fertilizers, organic waste) that is developed in the basin, and some point sources associated with industrial processes (slaughterhouses, tanneries and bleached cellulose production, etc.). On the other hand, in the sector of the mouth, where the wetland is, the main threats are associated with agricultural and forestry activities, as well as with the small settlement of La Pesca, which is located on the northern shore of the wetland (EGA 2006).

Due to the pressure exerted by anthropic activities on the coastal wetland and the Mataquito River, a number of studies have been carried out during the last 4 years to evaluate the quality of surface waters and to establish sediment, biotic and hydrological characteristics, especially in the wetland sector. In this chapter, a summary of the information obtained in relation to the aforementioned indicators is presented, in order to establish, from an integral perspective, the short-term dynamics of the Mataquito wetland by recognizing the variables that define it and its ranges Natural variation. Thus, with this synthesis we intend to provide a tool that allows planning, development and maintenance of strategies for conservation and recovery of this important coastal ecosystem.

[1] Unión de Ornitólogos de Chile.

[2] Servicio Agrícola y Ganadero.

[3] Comité nacional Pro Defensa de la Flora y Fauna.

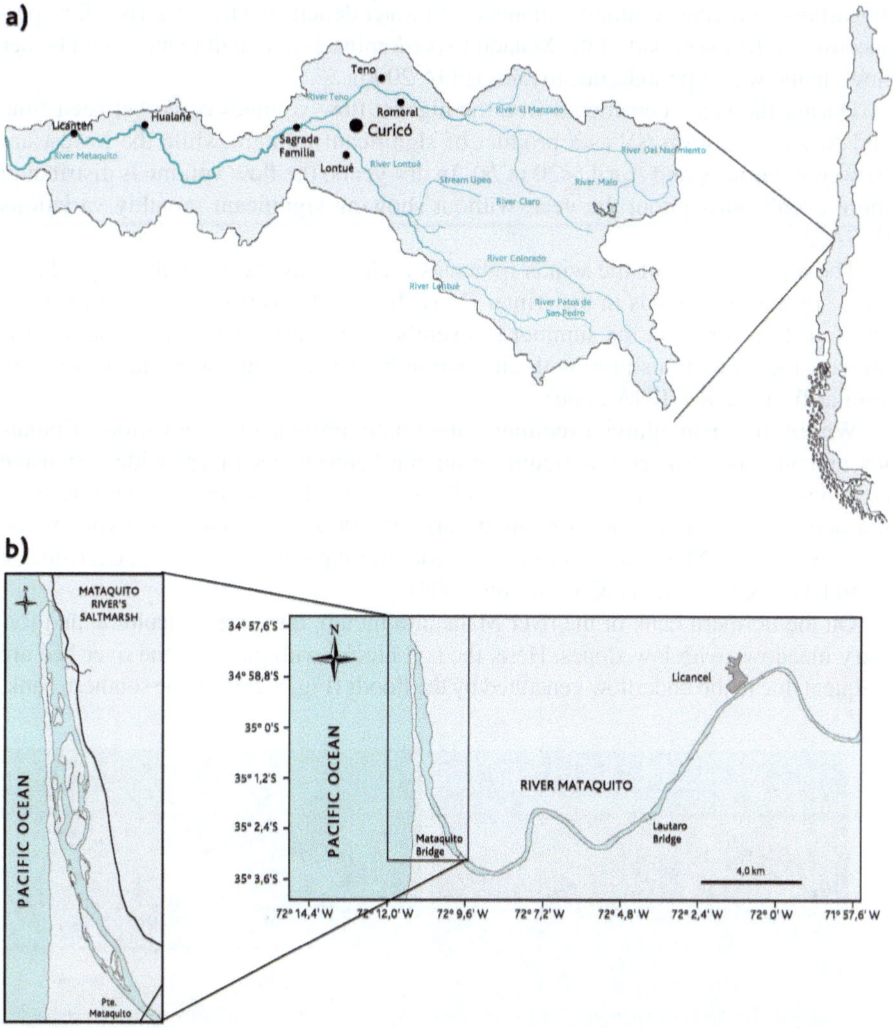

Fig. 8.1 Mataquito River basin (**a**) and detail of the river mouth where the wetland develops (**b**)

Study Area

The Mataquito River originates at the confluence of the Teno River, which drains the northern portion of the basin, and the Lontue River, which does the same in the southern portion (Fig. 8.1a). The Mataquito river basin crosses the provinces of Curico and Talca, in which the main human settlements of the Maule Region (INE 2002) are found.

This basin is under the influence of a pluviestational Mediterranean climate, which dominates the Coastal Mountain Range, the Central Valley and the lower floors of the Andes Mountain Range, under the 2000 msnm. This zone is characterized by presenting at least two consecutive summer months with water deficit, and from the river flow perspective, in the lower part of the Mataquito predominates a rainfall regime, with higher flows in the winter periods, due to rains (EGA 2006).

During the years considered wet, the highest flow volumes occur between June and August (>200 m^3/s), as a product of significant rainfall, while the lowest are between February and April (<20 m^3/s). In dry years, the flow volume is distributed more evenly throughout the year, without showing significant monthly variations (EGA 2006).

The annual cycle of the winds indicates a clear seasonality, with the predominance of northern winds in the winter (May, June and July), and winds of the third and fourth quadrant in the summer (November–February). The rest of the months are considered as transition, with alternation between south and north winds with similar frequencies (EGA 2006).

Within the basin, alluvial sediments dominate, provided by the various tributaries, and on which a layer of agricultural soil has been created that provides extensive crop fields. However, it is possible to identify some discontinuities resulting from the deposition of large lahars or mud currents 7000 years ago. This explains the scarcity of some Maulian soils from the agricultural point of view (e.g., Cerrillos de Teno DGA & Cade-Idepe Consultores 2004).

On the northern bank of the river Mataquito mouth, there are agricultural and forestry meadows with low slopes. Here, the soil blocks with plants in the river bed are frequent due to the underflow generated by the floods (Fig. 8.2a). On the southern bank,

Fig. 8.2 View of the river mouth of Mataquito River at the north (**a**) and south (**b**) bank

however, there is an extension of coastal dunes that separate the course of the river from the coastal edge of the Pacific Ocean. In the stabilized dunes is a varied flora. However, vegetation is rather sparse, except at certain points where favorable substrate conditions permit plant establishment (D. Arcos, personal observation, Fig. 8.2b). In general, coastal dunes are defined as mainly composed of silica and/or carbonate particles, ranging in size from fine to medium sand. The dune sand originates in the intertidal sector of the beach, from where it is mobilized mainly by the action of the wind. This displacement depends mainly on three factors: (**a**) the water content of the sand, (**b**) the velocity of the wind on the surface of the sand, and (**c**) the size of the particles. For the dune sand, the movement starts approximately with intensities of wind >4 m/s and there are no increases in the amount of particles mobilized at wind intensities >10 m/s (McLachlan and Brown 2006).

Mataquito Wetland Characterization

Physicochemical Characteristics of the Wetland Waters

To characterize the physical and chemical properties of the Mataquito wetland, different physical variables were evaluated (e.g. temperature, pH) and indicators of environmental quality, such as inorganic and organic compounds, brown essential and non-essential metal, pesticides and other relevant compounds. Because the wetland sector produces a mixture between inland (river) and marine waters, the characterization has also considered capturing variations in temperature and conductivity across the water column.

The historical characterization of the chemical quality of the surface waters of the wetland presented in this chapter is established using Gibbs (1970) models and the use of the surface water quality index (SWQI, DGA & Cade-Idepe Consultores 2003). The graphical representation of the Gibbs models shows how the processes of atmospheric precipitation, evaporation-crystallization and rock type affect the chemical characteristics of the water, based on variations in sodium, calcium and total dissolved solids (Gibbs 1970). On the other hand, the SWQI allows to integrate the information of certain environmental variables, finally obtaining a standardized value that is associated to a class or quality of the waters. In order to evaluate the quality of the wetland waters, the standardization of classes and the SWQI quality ranges proposed by the DGA (DGA & Cade-Idepe Consultores 2003) have been used. For this purpose, the following variables are considered: dissolved oxygen, pH, conductivity, DBO5 and total suspended solids, RAS index (NCh 1333 Of.78), chlorides and sulfates.

Table 8.1 presents a summary of the physicochemical variables monitored in the Mataquito river wetland sector between the summer of 2006 and November 2008. This shows the limits specified in the Chilean Standard NCh 1333 of 78 ("Water quality requirements for different uses "). Most of the variables presented wide ranges of variation during the analyzed period. This may be related to the recurrent

Table 8.1 Physicochemical parameters monitored in superficial waters of the Mataquito river wetland sector during the 2006 and 2008 period

Parameters	Range	NCh 1.333
Physico-chemical indicators		
Conductivity (μS/cm)	146–57.100	<750
DBO_5 (mg/L)	0.3–28.6	
DQO (mg/L)	<1–71.6	
Color (Pt-Co)	1–34	100
Oxygen (mg/L)	8.10–16.09	>5
pH	6.71–8.67	6.0–9.0
Suspended Solids (mg/L)	3.5–221.0	
Temperature (°C)	10.0–22.1	
Sediment Solids	<0.1–2.0	
Total Solids (mg/L)	120–401.390	
Total Alcalinity (mg/L)	34.7–111.7	>20
Hardness (mg/L)	58–6.600	
Inorganic components		
Ammonia (mg/L)	0.024–1.25	
Residual Chlorine (mg/L)	<0.02–0.06	
Total Chlorine (mg/L)	0.03–0.08	
Chlorides (mg/L)	9.8–19.478	200
Nitrogen Total (mg/L)	<0.1–83.0	
Nitrate (mg/L)	<0.002–2.8	
Nitrite (mg/L)	0.001–0.075	
Sulfates (mg/L)	10–2.727	250
Sulfurs (mg/L)	0.002–1.1	
Foskor (mg/L)	0.031–4.6	
Organic components		
Oils and fats (mg/L)	1.3–74.0	5
Detergents (mg/L)	0.003–0.14	
Foaming power (mm)	<2	
Phenol index (mg/L)	<0.001–0.047	
Chlorophenols (mg/L)	<0.0001	
Carbonates (mg/L)	0–34	
Bicarbonates (mg/L)	34.7–111.7	
Phosphates (mg/L)	0.06–14.1	
Metal essentials		
Sodium (mg/L)	3.20–71.450	
Copper (mg/L)	0.001–0.029	0.2
Total Chrome(mg/L)	0.00028–0.0014	0.1
Hexavalent Chrome (mg/L)	<0.03	
Iron (mg/L)	<0.03–3.7	5
Manganese (mg/L)	0.002–1.4	0.2
Nickel (mg/L)	<0.003–0.038	0.2

(continued)

Table 8.1 (continued)

Parameters	Range	NCh 1.333
Zinc (mg/L)	<0.0003–0.11	2
Potassium (mg/L)	<0.5–500	
Magnesium (mg/L)	<0.05–2.159	
Calcium (mg/L)	4.5–504	
Non-essential metals		
Aluminum (mg/L)	0.010–7.688	5
Cadmium (mg/L)	<0.0001–0.0025	0.01
Lead (mg/L)	<0.003–0.017	5
Mercury (mg/L)	<0.0002–0.0052	0.001
Pesticides		
2.4 D (μg/L)	<30	
4.4 DDD (μg/L)	<0.1	
4.4 DDE (μg/L)	<0.1	
4.4 DDT (μg/L)	<0.1	
Lindane (μg/L)	<0.1	
Metoxiclhlorin (μg/L)	<0.1	
Trichlorometane (μg/L)	<0.0001	
Pentachlorophenol (μg/L)	<0.1	
Other relevant elements		
AOX (μg/L)	4.63–280.0	
Fatty acids (mg/L)	<0.030–0.047	
Resinous acids(mg/L)	<0.2	
Lignin and Tannin (mg/L)	<0.060–0.52	
Lignins (mg/L)	0.063–21.80	
Tannins (mg/L)	<0.025–9.8	
Turpentine (mg/L)	<0.35	
Carbon Dioxide (mg/L)	0.3 – <10	
Totales Coliforms (n.M.P/100 ml)	2.200–5.000	
Fecal coliforms (n.M.P/100 ml)	700–1.700	1.000

inflow of marine waters into the wetland sector. The pH exhibits relatively more homogeneous temporal variations. The same happens for the variables power foam, chlorophenols, hexavalent chromium, pesticides, resin acids, turpentine and CO_2. All of these latter variables have also been recorded at concentrations centered or below the limits of analytical detection (Table 8.1).

In general, the range of variation of the parameters evaluated (Table 8.1) is within the limits specified in NCh 1333 Of 78 for aquatic life and irrigated waters. However, in some areas near the mouth, very specific levels of mercury, aluminum, magnesium, sulfates and chlorides were detected, which exceeded the established norm for irrigated water. Considering the secondary quality limits proposed by CONAMA (2000), in the wetland sector, water with an exceptional or good quality would dominate. Although there have also been identified periods in which surface waters have a lower environmental quality due to fluctuations in conductivity, total suspended solids, DBO_5, sulfates, chlorides, aluminum, oils and fats and phenol index. Finally, considering the international toxicological criteria CMC (criterion of maximum concentration) proposed by the EPA (2006), the waters of the wetland did not represent any acute toxicological risk for aquatic communities. During the study period, none of the variables considered in this index (zinc, mercury, pentachlorophenol, total chromium, nickel, cadmium and lead) exceeded the specified limits.

During the monitoring period, seasonal variations in the physicochemical parameters of the water were observed. During the summer-autumn period, when there are low volume flows in the Mataquito River and the influence of marine waters is greater, the following variables are increased: temperature, conductivity, hardness, alkalinity, total solids, bicarbonates, sulfates, sulfides, nickel, calcium, magnesium, potassium and sodium; while during periods of higher flow (winter-spring), the influence of marine waters is less. Therefore, suspended solids total, oils and fats, iron, aluminum, phosphorus and phosphates increase. The increase in these parameters is related to the greater sedimentary contribution that the basin receives during the winter-spring period. Exceptionally, in June 2007 a massive fish mortality event was observed, possibly related to the timely recording of partially treated riles in the Mataquito wetland sector. The presence of rile would have altered the color levels and the concentration of detergents (SAAM), lignin, tannins and AOX. Among these latter compounds, there are some with mutagenic and carcinogenic properties (Noma et al. 2001, Kaczmarczyk and Niemirycz 2005). Although AOX can occur naturally in the terrestrial and/or aquatic environment as derived from biohalogenation processes (Asplund and Grimvall 1991, Pee 1996), industrial activities have been viewed as the main pathway to the environment (Cherif et al. 2006).

The established Gibbs models indicate that there is a strong influence of the marine waters on the surface layer of the wetland water during the studies developed in June 2007, August 2007 and March 2008. However, this influence is not evident in the station near the head of the wetland (Puente Mataquito sector). Precisely, in this season the waters presented characteristics of the continental waters during the whole period of monitoring. As shown in Fig. 8.3, in November 2007 there was a greater influence of the erosive processes of the rocks, while during July and November 2008, surface waters presented proportions of sodium, calcium and dissolved solids similar to the averages reported for the continental surface waters (Fig. 8.3).

The variation of the environmental quality of the surface waters of the wetland sector measured through the SWQI, suggests that the quality decreases toward the mouth sector during all periods of study (Fig. 8.4). This is due to the marine influence. The effect of marine waters modifies the conductivity, pH, chloride and sulfate concentration and the adsorption ratio of sodium (RAS) in the continental waters of the sector associated.

with the main (head) water of the wetland (Puente Mataquito). In addition, the effect of marine influence on water quality is reflected in the standards obtained for August 2007 and March 2008 (good to fair), as well as for July 2008, November 2007, and 2008 (Excellent to good, Fig. 8.4).

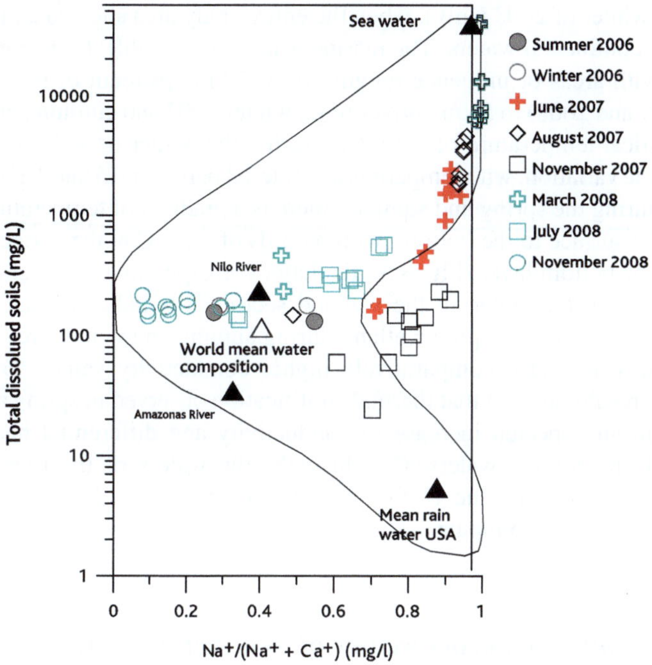

Fig. 8.3 Variation in $Na^+/Na^+ + Ca^+$ proportion with regards to concentration of total dissolved solids (Gibbs 1970)

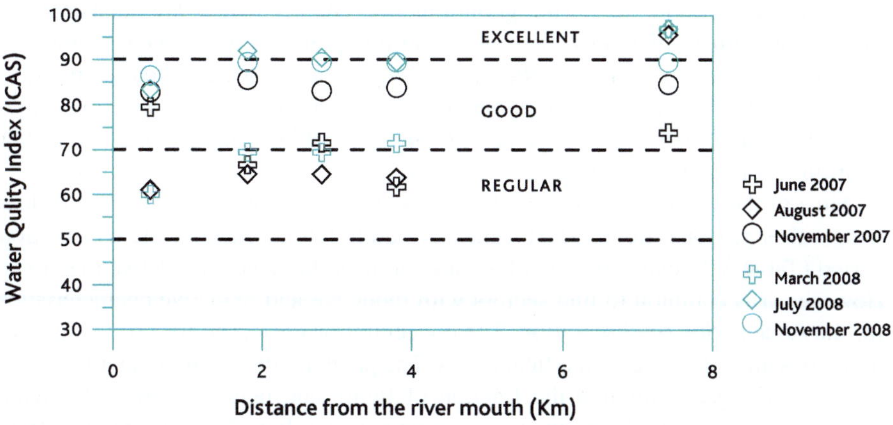

Fig. 8.4 Temporal evolution of the superficial water quality index (SWQI) in the Mataquito river wetland

Regarding the variations of temperature and conductivity obtained through the analysis of the water column, it is important to point out that throughout the year it is frequent to record the entrance of marine waters to the sector of the Mataquito river wetland through the bottom stratum. An exception to this pattern occurred during the winter of 2008 (July), when the entire study area showed a marked influence of the continental waters. The marine waters are variable, both temporally and spatially, with areas of influence ranging from 2 km upstream of the mouth (e.g., spring 2007 and 2008) to 5 km upriver (e.g., winter 2007 and autumn periods 2008).

The vertical temperature of the water during the winter periods (June–August) presents little variation, with temperatures centered between 9.5 and 12 °C (Fig. 8.5). You find, during the spring and summer, there is a gradient of temperature descending from the surface to the bottom, with records of >21 °C at the surface level, and <16 °C at the bottom layer (Fig. 8.5), the latter corresponding to the characteristics of the marine waters entering the wetland sector. This indicates that the marine waters that enter the Mataquito wetland during autumn and winter are warmer than inland waters and have comparatively higher conductivity values. On the other hand, these results suggest that thermal stratification observed in spring and summer results from an expected increase in conductivity and differential heating in the continental and marine waters. Conclusively, the waters of the river undergo a marked surface heating due to the greater solar radiation, being comparatively warmer than the marine inputs.

Physicochemical Characteristics of Sediments in the Wetland

The physicochemical characterization is based on the analysis of the spatial-temporal variations that presented the granulometric characteristics and the variables listed in Table 8.2. The granulometric characteristics are analyzed by interpreting ternary diagrams. Because direct interpretation from an environmental database is usually a complex situation (Zhang et al. 2008), exploratory multidimensional analysis of major components has been performed (Zhou et al. 2007, Narendra et al. 2010). We considered only those variables that were quantified over their limits of analytical detection.

Most of the variables present wide ranges of variation, although without a clear temporal or seasonal trend. MOT values are quite low throughout the observation period (Table 8.2), characteristic of an environment that favors oxidative reactions. However, it is common to find sectors with reductive and oxidative characteristics for the same period of observation. The concentrations recorded for resin acids and pesticides are below their analytical limits. The phenols show a significant variation (<0.01–5.33 μg/g), although the fraction of lignin and tannins presented a wide oscillation, these were determined as a whole between June 2007 and March 2008. However, between July and November 2008, these fractions were determined separately (Table 8.2).

Throughout the observation period, trace elements of terrestrial origin were predominant (iron and aluminum, Table 8.2). High levels of zinc and copper are also highlighted, while the secondary concentrations of lead and cadmium are minor. Considering the geo-accumulation index proposed by Müller (1981), the characteristics that present wetland floods would indicate the existence of an uncontaminated

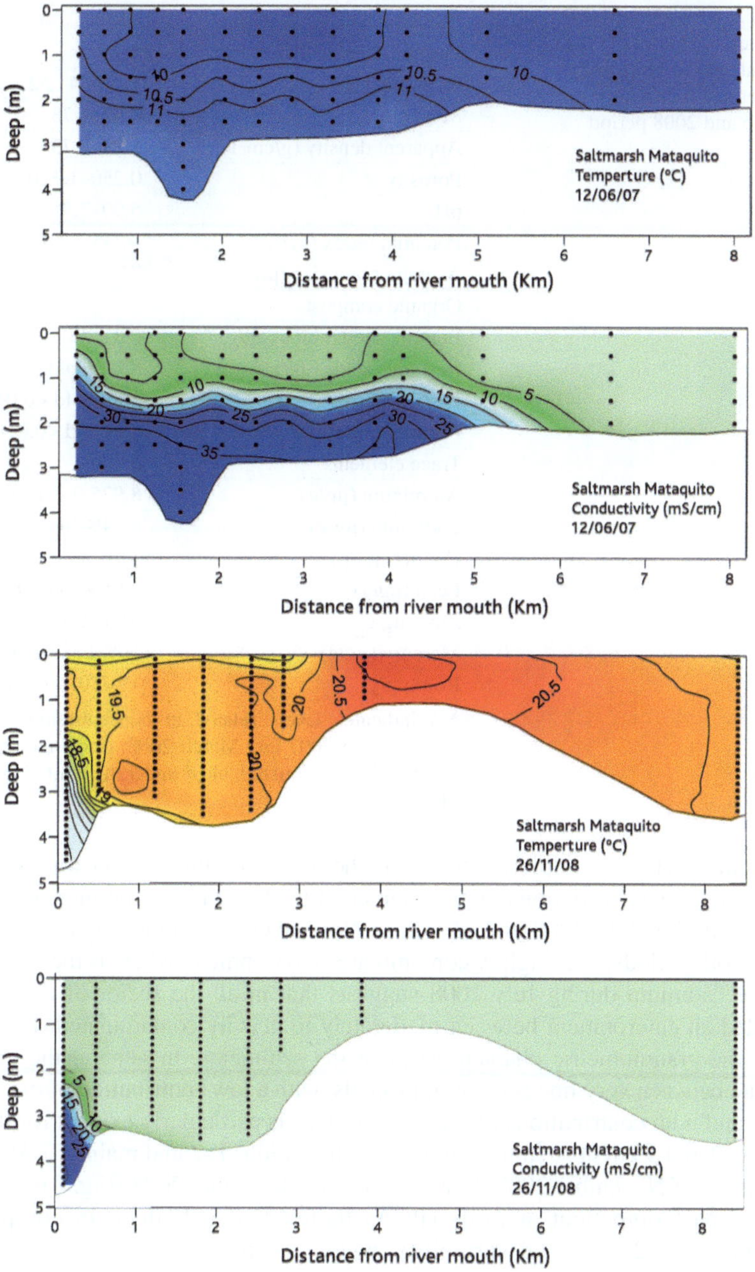

Fig. 8.5 Hydrographic sections of temperature (°C) and conductivity (mS/cm) in the Mataquito river wetland, Maule Region

Table 8.2 Physicochemical parameters monitored in the sediments of the Mataquito river wetland sector during the 2006 and 2008 period

Parameters	Range
Physico-chemical characteristics	
Humidity (%)	19.10–41.52
MOT (%)	0.62–4.78
Apparent density (g/cm^3)	1.29–2.75
Porosity	0.250–1.320
pH	5.90–7.75
Potential redox (mV)	−235 – 310
Average grain size (μm)	71.41–357.80
Organic compost	
Lignins and tannins (μg/g) **	0.31–53.2
Resinous acids (μg/g)	83.3% bld <0.1–7.31
Pesticides (μg/g)	100% bld <0.04
Phenols (μg/g)	25% bld <0.01–5.33
Trace elements	
Aluminum (μg/g)	8.975.9–30.381.8
Cadmium (μg/g)	82.4% bld <0.2–4.9
Copper (μg/g)	11.6–49.7
Lead (μg/g)	73.5% bld <0.2–55.0
Zinc (μg/g)	31.1–189.8
Mercury (μg/g)	10% bld <0.06–0.080
Iron (μg/g)	19.750–69.611

(**) indicates levels related to concentrations reported between June 2007 and March 2008. (mot) indicates percentage of organic matter, uld = under the limit of analytical detection

environment. However, timely increases in the concentrations of some trace elements suggest significant variations in environmental quality. In fact, the increase of lead and zinc during July 2008 and of iron in November 2008 indicates the existence, during both periods, of a slightly contaminated environment, whereas the determinations of cadmium during July 2008 suggests that in all the sector of the wetland prevailed an environment between moderately to heavily contaminated. Figure 8.5 shows the granulometric characteristics of the sediments. In general the granules oscillate between very fine and medium sands, with a low contribution of the gravels (<5%) and with contributions of mud >20%. This is particularly evident for the area between 2 and 3 km upstream from the mouth (stations E-2 and mainly E-3), during the periods of November 2007, March 2008 and November 2008 (Fig. 8.6).

Principal component analysis (PCA) indicates that both main components account for 62.2% of total variability, where components X and Y account for 47.4% and 14.8%, respectively (Fig. 8.7). Disintegration of the points on the axis of the first main component is observed. This is related to the recurrent record of comparatively higher levels of copper, MOT, moisture, aluminum and porosity of sediments at stations E-2 (2.8 km from the mouth) and mainly E-3

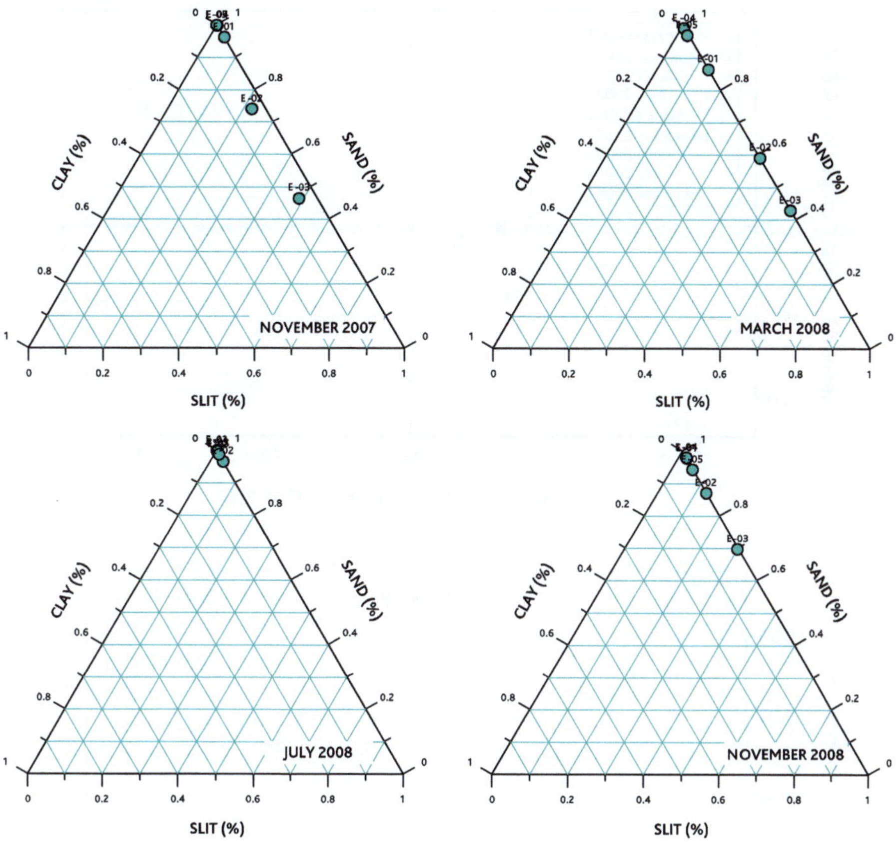

Fig. 8.6 Ternary diagrams for sediments in the Mataquito river wetland

(1.8 km from the mouth) . The rest of the sampling points presented magnitudes around the historical averages, with no greater dispersion of the information on the axis of the second main component (Fig. 8.7). It should be noted that for this analysis, there were only considered those variables that were evaluated in all the studies carried out in the wetland sector and that, in addition, they repeatedly presented numerically quantifiable information; which is to say, on the limits of analytical detection.

Biotic Characterization of the Wetland Sector

In this chapter, the results of the characterization of the aquatic and terrestrial biota cadastral survey during the period 2006–2009 are presented. Table 8.3 details the items characterized as well as the number of monitoring stations. For the

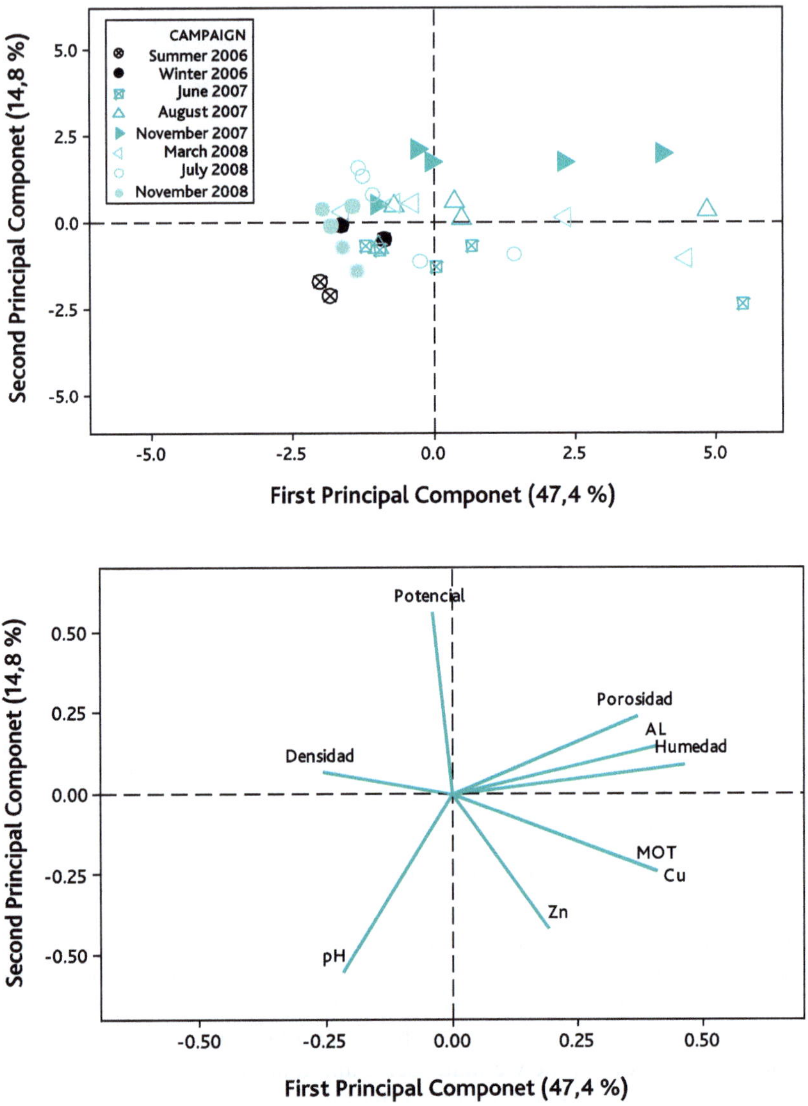

Fig. 8.7 Bi-dimensional representation of the significant two principal components obtained in the historical explorative multivariate analysis for sediments in the wetland

macrobenthos, zooplankton, zoobenthos, phytoplankton and phytobenthos the number of monitoring stations made for the period is given. The terrestrial fauna and flora (Phanerogams) is indicated with the symbol "+" for those periods for which information is available. As for the ichthyofauna, the number of sets made to characterize this item is shown (Table 8.3).

Table 8.3 Number of sampling points during the study period for aquatic and terrestrial biota in the Mataquito river wetland sector

	Periods								
	Summer 2006	Winter 2006	June 2007	August 2007	Nov 2007	March 2008	July 2008	Nov 2008	Jan 2009
Macrobenthos	–	–	–	–	4	4	4	4	–
Zooplankton	2	2	–	–	4	4	4	4	–
Zoobenthos	2	2	–	–	4	4	4	4	–
Phytoplankton	2	2	–	–	4	4	4	4	–
Phytobenthos	2	2	–	–	4	4	4	4	–
Terr. Flora	+	+	+	–	–	–	–	–	–
Terr. Fauna	+	+	–	–	+	+	+	+	–
Ichtyofauna	2	2 (*)	–	–	–	–	8	–	10

(–) indicates period without information. (*) indicates period without positive events recorded due to bad meteorological conditions

Water Communities

Table 8.4 presents each of the items identified in the sediments of the wetland sector during November 2007, March 2008, July 2008 and November 2008. In these campaigns, 19 macrobenthic species were identified. However, this specific richness presents important temporal variations. In November 2007, 6 species were recorded, while in the campaigns carried out in July and November 2008 the presence of 5 taxa was verified (Table 8.4). The highest specific richness (n = 10 sp.; Table 8.4) was recorded during March 2008. The predominant groups correspond to Crustaceans and Annelids; within the latter, the Polychaetes are particularly important (Table 8.4, Fig. 8.8a). The only species of Oligochaetes present at the study site was recorded in a timely manner towards the section of the head of the wetland during November 2008 (Table 8.4, Fig. 8.8a). Both Polychaetes and Crustaceans are distributed throughout the wetland area, although their diversity and richness tends to increase towards the sector of the mouth. Mollusks are recorded only in the vicinity of the mouth during November 2007 and March 2008 (Table 8.4, Fig. 8.8a).

The different taxa identified in the wetland show slight temporal variations in the relative abundances. In general, the Polychaetes group is the most abundant in all sampling campaigns (Fig. 8.8a), while Crustaceans are second in abundance (40–10%, Fig. 8.8a). However, in July 2008 this taxon experienced a significant increase in abundance (~78%, Fig. 8.8a). The greatest abundance of Crustaceans during this period is due to the unusual increase of Amphipod species.

In the intervals in which Mollusks and Oligochaetes are recorded, both present relatively marginal abundance, ranging from 0.02% to 2.06% (Fig. 8.8a).

Compared to the relative abundance patterns, the biomass of the macrobenthic taxa presented significant temporal variations (Fig. 8.8b). During November 2007 and March 2008 the Polychaetes had the highest biomass, with percentages of

Table 8.4 Macrobenthic community recorded in sediments of the Mataquito river wetland sector during the November 2007 (1), March 2008 (2), July 2008 (3) and November 2008 (4) campaigns

Phylum	Class/order or family	Specie	1	2	3	4
Annelida						
	Polychaeta/Spionidae	*Paraprionospio* sp.	–	–	+	+
	Polychaeta/Spionidae	*Prionospio* sp.	–	–	+	–
	Polychaeta/Spionidae	*Microspio* sp.	+	–	–	–
	Polychaeta/Spionidae	*Spiophanes* sp.	–	+	–	–
	Polychaeta/Nereidae	*Platynereis* sp.	+	–	–	–
	Polychaeta/Nereidae	*Nereis* sp	–	+	+	+
	Polychaeta/Capitellidae	*Capitella* sp.	+	+	–	–
	Polychaeta/Cossuridae	*Cossura chilensis*	–	+	–	–
	Polychaeta/Glyceridae	*Glycera americana*	–	+	–	–
	Olygochaeta	*Oligochaeta* sp	–	–	–	+
Crustaceans						
	Amphipoda	*Paracorophium* sp	+	–	–	+
	Amphipoda	*Lysianassidae* sp	+	+	–	–
	Amphipoda	*Gammarus* sp	–	+	+	–
	Isopoda	*Excirolana* sp	+	–	+	–
	Decapoda	Megalopa	–	+	–	–
	Decapoda	*Hemigrapsus crenulatus*	–	–	–	+
Mollusks						
	Carditidae	*Cyclocardia compressa*	+	–	–	–
	Mytilidae	*Mytilus* sp	–	+	–	–
	Mactridae	*Mulinia* sp	–	+	–	–

The symbols (+) and (–) indicate presence and absence, respectively

83.8% and 58.1%, respectively. According to the increase in relative abundance observed in July 2008, Crustaceans correspond to the item with the highest biomass (63.4%; Fig. 8.8b). In November 2008, this group contributed the highest biomass (87.9%, Fig. 8.8b). This is due to the fact that during this campaign two specimens of *Hemigrapsus crenulatus* were collected, decapod crabs that in comparison to the rest of the identified items present large size and body mass. Consistent with the low abundance of Mollusks and Oligochaetes, their contribution to total biomass is quite poor during the periods in which they are recorded.

In all of the studies, the spatial variations are observed, both in the specific richness and relative abundance of macrobenthic organisms through the wetland. Both descriptors tend to increase from the sector of the mouth to the interior of the wetland, being a maximum in abundances and wealth between 1.5 and 3 km upstream. Towards the sector of the head of the wetland, the wealth and abundances decrease, especially the latter. The lowest densities occur at the monitoring station located towards the terminal end of the wetland sector, near the Mataquito Bridge.

In the sampling campaigns, 12 different classes of zooplankton organisms were identified (Fig. 8.9), as well as three types of organisms that could not be assigned to this taxonomic category (categorized as indeterminate in Fig. 8.9). Both the representation just as the relative abundance of these entities shows a significant

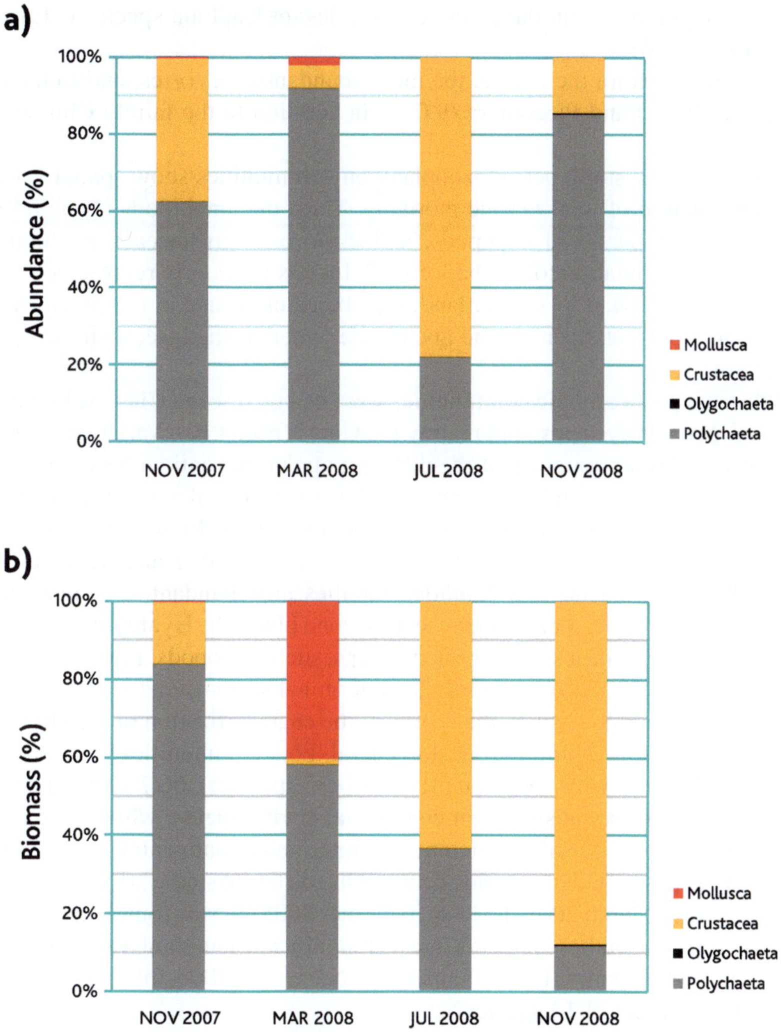

Fig. 8.8 Temporal distribution of relative abundance (**a**) and biomass (**b**) for the macrobenthic groups registered in Mataquito wetland sediments

temporal variability (Fig. 8.9). For example, the *Clitellata Class* was the predominant group in March 2006 (relative abundance = 38.7%), but in July 2008 (22.1%) and November 2008 (13.4%) its relative abundance was significantly reduced. Meanwhile, in August 2006 and March 2008 this group practically disappeared from the study area. Similarly, the abundant point maxima exhibited by the Maxilipoda, Insecta, Malacostraca and Branchiopod Classes indicate that the structure of the zooplankton communities of the wetland is highly dynamic. Within the group of Malacostraca are Amphipods of the Hyalellidae Family, which represent 17.8% of the total abundance. In the Maxillopoda group, the copepods of the orders Cyclopoida (12.2%) and Calanoidae (9.7%) are abound. Meanwhile, within the Branchiopod and Clitellata Classes are relevant Daphnia species (11.8%) and Oligochaetes (8.6%).

For its part, within the insects the most abundant taxa correspond to the orders Hemiptera (9.7%) and Plecoptera (9.0%), in addition to the family Chironomidae (5.9%).

Throughout the study period, zooplankton communities show spatial variations in the distribution of total marine mortality. Thus, the amphipods Hyalellidae, the calanoid and harpacticoid copepods, and cladoceran crustaceans predominantly dominate the terminal sector of the wetland. Insects are poorly represented. In contrast, towards the head of the wetland, the abundant groups in the previous sector decrease in representativeness and abundance, whereas the insects show the opposite pattern.

Within the zoobenthonic communities were detected seven Classes, highlighting the Maxilipods, Polycaetes, and mainly the Clites (Fig. 8.10). Variations in relative abundance and representation of the different taxonomic entities recognized in the wetland during the six studies suggest that the structuring of zoobenthonic communities is dynamic over time. In most studies, except for March 2006, vermiform forms predominate in all sectors of the wetland (Fig. 8.10). Thus, organisms of the Nepthyidae (Polychaeta) and Naididae families are abundant, that in their total amount to 61.2% of the total abundance. In second place, the Hyalellidae Amphipods stand out (17.7%). The less represented groups, such as isopods, mollusks and polychaetes, are restricted in distribution to the terminal sector of the wetland.

Figures 8.11 and 8.12 show the results of the characterization of the phytoplankton and phytobenthic communities. In general, both communities have low values of specific richness, with 45 in total species (<35 sp. each). 66.7% of the total species correspond to organisms from continental environments, while the remaining 33.3% corresponds to a saline-tolerant taxa (marine or continental). The number of species in the phytoplankton varies between the different studies (Fig. 8.11a), ranging from 34 sp. (March 2008) to 15 sp. (summer 2005). It was found that the species richness in phytobenthos does not present significant temporal variations (20–26 sp., Fig. 8.12a). However, in the campaign of November 2008, this decreases exceptionally to a number of 14 species.

Usually, phytobenthic species are recorded in the water column, forming part of the planktonic component (e.g. *Fragilaria ulna, Navicula sp., Melosira varians and Diatoma vulgare*). In all the studies carried out within both communities, the *Bacillariophyceae* Class is mostly dominant (Figs. 8.11 and 8.12), especially in phytobenthos. In this group, diatoms present abundance of >80%, and reach a maximum of 100% during November 2008 (Fig. 8.12g). Within the phytoplankton communities, their abundance is comparatively smaller, ranging from 60% to 83.3% (Fig. 8.11). Among the most abundant and representative species of the Bacillariophyceae Class are: *F. ulna, M. varians, Gomphoneis minuta, Frustilia vulgaris, Navicula and D. vulgare, Chaetoceros and Leptocylindrus*.

In both phytoplankton and phytobenthos, the Chlorophyceae Class constitutes the second most important item, with *Scenedesmus quadricauda* being the most representative species. However, as of July 2008 the registration of green algae within the phytobenthic group has disappeared (Fig. 8.12f, g). The abundance of

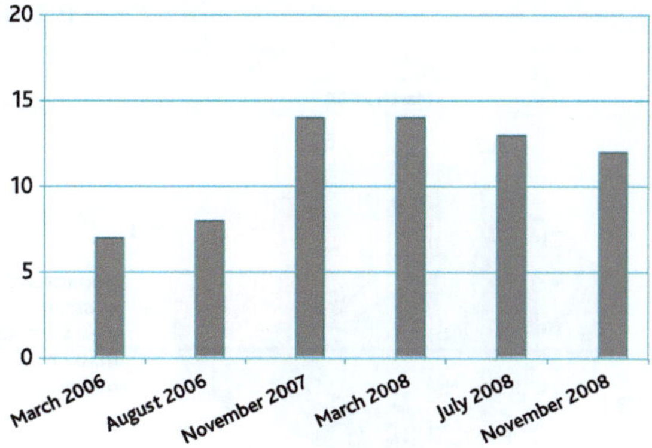

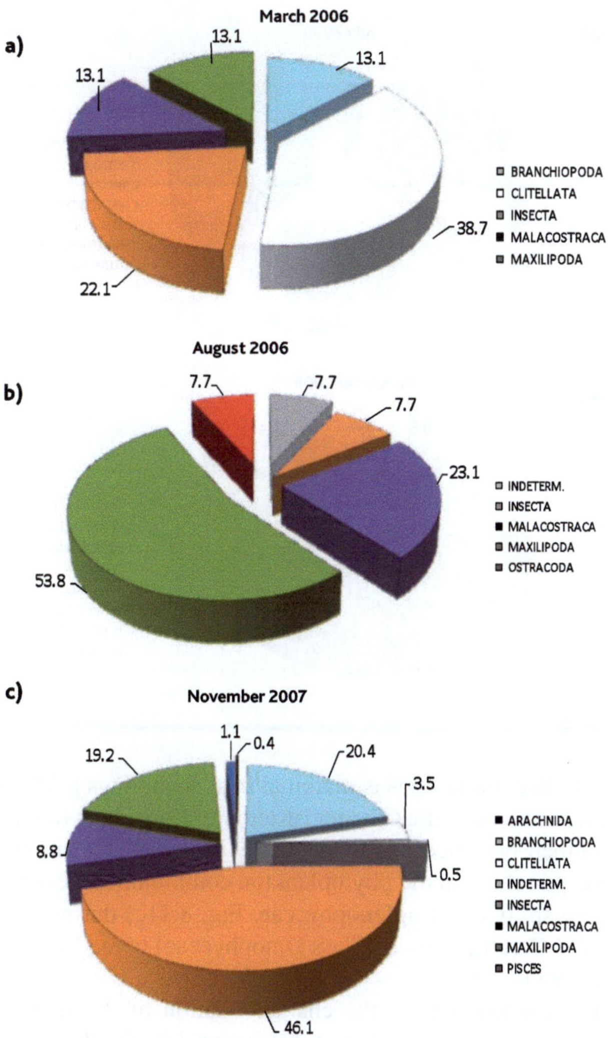

Fig. 8.9 Relative abundance for Classes identified within the zooplankton communities in the Mataquito wetland during the six sampling campaigns

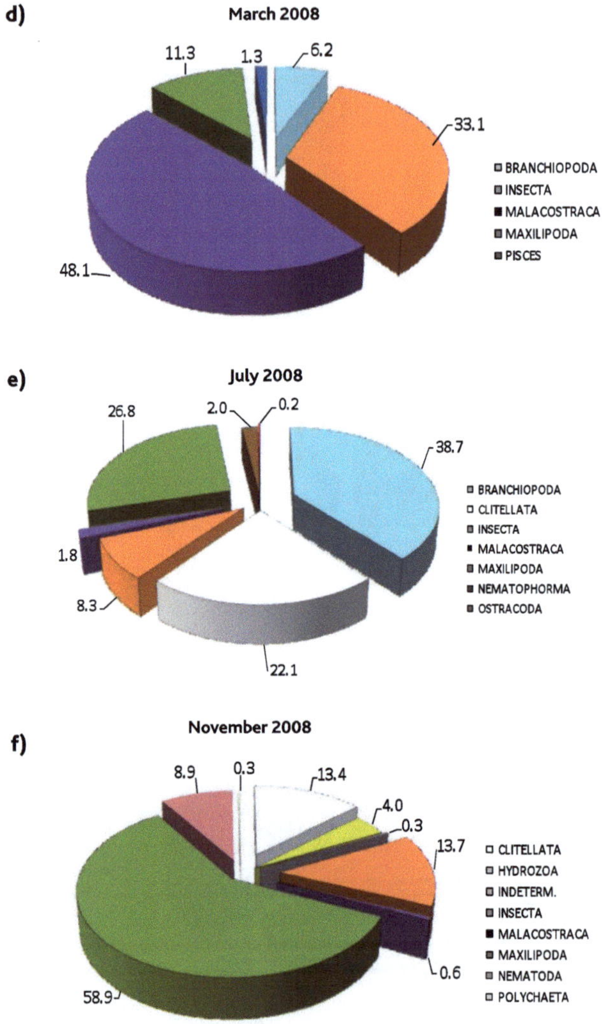

Fig. 8.9 (continued)

cyanophytes in both communities is marginal in all samplings (<7%, Figs. 8.11 and 8.12). It is important to note that the registration of blue algae in the phytobenthos is irregular, and their presence was verified only in 3 of the 6 campaigns carried out (Fig. 8.12). Exceptionally, in the phytoplankton communities were identified specimens of *Euglena* sp. (Class Euglenophyceae, Fig. 8.11c) during November 2007, and individuals of Ceratium furca (Class Dinophyceae) in the March and July 2008 campaigns (Fig. 8.11d, e).

In the studies carried out for the characterization of the Ichthyofauna of the Mataquito river wetland, a total of nine species were identified. Among the different studies, the specific richness showed significant variations, ranging from four species (summer 2006) to a maximum of seven (winter 2008 and summer 2009,

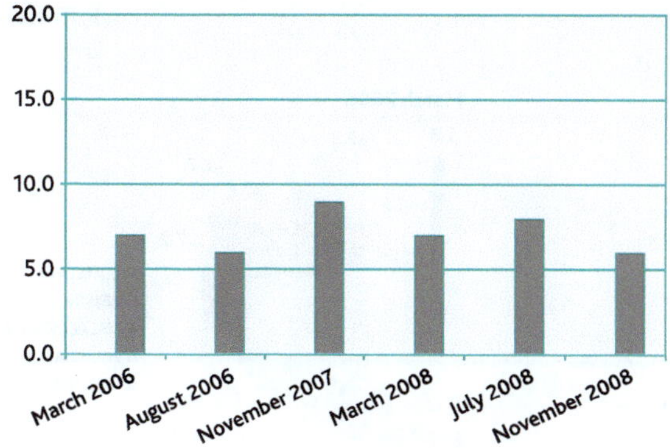

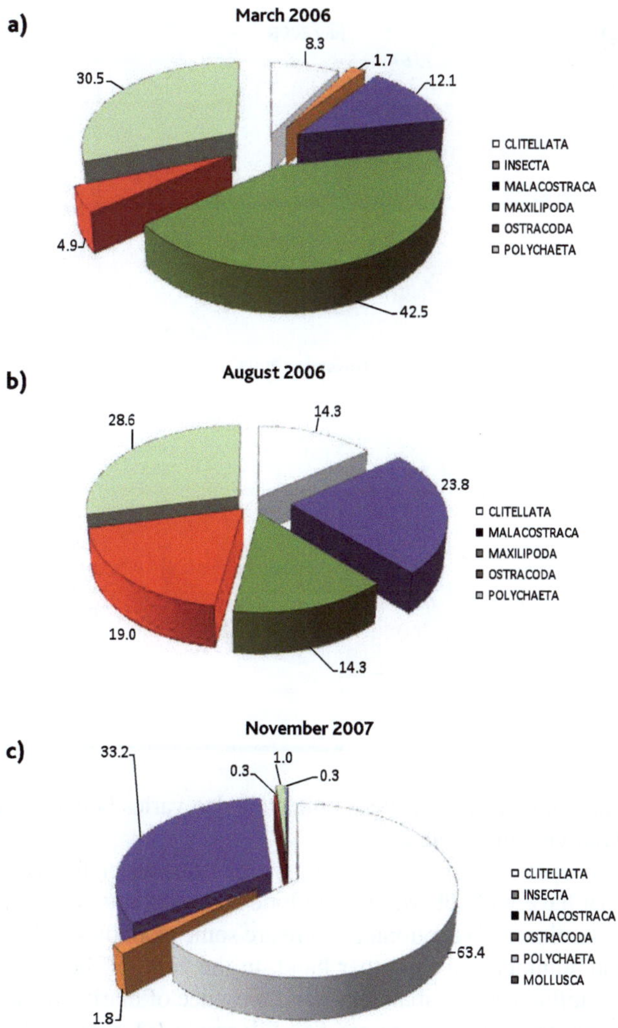

Fig. 8.10 Relative abundance for Classes identified within the zoobenthic communities in the Mataquito wetland during the six sampling campaigns

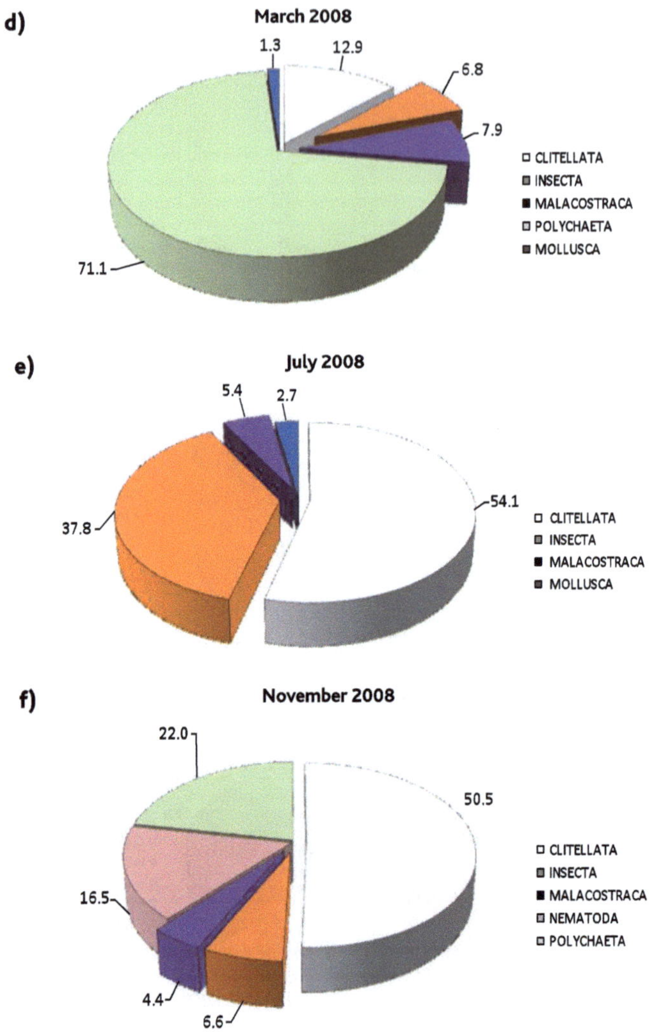

Fig. 8.10 (continued)

Table 8.5). The number of specimens collected also varies between sampling periods (85–150 individuals, Table 8.5).

In the sector upstream of the wetland, typical freshwater fish (*Percilia gillissi*) and/or taxa that have wide distribution in lotic environments (*C. carpio*, *O. mykiss* and *S. trutta*) dominate. Exceptionally there are some examples of M. cephalus during winter and summer. On the other hand, in the zone of the mouth, due to the greater marine influence, the abundance and presence of marine fishes is increased. Among the most frequent species are Sea Silverside (*O. regia*), *Pejerrey* Chilean Silverside (*B. australis*), Flathead Grey Mullet (*M. cephalus*), Patagonian Blenny (*Eleginops maclovinus*), Pacific Manhaden (*Ethmidium maculatum*) and Pampanito (*Trachinotus paitensis*). In general, Silverside/*Pejerrey* (*B. australis*) is the most

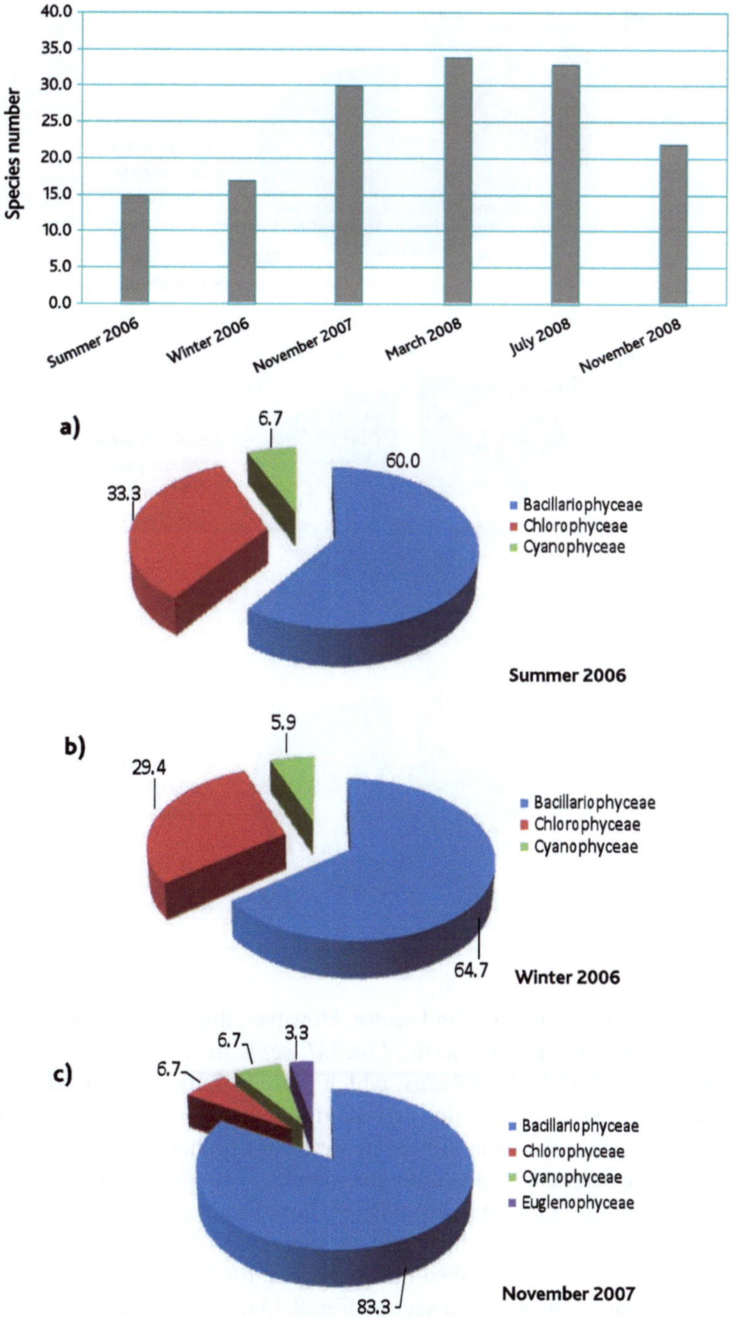

Fig. 8.11 (**a**) Temporal evolution of specific richness of phytoplankton communities in the Mataquito wetland. (**b–g**) Relative abundances for classes identified within the phytoplankton communities during the six sampling campaigns

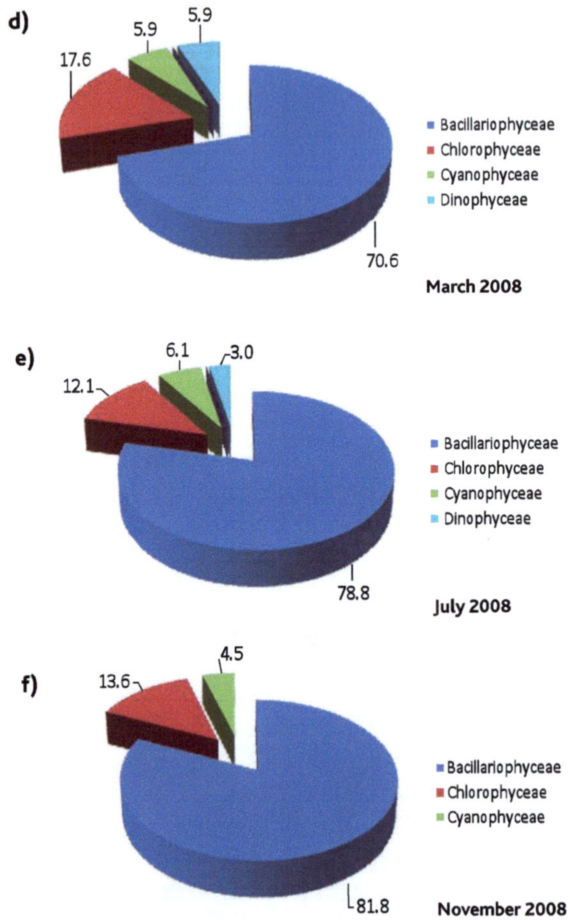

Fig. 8.11 (continued)

abundant species within the wetland sector. However, during summer 2009 the most important species corresponded to the *Lisa (M. cephalus).*

With the exception of *M. cephalus*, which presents important abundance only in the summers, the rest of the species identified in the wetland tend to exhibit similar abundance throughout the year. However, variations in their densities occur at spatial scale, with differences in absolute abundances between the mouth and the sectors located upstream of the Mataquito River. For example, *C. carpio* presents mean equivalent densities between winter and summer (0.0007 ind./m^2 and 0.0008 ind./m^2, respectively). However, its absolute abundance for both seasons decreases from the upstream areas to the wetland sector (Fig. 8.13a, b). In the case of *B. australis*, the spatial distribution of maximum abundances varies seasonally. In fact, Silverside density is lighter in summer than in winter, but the seasonal peak of maximum density occurs in upstream areas (Fig. 8.13c). During winter, the maximum in absolute abundance is recorded towards the wetland sector (Fig. 8.13e).

Table 8.5 Fish species identified and total number of specimens (total n°) collected in the Mataquito river wetland sector

Campaign	Scientific name	Common name	n° total
Summer 2006	*Basilichthys australis*	Silverside	
	Odontesthes regia	Sea Silverside	
	Cyprinus carpio	Common Carp	
	Mugil cephalus	Flathead Grey Mullet	85 (*)
Winter 2008	*Basilichthys australis*	Silverside	
	Odontesthes regia	Sea Silverside	
	Cyprinus carpio	Common Carp	
	Mugil cephalus	Flathead Grey Mullet	
	Eleginops maclovinus	Patagonian Blenny	
	Oncorhynchus mykiss	Rainbow Trout Arcoíris	
	Salmo trutta	Brown Trout	150
Summer 2009	*Basilichthys australis*	Silverside	
	Odontesthes regia	Sea Silverside	
	Eleginops maclovinus	Patagonian Blenny	
	Ethmidium maculatum	Pacific Manhaden	
	Trachinotus paitensis	Paloma Pampanito	
	Cyprinus carpio	Common Carp	
	Mugil cephalus	Flathead Grey Mullet	102

(*) total of individuals collected includes only pejerrey individuals. For the rest of the periods, the sum of individuals of all species collected is shown

Land Communities

The Mataquito River is located within the biogeographic region of the Maquino Deciduous Forest (Gajardo 1994), in which *Nothofagus glauca, Azara petiolaris* are dominant, in addition to the relic species *Nothofagus alessandrii and Gomortegakeule* in the humid sectors. Within this geographical area there are three distinct vegetative formations (Luebert and Pliscoff 2004): (**a-**) Mediterranean coastal *sclerophyllous* forest of *Lithrea caustica* and *Azara integrifolia,* (**b-**) "Mediterranean coastal deciduous forest *of Nothofagus glauca and Persea lingue*" and (**c-**) "Mediterranean coastal dunes of *Ambrosia chamissonis and Distichlis spicata.*" Both the deciduous and sclerophyllous forests present in this region are highly affected by the replacement of native vegetation by *Pinus radiata* plantations (Luebert and Pliscoff 2004).

The EGA study (2006) indicates that differences in the soil characteristics present on the north and south sides of the wetland determine that the land communities that develop in both sectors differ in structure (Fig. 8.14). On the north bank, the forest plantations are located in the sector furthest from the body of water. While in the intermediate sectors of vegetation formation native trees and shrubs are mixed, towards the wetland, on the edge of the body of water, shrubs and grazing formations are dominant (Fig. 8.14). Here, the dominant species are *Bromus, Medicago hispidata, Trifolium repens, Rumex acetosella and Hypochaeris radicata.* Moreover, a

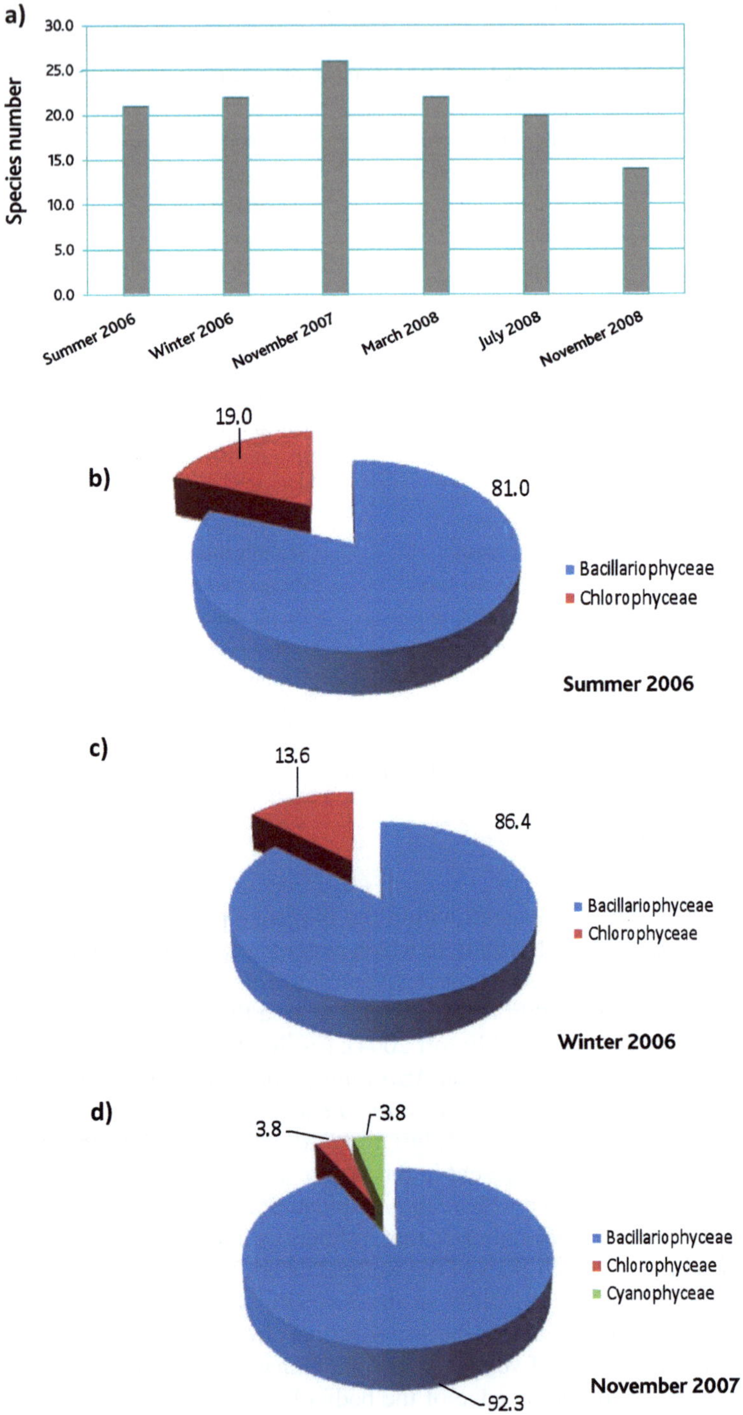

Fig. 8.12 (**a**) Temporal evolution of specific richness of phytobenthic communities in the Mataquito wetland. (**b–g**) Relative abundances for Classes identified within the phytobenthic communities during the six sampling campaigns

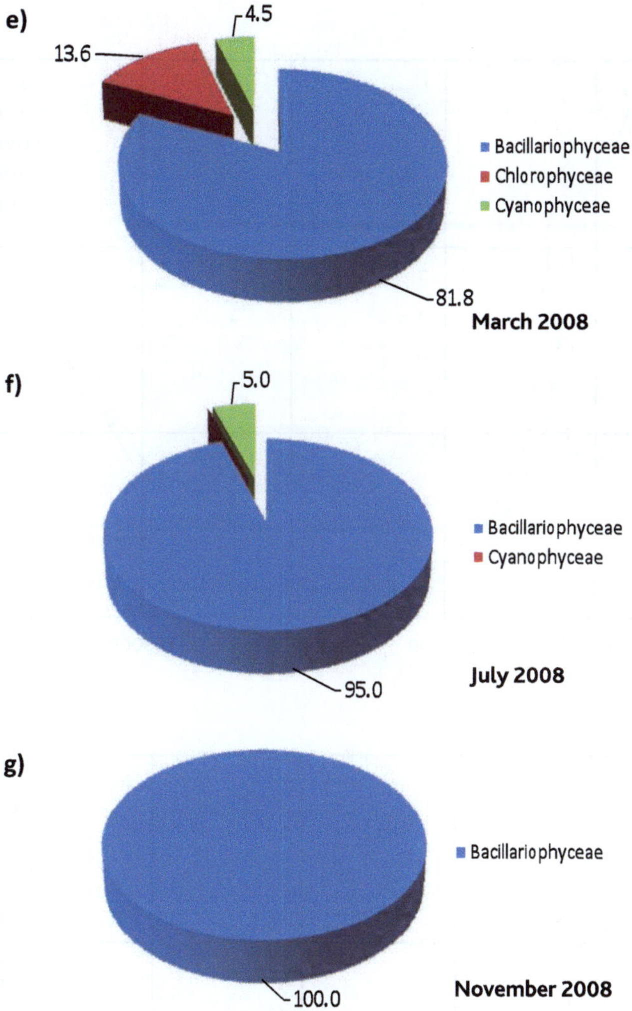

Fig. 8.12 (continued)

very diverse coastal dune system is developed on the southern shore, especially during summer periods. In this zone, the shrub and herbaceous forms abound, emphasizing the species *Distichlis spicata, Scirpus americanus, Paspalum distichum and Polygonum persicaria* (Fig. 8.14).

Table 8.6 shows the list of plant species identified on both shores of the wetland during summer and winter, respectively. Here, it can be appreciated that the horticulture (floristic) spectrum in both seasons is dominated widely by introduced species. The representation of native species is low on shores, as well as during seasons. In fact, these are equivalent to 23.1% (winter period) and 26.1% (summer period) of the assemblies present on both banks (Table 8.6).

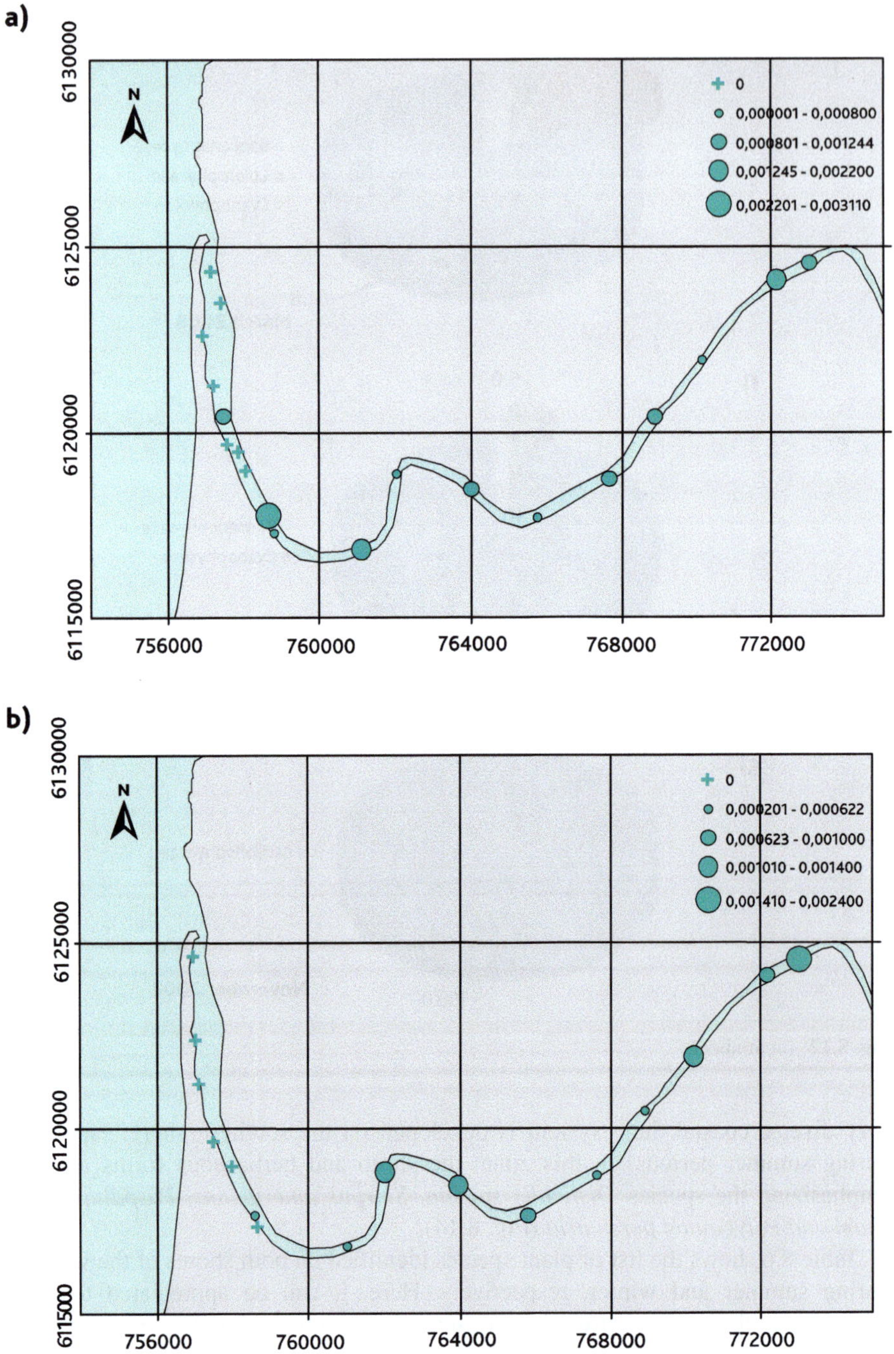

Fig. 8.13 Spatial distribution of absolute abundance (ind/m^2) for the species *Cyprinus carpio* [summer (**a**) and winter (**b**)] and *Basilicthys australis* [summer (**c**) and winter (**d**)]

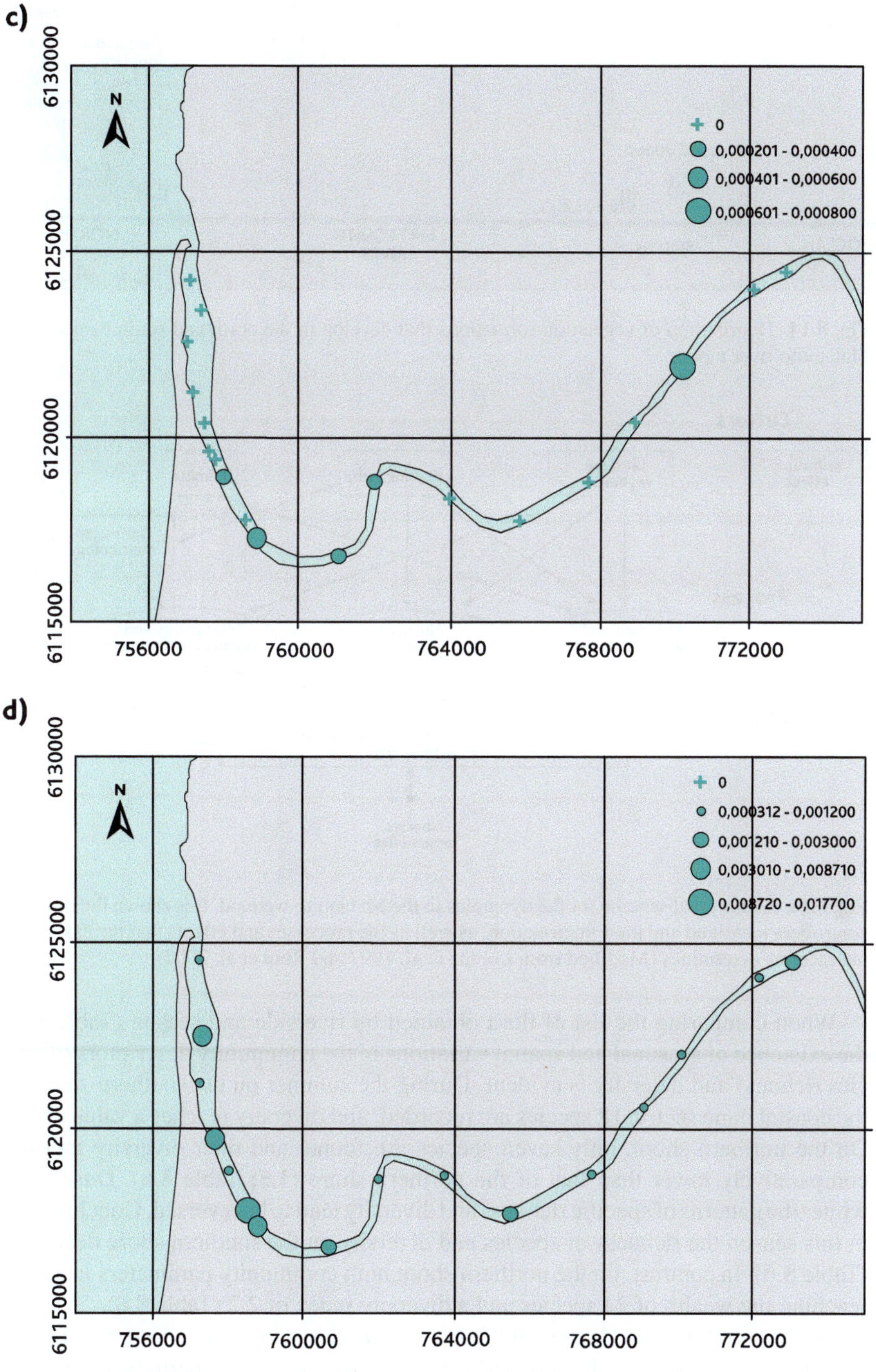

Fig. 8.13 (continued)

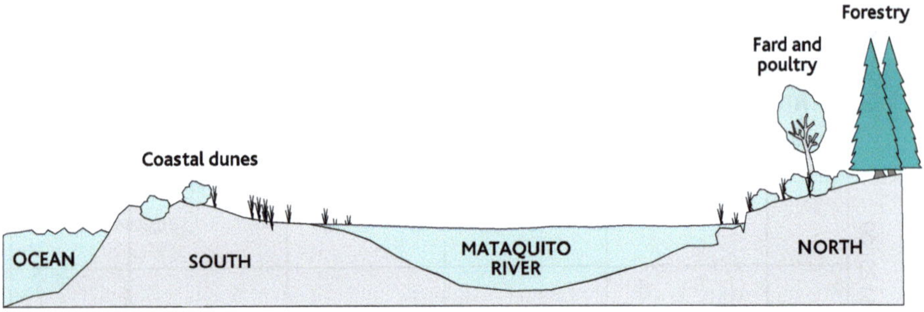

Fig. 8.14 Distribution of vegetation formations that develop in the north and south banks of the Mataquito river mouth

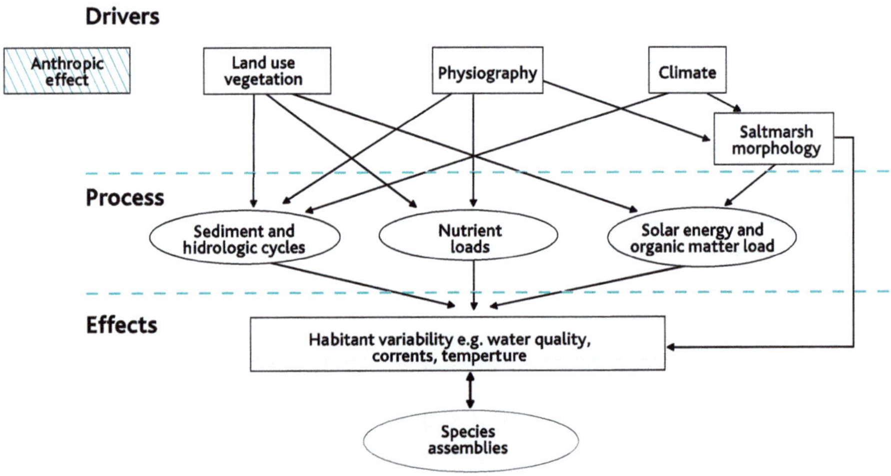

Fig. 8.15 Conceptual scheme for the dynamics in the Mataquito wetland. It is shown the different controllers identified and their interrelation, as well as the processes and effects that the cascade of interactions determines (Modified from Lorenz et al. 1997 and Roni et al. 2002)

When comparing the list of flora obtained by riverside and season (Table 8.6), the existence of seasonal and spatial variations in the community descriptors of species richness and diversity is evident. During the summer on the southern shore, in the coastal dune sector, 18 species are recorded, and diversity reaches a value of 2.5. On the northern shore, only seven species are found, and their diversity index is comparatively lower than that of the northern shore (**1.5;** Table 8.6). During the winter the patterns of specific richness and diversity tend to be reversed. Conclusively, in this season the richness of species and diversity on the southern shore decreases (Table 8.6). In contrast, on the northern shore both community parameters increase, reaching the wealth of 24 species and a diversity index of 2.2 (Table 8.6).

In the sampling periods the land fauna was quite scarce. There were no positive sightings in the summer of 2006, and July and November 2008 periods. During the winter of 2006, the only sightings reported was that of Parastacus pugnax.

Table 8.6 Plant species identified in the wetland sector during the summer and winter periods

Specie	Family	Life form	Summer	Summer	Winter	Winter
*Acacia caven**	Mimosaceae	Tree	–	–	+	–
*Ambrosia chamissonis**	Asteraceae	Herb	–	+	–	+
*Anthemis arvensis**	Asteraceae	Herb	+	–	–	–
*Baccharis linearis**	Asteraceae	Shrub	–	+	–	–
*Baccharis salicifolia**	Asteraceae	Shrub	–	–	+	–
Brassica rapa	Brassicaceae	Herb	–	+	+	–
Bromus sp.	Poaceae	Herb	–	–	+	–
Carduus pycnocephalus	Asteraceae	Herb	–	–	+	–
Cerastium fontanum	Caryophyllaceae	Herb	–	+	–	–
Chenopodium album	Chenopodiaceae	Herb	–	+	–	–
Conium maculatum	Apiaceae	Herb	–	–	+	–
Cynara cardunculus	Asteraceae	Herb	–	–	+	–
Dactilys glomerata	Poaceae	Herb	–	–	–	+
*Distichlis spicata**	Poaceae	Herb	–	+	–	–
*Equisetum bogotense**	Equisetaceae	Herb	–	–	+	–
Galega officinalis	Fabaceae	Herb	–	–	+	–
Hypochaeris radicata	Asteraceae	Herb	+	–	+	–
*Juncus bufonius**	Juncaceae	Herb	–	–	+	–
Lupinus arboreus	Fabaceae	Shrub	–	–	+	–
Lythrum hyssopifolia	Lythraceae	Herb	–	–	+	–
*Margyricarpus pinnatus**	Rosaceae	Herb	–	–	+	–
Medicago hispidata	Fabaceae	Herb	–	–	+	–
Medicago sativa	Fabaceae	Herb	–	+	+	–
Melilotus alba	Fabaceae	Herb	–	+	–	–
Paspalum distichum	Poaceae	Herb	+	+	–	–
Plantago lanceolata	Plantaginaceae	Herb	+	+	+	–
Polygonum aviculare	Polygonaceae	Herb	–	–	+	–
Polygonum persicaria	Polygonaceae	Herb	–	+	–	–
Raphanus sativus	Brassicaceae	Herb	+	–	+	–
Rubus ulmifolius	Rosaceae	Shrub	–	–	+	–
Rumex acetosella	Polygonaceae	Herb	+	–	+	–
Rumex crispus	Polygonaceae	Herb	–	–	+	+

(continued)

Table 8.6 (continued)

Specie	Family	Life form	Summer	Summer	Winter	Winter
Rumex sp.(?)	Polygonaceae	Herb	–	+	–	–
*Salix humboldtiana**	Salicaceae	Herb	–	+	–	–
Salsola kali	Chenopodiaceae	Herb	–	+	–	–
*Scirpus americanus**	Cyperaceae	Herb	–	+	–	–
Taraxacum officinalis	Asteraceae	Herb	–	–	+	–
Trifolium repens	Fabaceae	Herb	+	–	–	–
Vicia hirsuta	Fabaceae	Herb	–	+	–	–
Vicia sp.	Fabaceae	Herb	–	–	+	–
Xanthium cavanillesii	Asteraceae	Herb	–	+	–	–
Xanthium spinosum	Asteraceae	Shrub	–	+	–	–
índice de diversidad			1.5	2.5	2.2	1.1

n and s indicate plant species for the north and south banks, respectively. (+) indicates presence, (–) indicates absence, (*) native taxa, (?) indicates that for the taxa, it was not possible to determine its biogeographic origin

Meanwhile, in the campaigns carried out in November 2007 and March 2008 the presence of the common rabbit (*Oryctolagus cuniculus*) was detected.

Toxicological Characterization of the Waters of the Wetland

The Mataquito river basin receives the discharge of domestic wastewater, contributions of public agricultural activities and/or wastes derived from industrial processes. Due to the anthropic intervention one could expect a marked deterioration in the quality of its waters, especially towards the terminal portion of the basin. Thus, determining the state of the water resources of the basin is fundamental to know the effect of pollutants on the persistence and health of organisms present in this ecosystem. In general, chemical determinations of water usually reveal the presence of compounds and elements toxic to biota. However, to know these effects adequately, it is necessary to perform toxicity bioassays. With this approach, it is possible to quantify and verify the interaction between the matrix to be tested (water and/or sediment) and the survival and/or reproduction of test organisms (Larrain 1995), and then, on the basis of the results obtained in these tests, generate predictions referring to the state of a natural community permanently exposed to the evaluated matrix.

In this section we present the results of the acute and chronic bioassay experiments performed in five occasions: June and November 2007, March, July and November 2008. In June 2007, insightful experiments were carried out. The Daphnia magna test organism was used for these tests and the recommendations for the development and evaluation proposed in the Chilean standard NCh 2083.Of 1999 were followed according to the conditions presented in Table 8.7.

Field work to evaluate the quality of the surface water of the wetland show that there is no acute toxic effect on *D. magna* in any period of the year. Thus, an LC_{50} value associated with these experiments was not recorded, in which the *D. magna* specimens were exposed by 24 and 48 h. The low toxicity of the surface water of the wetland is also evident when considering the results of the chronic trials (Table 8.8).

Table 8.7 Summary of the conditions of acute toxicity and criteria for acceptability of the bioassay performed with *D. magna*

Test type	Acute toxicity with no refill	Chronic toxicity with refill
Temperature	22° ± 1 °C	22° ± 1 °C
Photoperiod	16 h luz- 8 darkness	16 h luz- 8 darkness
Organism age	Neonates <24 h	Neonates <24 h
n° of replica	4	10
n° of organisms	5	1
n° of organisms per concentration	20	10
Test solution Airing	80% of saturation in the diluted water	80% saturation in the diluted water
Replacement of test solutions	n/a	Every 48 h
Type of packaging	Plastic	n/a
Solution Volume (ml)	20	50
Feeding regime	No feed	Every 48 h
Dilution Water (reconstituited) from	Hardness: 250 ± 25 mg/L	Hardness: 250 ± 25 mg/L de
	CaCO3; pH: 7,8 ± 0,2	CaCO3; pH: 7,8 ± 0,2
Duration of biotest	48 h	21 days
Average Effect	Mortality	Reproduction
Result Expresion	LC_{50}	NOEC y LOEC
Positive Control	Potassium dichromate	n/a
Criterion of acceptability	>80% of survival in controls	>80% of survival in controls

(n/a): does not apply, (LC_{50}): lethal concentration for 50% of the organisms exposed to treatment; (NOEC): in the experiment, effects of concentration are not observed, and (LOEC): lowest concentration at which the effect on survival and/or reproduction of the organism tested is observed

Table 8.8 Statistical data for the number of newborn counted per treatment and period after 21 days exposed. The control treatment refers to one done based on reconstituted water. The probability value obtained when contrasting control and superficial water treatments are presented. Numbers in parenthesis correspond to one standard deviation

	Control		Agua superficial		
Period	Average	Range	Average	Range	Value-p
November 2007	74.5 (10.23)	62–92	74.8 (15.94)	41–97	1.000
March 2008	57.7 (5.20)	49–65	63.7 (6.09)	51–70	0.285
July 2008	50.7 (6.05)	42–60	46.4 (2.91)	41–50	0.421
November 2008	57.4 (7.4)	44–69	54.6 (10.1)	40–67	0.992

In these experiments the neonatal count varied both in laboratory controls and in bioassays. However, in none of the studies were the differences in mean number of infants between the control and the experiment statistically significant (Table 8.8). This implies that the waters of the wetland had no sublethal effect on the test organisms after 21 days of exposure.

Discussion and Connections

The results presented here on the characterization of the waters of the wetland of the river Mataquito, indicate that in this sector is very frequent for the marine waters to enter through the bottom stratum. In periods when the marine influence is greater, specifically when the river has a lower flow, brackish waters are able to reach 5 kilometers upstream from the mouth. This situation is in contrast to the results derived from previous studies in the Mataquito river basin (EGA 2006, CREA 2007a, b, 2008a, b, c). These studies show that, between the locality of Licanteny and the Mataquito Bridge, the conductivity is homogeneously distributed and that the entire water column has conductivities attribu to continental waters (<5 mS/cm; EGA 2006, CREA 2007a, b, 2008a, b, c). This way, based on the hydrographic records presented here, it is possible to characterize spatial stratification of the waters of the Mataquito river wetland, as follows: a) in its first 3 km, the wetland zone would behave like a "salt wedge" estuary, where the salt wedge entering from the sea penetrates the stratum (B) within the next 2 km, the mixing processes associated with tides and wind transform the estuary into a "partially homogeneous" one, and (c) while in the subsequent 2 kilometers, wind action tends to further dilute the influence of the marine waters, giving way to a "vertically homogeneous" estuary, where the conductivity presents characteristic fluvial values (CREA 2007a, b).With regard to the quality of the surface water of the wetland, the evaluation of the physical and chemical parameters suggests that these resources are generally within the limits specified in NCh 1333 Of 78 for irrigated water and aquatic life. Moreover, considering the secondary quality limits proposed by CONAMA (2000), the wetland is dominated by quality water, between exceptional and good. The toxicological results obtained in the experiments with D. magna are consistent with the inferred quality for the surface waters of the wetland. This implies that water in the mouth sector has no toxic effect (acute and/or chronic). However, in concentrations closer to the mouth, there are some elements (e.g. mercury) that have been detected, and that exceed the quality limits imposed by the norm NCh 1333 Of 78. In addition, there have been some periods in which the environmental quality is lower. It was precisely during the fish mortality event observed in June 2007 that the surface waters presented "regular quality". In this period, there were increases in color levels, detergents, lignin, tannins and AOX due to the punctual shedding of riles derived from the partially treated cellulose bleaching processes. However, the June 2007 conditions constitute an isolated situation, where the values obtained for the chemical variables conform to the range of variation classified as normal for the basin in the studies developed later.

When comparing the results of the physical-chemical characterization presented here with those previously recorded in the Mataquito basin (DGA & Cade-Idepe Consultores 2004), the volume magnitudes obtained for conductivity, total suspended solids, DBO_5, sulfates, sulphides, nitrites and manganese are comparatively poor. The levels of aluminum, total chrome, copper and iron are lower. These differences are explained by the greater influence exerted by the drag by runoff in the upper sectors of the Mataquito river. As a matter of fact, it has been identified that this process is capable of increasing the levels of aluminum silicate compounds. Observations indicate that changes in soil conditions mediated by the melting of acid snows and/or slightly acidic rains mobilize the insoluble aluminum accumulated in the soils. This phenomenon increases its concentration in the natural waters of the river (DGA & Cade-Idepe Consultores 2004).

The comparison between the results of the physicochemical characterization for the period 2006–2008 and those published for other basins in southern central Chile (Table 8.9), reveals the peculiarities of the system studied. Unlike other basins, in the Mataquito wetland the concentrations of total solids, bicarbonates, sulfates, sulfides, chlorides and essential metals are modulated mainly by the presence of subsurface marine waters and the mixing and diffusion processes. The same happens for the variables of conductivity, hardness and alkalinity. However, the variations in the physicochemical variables in the wetland seem to be explained by the oscillations in the flow, which is modulated mainly by the precipitation patterns and, secondarily, by the melting of the snow. This modifies the process of runoff and trawling of material throughout this basin. Indeed, during periods of low flow (summer-autumn), when the influence of marine waters is greater in the wetland sector, all previous variables increase significantly. On the contrary, in winter and spring, when the flow is higher and the marine influence decreases, increases in the concentration of total suspended solids, organic compounds, aluminum and phosphorus are observed.

In general, fine sands predominate in the wetland sector, although there are seasonal variations in sediment granulometry. Thus, during the winter and spring, the middle size sand dominates in the sediments. It is precisely because of the higher flow rates in the Mataquito River during this period that a very energetic environment persists that by the action of trawling, prevents the depositing of the finer fractions (Friedman 1961). In contrast, towards the end of summer and early fall, the lower flows experienced by the basin, determine less energetic environments favoring the deposition of fine sediments (Duane 1964).

Table 8.9 Basins in central-south Chile with which results from the Mataquito river wetland characterization have been compared

Basin	Reference
River Maule	CREA (2009)
River Chillán	Debels et al. (2005)
River Quilque	Habit et al. (2005)
River Damas	Figueroa et al. (2003)
River Traiguén	Rivera et al. (2009)
River Cruces	Schaefer & Einax (2010)

With respect to the sediments that are developed in the Mataquito wetland sector, these are characterized by the predominance of neutral pH and Redox potentials associated to oxidative processes. However, in most of the studies carried out, at least one sector of the wetland is identified with reducing characteristics. Redox potential is a factor that determines the stability and biochemical transformation of organic matter, as well as the distribution, type and physiological activity of the microorganisms that inhabit the soil (Teasdale et al. 1998). Changes in this potential may eventually cause variations in the solubility and speciation of trace elements, causing alterations in flows from sediments to adjacent water (Kerner and Wallman 1992, Pardue & Patrick 1995). Hence, the predominance of values associated with oxidative environments is in line with the low values of MOT (<5%) recorded in wetland sediments and high levels of dissolved oxygen in the adjacent water column (>5 mg/L). In all of the studies, an inverse correlation was found between the particle size of the sediment and the concentrations of MOT, aluminum, copper, iron and zinc. The organic compounds evaluated in the sediments generally have low concentrations and do not show spatial or temporal tendencies that suggest differential accumulation. The concentrations of trace elements measured in the sediments of the wetland between 2006 and 2008 are within the natural ranges described for the terrestrial crust and other continental aquatic systems of our country (Table 8.10). When considering the toxicological limits "Threshold effects levels" (TEL) and "Probable effects levels" (PEL) established for sediments by FDEP (1994), practically all

Table 8.10 Levels for concentrations of trace elements measured in the Mataquito wetland, in the earth's crust and other aquatic systems in Chile. Units of elements expressed in μg/g

Element	Mataquito Wetland	Other Studies
Aluminium	8.975–30.381	69.300–83.200 Upper crust (Chester 1990)
Cadmium	<0.2–4.9	0.2–0.03 Upper crust (Chester 1990)
		0.2–0.5 Mapocho river(Segura et al. 2006) 0.31–1.41 Aysén river (Ahumada et al. 2007)
Copper	11.6–49.7	32–55 Upper crust (Chester 1990)
		16–24.900 Cachapoal river (Narváez et al. 2007) 30–6.789 Elqui river (Oyarzún et al. 2004) 209–2.850 Mapocho river (Segura et al. 2006) 37.3–67.0 Aysén river (Ahumada et al. 2007)
Lead	<0.2–55.0	12.5–16.0 Upper crust (Chester 1990)
		17–61 Mapocho river (Segura et al. 2006) 4.8–118 Cachapoal river (Narváez et al. 2007) 63.4–202.8 Urban lagoons in Concepción (González et al. 2009)
Zinc	31.1–189.8	70–127 Upper crust (Chester 1990)
		5.9–768 Cachapoal river (Narváez et al. 2007) 12.70–108.22 Aysén river (Ahumada et al. 2007) 577–1.290 Mapocho river (Segura et al. 2006)
Mercury	<0.06–0.080	0.056 Upper crust (Wedepohl 1991)
		0.00–5.300 Punitaqui stream (Higueras et al. 2004) 0.036–0.080 Chile Andean lakes (Urrutia et al. 2002)

determinations for trace elements are under the PEL index, with the exception of July Of 2008. During this period the levels of lead and zinc surpass the lower index TEL. Meanwhile, the concentrations of cadmium detected in the wetland and head of the Mataquito in July 2008 are above the categories.

TEL and PEL, respectively. The quality obtained for the sediments during this interval is related to the increase of the precipitations. Thus, the greater deposit of lead, zinc, mercury and cadmium would be explained by the concomitant increase in the Mataquito flow and its greater capacity to transport materials through the basin.

At Phyla level, the composition of the macrobenthic communities present in the wetland does not present seasonal variations among the different studies carried out, being dominated by annelids and crustaceans. The most representative classes include the most frequent groups in the soft bottoms of southern Chilean estuarine environments such as polychaetes (9 sp), amphipods (3 sp), bivalves (3 sp) and decapods (2 sp) (Bertrán et al. 2001, 2006, Donoso 1991, Jaramillo et al. 1995, 2008, Moscoso et al. 2006; Quijón et al. 1996; Richter 1985). The salinity gradient that develops in the wetland, characteristic of estuarine environment, determines the spatial variations in the distribution of macrobenthic taxa, as well as spatial patterns observed in the relative abundance and species richness of these associations. In fact, the higher incidence of ocean water penetration in the mouth sector determines the predominance of salinity tolerant species (e.g. bivalve mollusks). Meanwhile, towards the sector of the head of the wetland, where the marine influence is smaller, the representation and abundance of organisms characteristic of inland waters and intolerant to salinity increases (e.g. Oligochaetes). The highest density and specific richness recorded in the middle sector of the wetland would be related to the development of ecotonal communities given both by the transition between two environments with differential influences of the marine waters, and by the predominance of sediments constituted by fine to very thin sands. For other estuaries in the south central region of Chile, it has been reported that both salinity and sediment characteristics determine the dynamics and structuring of macrobenthic communities (Jaramillo et al. 1995, Quijón and Jaramillo 1993, Quijón et al. 1996).

Analysis of the community structure of zooplankton and zoobenthos indicate that both communities present temporal variations in the abundance and representativeness of their dominant groups. This is because the taxa that compose these communities are especially sensitive to fluctuations in environmental parameters and intensity/presence of biological interactions, such as competition and depredation (Jeffries 1967, Soetaert and Van Rijswijk 1993, Uriarte and Villate 2004). As for macrobenthic communities, the observed rate distribution patterns are related to the salinity gradient that exists within the estuary system. In the particular case of zoobenthos, the great abundance of vermiform forms registered in the wetland is due to the dominance of soft bottoms throughout the system.

The wide dominance of salinity tolerant species (>66%) within phytobenthic and phytoplankton associations is consistent with the estuarine characteristics of the wetland. In fact, the genera Chaetoceros and Leptocylindrus, both very abundant in phytoplankton during the study period, acquire importance in environments with strong marine influence. These species are considered typical of coastal environments and are abundant in the mouth (Wong and Townsend 1999). Although fluctuations in relative abundance are evident and specific richness, especially in the phytoplankton, in the wetland the preponderant species are maintained through the time. Thus, the temporal variations in these phytoplankton community descriptors will be given mainly by changes in the representation and density of the less dominant taxa. These fluctuations would be related to inherent processes of wetland dynamics, such as: (**a**) morphometric changes, (**b**) variations in temperature and salinity, (**c**) water circulation (tides, waves, flow), (**d**) variations in the chemical characteristics of the medium, (**e**) fluctuations in light intensity and (**f**) predation (Ketchum 1954, Wood 1967, Lucas et al. 1999, May et al. 2003, Tett et al. 2003, Ferreira et al. 2005).

The fish communities of the wetland show low species richness (n = 9). This is not a peculiarity for the system, since the freshwater aquaculture of Chile is little diversified due to particular evolutionary and biogeographical processes (Campos et al. 1993a, b, Vila et al. 1999, Dyer 2000). However, considering that the ichthyofauna of the wetland develops in a typical wetland environment (fine substrates, low flow velocities and moderate summer temperatures), species richness would be expected in comparison with the upstream sector (Campos et al. 1993a, b, Vila et al. 1999, Dyer 2000).

Spatial variations in the distribution of species observed in the wetland and upstream sectors are related to the influence of marine waters. In fact, as a result of the salinity gradient that develops through the estuaries, freshwater fish expand their range of distribution towards the upper and middle sectors of the estuary, while marine fish, mainly in their juvenile stages, follow the masses of Water entering the river basins (Campos and Moreno 1985). Occasional records of *M. cephalus* in the upstream sectors of the Mataquito wetland are related to the latter strategy. Due to its great osmo-regulatory capacity (Riede 2004), this species penetrates into inland watercourses to complete its life cycle (Figueroa et al. 2009), and can reach up to 50 kilometers upstream (Habit 1988).

As for the seasonal variations in absolute abundances, the fish do not show significant changes, except for *M. cephalus*. This species tends to exhibit higher densities during the summer. The observed changes in absolute abundances at spatial scale for *C. carpio* and *B. australis* obey the gradient in salinity established in estuarine environments. In fact, the maximum densities for these freshwater taxa occur in sectors with lower salinity. In the case of *C. carpio* the major absolute abundance is concentrated upstream of the wetland throughout the year. In contrast, the spatial location of the maximum densities for *B. australis* oscillates according to the seasonal variations in the influence of the marine waters through the Mataquito. Thus, the greater influence of marine waters during periods of low flow (summer-autumn) determines that Pejerrey's maximum densities are concentrated in the upstream sector. On the contrary, the lower influence of the marine waters during the winter allows the maximum abundances for this species to occur around the wetland. The strong predominance of introduced species (>76%) within the terrestrial flora spectrum of the Mataquito wetland indicates a high degree of anthropic intervention in the area. These results are consistent with the findings of previous research conducted upstream (EGA 2006) and other sectors of central Chile (San Martín et al. 2001). In fact, the important proportion of introduced species that characterizes riparian zones has been interpreted as the result of anthropic activities since colonial times, specifically those related to water use, agriculture, and forest plantations. Seasonal and spatial variations in community descriptors of species richness and diversity of vegetation communities are mainly due to changes in the representativeness and relative abundance of herbaceous species. These forms of life present annual cycles, manifesting themselves and increasing in coverage only during the rainy season on the north bank (winter). However, in the case of the southern shore, the decrease in diversity and representation of the observed rate in the winter may be due to the instability that the substrate acquires during the rainy season (San Martín et al. 1992).

The faunal spectrum obtained for the wetland contrasts with the results obtained in previous studies in the wetland sector (EGA 2006, Estades et al. 2007, Parada 2007, CREA 2007a, b, 2008a, b, c), which have emphasized In the importance of the wetland as a site of biological conservation, for its great diversity and wealth of birds. However, it is likely that during the monitoring periods some species that are highly mobile will not be detected and have been recorded in the upstream areas of the wetland. Thus, it is not ruled out that occasionally, towards the sector of the mouth, are seen species of *Abrothrix olivaceus*, the Chilean frog (*Caudiverbera caudiverbera*), the four-toed toad (*Pleuroderma thaul*), *Bufo chilensis* (arunco), the tenuous lizard (*Liolaemus tenuis*) and the Chilean lizard (*Liolaemus chilensis*).

The results show that both the biotic and abiotic components of the Mataquito wetland experience variations at intra and interannual scales. However, generally the greatest abundance and species richness for the different groups studied occurs between 1.5 and 3 km upstream of the mouth of the river. Indeed, in this sector there is a dilution of the salinity product of the mixture between continental and marine waters, which allows the development of ecotonal communities. Thus, the main factors that explain the dynamics of the system involve variations in the hydrological regime and the quality of its waters, both determined by seasonal fluctuations in the degree of penetration of the ocean waters and the volumes of the maximum and minimum flows. The latter, controlled in turn by the regional climate, mainly precipitation and thawing.

The performance and inferred dynamics for the Mataquito wetland is consistent with the conceptual model for the functioning of the aquatic systems proposed by Lorenz et al. (1997) and Roni et al. (2002). In this study, one recognizes a series of drivers that interact with each other and that modify the different environmental processes. Such changes in turn modify habitat variability. Finally, the effects are transmitted to the biotic component of the system. Figure. 8.15 shows this conceptual model adapted for the Mataquito wetland. The conceptualization of wetland dynamics implies that for this system it is possible to generate specific predictions about the biological consequences of changes in controllers. Basically, for the wetland the most relevant controllers include climate, physiography, vegetation/soil and anthropic processes. These last ones constitute the preponderant controller, since they have the potential to modify the other natural controllers, besides the processes that occur in the system.

The applicability of the previous model in the wetland is subject to the consideration of the most important features that ignore our research on the functioning and dynamics of the ecosystem. The first is the high degree of resilience that the wetland communities present to natural and/or anthropogenic disturbances. While the second corresponds to the rapid response times that the ecosystem exhibits in the face of changes in its controllers and the concomitant modification of the processes that govern its operation.

The characteristics influence that underlie the dynamics of the wetland is evident when considering the previous environmental conditions, during and after the event of contamination by industrial riles of June of 2007. Precisely, after this event the ecosystem was able to persist through time under very similar environmental characteristics prior to June 2007, even though there were later fluctuations in the community descriptors. Thus, these antecedents would indicate that the ecosystem of the coastal wetland of the river Mataquito has been adjusted to the action of the diverse controllers identified in this study. Specifically, those that modify the balance and mixing of continental and oceanic waters entering through the bottom stratum. Hence, on the basis of the conceptualization presented here, it is possible to hypothesize about the effects that can generate other disturbances in the system. For example, field observations indicate that the tsunami that affected the coast of central Chile on February 27, 2010 altered the geomorphology of the Mataquito basin due to the disappearance of the bar in the terminal sector of the river.

This would have led to a greater entrance of oceanic waters to the interior, immediately modifying the physical and chemical factors. This, in turn, would have altered patterns of species distribution, as well as specific abundances and wealth of them. However, the high resilience and rapid response times that characterize the communities present in the wetland would determine that there are now clear signs of ecosystem recovery.

Acknowledgments The authors thank ARAUCO S. A. for the funding given for the development of the various studies carried out in the Mataquito river wetland.

References

Ahumada, R., E. González, V. Medina, and A. Rudolph. 2007. Analysis of mayor zinc sources in marine sediments of the estuarine zone in Aysén fiord, Chile. *Investigaciones Marinas* 35 (2): 5–14.

Asplund, G., and A. Grimvall. 1991. Organohalogens in nature. *Environmental Science & Technology* 25: 1346–1350.

Bertrán, C., J. Arenas, and O. Parra. 2001. Macrofauna del curso inferior y estuario del río Biobío (Chile): cambios asociados a la variabilidad estacional del caudal hídrico. *Revista Chilena De Historia Natural* 74: 331–340.

Bertrán, C., L. Vargas, F. Peña, s. Mulsow, J. Tapia, e. Hauenstein, R. Schaletter, and A. Bravo. 2006. Macrofauna bentónica de los humedales de tres lagos salinos en el borde costero del sur de Chile. *Ciencias Marinas* 32 (3): 589–596.

Campos, H., and C. Moreno. 1985. Asociaciones de peces en estuarios chilenos, Pacífico Sur Americano. In *Fish community ecology in estuaries and coastal lagoons: towards an ecosystem integration (En. L. A. Yáñez-Arancibia)*, 407–414. México: UNAM Press.

Campos, H., F. Alay, V. Ruiz, and J. Gavilán. 1993a. Antecedentes biológicos de la fauna íctica presente en la hoya hidrográfica del río Biobío. In *Evaluación de la calidad del agua y ecología del sistema limnético y fluvial del río Biobío*, ed. F. Faranda and O. Parra. Concepción: Serie Monografías Científicas Eula.

Campos, H., V. Ruiz, J. Gavilán, and F. Alay. 1993b. *Peces del río Biobío*. Serie Publicaciones de divulgación. Centro Eula. Chile: Universidad de Concepción.

Cherif, S., R.B. Fradj, and A. Jrad. 2006. Quality of treated wastewater: method validation of AOX. *Accreditation and Quality Assurance* 11: 632–637.

Chester, R. 1990. Trace elements in the oceans. In *Marine geochemistry*, ed. R. Chester, 346–421. New York: Chapman & Hall.

CONAMA. 2000. *Diagnóstico y propuesta de manejo de los humedales de la región del Maule*. Santiago: Chile Ambiente.

CREA. 2007a. *Caracterización ambiental porción terminal del río Mataquito, Región del Maule. Planta Celulosa Licancel. Julio 2007*, 133. Concepción: Centro Regional de Estudios Ambientales. Universidad Católica de la Santísima Concepción.

———. 2007b. *Monitoreo integral de la sección del río Mataquito, Región del Maule. Planta Celulosa Licancel. Noviembre 2007*, 192. Concepción: Centro Regional de Estudios Ambientales. Universidad Católica de la Santísima Concepción.

———. 2008a. *Monitoreo integral de la sección del río Mataquito, Región del Maule. Planta Celulosa Licancel. Marzo 2008*, 231. Concepción: Centro Regional de Estudios Ambientales. Universidad Católica de la Santísima Concepción.

———. 2008b. *Monitoreo integral de la sección del río Mataquito, Región del Maule. Planta Celulosa Licancel. Julio 2008*, 237. Concepción: Centro Regional de Estudios Ambientales. Universidad Católica de la Santísima Concepción.

———. 2008c. *Monitoreo integral de la sección del río Mataquito, Región del Maule. Planta Celulosa Licancel. Noviembre 2008*, 245. Concepción: Centro Regional de Estudios Ambientales. Universidad Católica de la Santísima Concepción.

———. 2009. *Informe Final: "Estudio de la calidad de las aguas, sedimentos y comunidades bentónicas estuarinas del río Maule"*, 85. Concepción: Centro Regional de Estudios Ambientales. Universidad Católica de la Santísima Concepción.

Debels, P., R. Figueroa, R. Urrutia, R. Barra, and X. Niell. 2005. Evaluation of water quality in the Chillán river (central Chile) using physicochemical parameters and a modified water quality index. *Environmental Monitoring and Assessment* 110: 301–322.

Dga & Cade-Idepe Consultores. 2003. *Diagnóstico y clasificación de los cursos de agua según objetivos de calidad. Informe Final Volumen I. Dirección General de Aguas*. Gobierno de Chile: Ministerio de Obras Públicas.

———. 2004. *Diagnóstico y clasificación de los cursos de agua según objetivos de calidad. Cuenca del río Mataquito. Dirección General de Aguas*, 106. Gobierno de Chile: Ministerio de Obras Públicas.

Donoso, E. 1991. Distribución y estructura comunitaria de la macroinfauna intermareal en dos estuarios del sur de Chile. Tesis para optar al título de Profesor de Biología, Química y Ciencias Naturales. Universidad Austral de Chile. 33. + anexos.

Duane, D.B. 1964. Significance of skewness in recent sediments, western Pamlico Sound, North Carolina. *Journal of Sedimentary Petrology* 34: 864–874.

Dyer, B. 2000. Systematic review and biogeography of the freshwater fishes of Chile. *Estudios Oceanológicos (Chile)* 19: 77–98.

EGA. 2006. *Informe Final. Línea Base Ambiental río Mataquito. Planta Licancel*, 255. Santiago: Celulosa Arauco y Constitución S. A.

EPA. 2006. *National Recommended Water Quality Criteria*. Washington, D.C.: Office of Water, Office of Sciense and Technology.

Estades, C., M.A. Vukasovic, and R. Thomson. 2007. Evaluación de la avifauna del estuario del río Mataquito los días 09, 18 y 25 de Junio del 2007. In Caracterización ambiental porción terminal del río Mataquito, Región del Maule. Planta Celulosa Licancel. Julio 2007, eds. D. Arcos., J. Ortiz, A. Silva and P. Torres, 133. Centro Regionalde Estudios Ambientales. Universidad Católica de la Santísima Concepción.

FDEP. 1994. Approach to assessment of sediment quality in Florida coastal waters, Development and evaluation of sediment quality assessment guidelines Vol. I, macdonald Environmental Sciences Ltd., Ladysmith, British Columbia (1994). Prepared for Florida Department of Environmental Protection, Office of Water Policy, Tallahasee, Florida.

Ferreira, J.G., W.J. Wolff, T.C. Simas, and S.B. Bricker. 2005. Does biodiversity of estuarinephytoplanktondependonhydrology? *Ecological Modelling* 187: 513–523.

Figueroa, R., C. Valdovinos, E. Araya, and O. Parra. 2003. Macroinvertebrados bentónicos como indicadores de calidad de agua de ríos del sur de Chile. *Revista Chilena De Historia Natural* 76: 275–285.

Figueroa, R., M.L. suárez, A. Andreu, V.H. Ruiz, and M.R. Vidal. 2009. Caracterización ecológicadehumedalesdelazonasemi-áridaen Chilecentral. *Gayana* 73 (1): 76–94.

Friedman, G. 1961. Distinction between dune, beach, and sands from their textural characteristics. *Journal of Sedimentary Petrology* 31 (4): 514–529.

Gajardo, R. 1994. *La vegetación natural de Chile*, 165. Santiago: Clasificación y distribución geográfica.

Gibbs, R.J. 1970. Mechanismcontrollingworldwaterchemistry. *Science* 170: 1088–1090.

González, E., M. Retamal, V. Medina, and R. Ahumada. 2009. Enriquecimiento, disponibilidad y contaminación de metales traza (Cd, Cu, Pb y Zn) en sedimentos de lagunas urbanas de Concepción-Chile. *Quimica Nova* 32 (4): 902–907.

Habit, E. 1988. Peces. In *Estudiodelínea basepara la evaluacióndel impactoambiental del complejo forestal industrial Itata*, ed. O. Parra and E. Habit, 79–86. Concepción: Ediciones Centro Eula-Chile, Universidad de Concepción.

Habit, E., O. Parra, and C. Valdovinos. 2005. Ictiofauna de un sistema fluvial receptor de aguas servidas: respuestas a una nueva planta de tratamiento (Río Quilque, Chile Central). *Gayana* 69 (1): 94–103.

Higueras, P., R. Oyarzún, J. Oyarzún, H. Maturana, J. Lillo, and D. Morata. 2004. Environmental assessment of copper–gold–mercury mining in the Andacollo and Punitaqui districts, northern Chile. *Applied Geochemistry* 19: 1855–1864.

INE. 2002. *Censo 2002, Vol I: Resultados Población, País – Región*, 356. Chile: Instituto Nacional de Estadísticas.

Jaramillo, E., A. Mclachlan, and J. Dugan. 1995. Total simple area and estimates of species richness in exposed sandy beaches. *Marine Ecology Progress Series* 119: 311–314.

Jaramillo, M., P. Muñoz, V. Delgado, and C. Bertrán. 2008. Spatio-temporal distribution of spionids (Polychaeta-Spiondae) in an estuarine system in south central Chile. *Revista Chilena De Historia Natural* 81: 501–514.

Jeffries, H.P. 1967. Saturation of estuarine zooplankton by cogeneric associates. In *Estuaries*, In: Publications of the American Associations for the Advancement of Science, ed. G.H. Lauff, vol. 83, 500–508.

Kaczmarczyk, A., and E. Niemirycz. 2005. Adsorbable organic halogens (AOX) in Polish Rivers – levels and changes. *Acta Hydrochimica et Hydrobiologica* 33 (4): 324–336.

Kerner, M., and K. Wallman. 1992. Remobilization events involving Cd and Zn from intertidal mud flat sediments in the Elbe Estuary during the tidal cycle. *Estuarine, Coastal and Shelf Science* 35: 371–393.

Ketchum, B. 1954. Relation between circulation and planktonic populations in estuaries. *Ecology* 35 (2): 191–200.

Larrain, A. 1995. Criterios ecotoxicológicos para evaluar alteraciones ambientales y establecer parámetros de control: importancia de los bioensayos de toxicidad. Cienc. Tec. Mar, CONA (N° especial):39–47.

Lorenz, C.M., G.M. van Dijk, A.G.M. van Hattum, and W.P. Cofino. 1997. Concepts in river ecology: implications for indicator development. *Regulated Rivers: Research & Management* 13: 501–516.

Lucas, L.V., J.R. Koseff, G. Monismith, J.E. Cloern, and J.K. Thompson. 1999. Processes governing phytoplankton blooms in estuaries II: The role of horizontal transport. *Marine Ecology Progress Series* 187: 17–30.

Luebert, F., and P. Pliscoff. 2004. *Clasificación de pisos de vegetación y análisis de representatividad ecológica para áreas propuestas para protecciónde la ecorregión*, 174. Valdivia: Serie de Publicaciones WWF programa Ecorregión Valdiviana.

May, C.l., J.R. Koseff, L.V. Lucas, J.E. Cloern and Schoellhamer D.H.. 2003. Effects of spatial and temporal variability of turbidity on phytoplankton blooms. *Marine Ecology Progress Series* 254: 111–128.

Mclachlan, A., and A.C. Brown. 2006. *The ecology of sandy shores*. 2nd ed, 373. New York: Academic Press.

Moscoso, J., A. Rudolph, R.D. Sepúlveda, and C. Suárez. 2006. Effect of temporary closure of the mouht of an estuary on the benthic macrofauna: Lenga – Chile, a case study. *Bulletin of Environmental Contamination and Toxicology* 77: 484–491.

Müller, G. 1981. The Heavy Metal Pollution of the Sediments of Neckars and its Tributary. *A Stocktaking Chemische Zeit* 150: 157–164.

Narendra, K., P.K. Omana, and M. Mohan. 2010. Water and sediment quality of Ashtamudi estuary, a Ramsar site, southwest coast of India – a statistical appraisal. *Environmental Monitoring and Assessment* 165: 307–319.

Narváez, J., P. Richter, and I. Toral. 2007. Preliminary physical and chemical characterization of river waters and sediments affected by copper mining activity in central Chile. Application of multivariate analysis. *Journal of the Chilean Chemical Society* 52 (3): 1261–1265.

Noma, Y., S. Yamane, and A. Kida. 2001. Adsorbable organic halides (AOX), AOX formation potential, and PCDD's/DF's in landfill leachate and their removal in water treatment processes. *Journal of Material Cycles and Waste Management* 3: 126–134.

Olivares, P. 2007. Diseño e implementación de una red de humedales protegidos en la región del Maule, zona mediterránea de Chile. In *En: Gestión sostenible de humedales*, eds. M. Castro and L. Fernández. Santiago: CYTED & Programa internacional de interculturalidad.

Oyarzún, R., J. Lillo, P. Higueras, J. Oyarzún, and H. Maturana. 2004. Strong arsenic enrichment in sediments from the Elqui watershed, northern Chile: industrial gold mining at El Indio–Tambo district versus geologic processes. *Journal of Geochemical Exploration* 84: 53–64.

Parada, M. 2007. *Recuento de flamencos en la desembocadura del río Mataquito (12 de Junio al 13 de Agosto del 2007)*. Informes Flamencos Mataquito. Bio-Rnat. Ltda. 11 pp.

Pardue, J.H., and W.H. Patrick Jr. 1995. Changes in metal speciation following alteration of sediment redox status. In *Metal contaminated aquatic sediments*, ed. Allen E. Herbert, 169–185. Chelsea: Ann Arbor Press.

Pee, K.H. 1996. Biosynthesis of halogenated metabolites by bacteria. *Annual Reviews in Microbiology* 50: 375–395.

Quijón, P., and E. Jaramillo. 1993. Temporal variability in the intertidal macroinfauna in the Queule river estuary, south central Chile. *Estuarine Coastal Shelf Science* 37: 655–667.

Quijón, P., E. Jaramillo, and M. Pino. 1996. Macroinfaunal assemblages associated with mussel and clam beds in an estuary of southern Chile. *Estuaries* 19 (1): 62–74.

Richter, W. 1985. Distribution of the soft-bottom macrofauna in an estuary of southern Chile. *Marine Biology* 86: 96–100.

Riede, K. 2004. *Global register of migratory species from global to regional scales. Final Report of the R & D Projekt. 808.05.081*, 329. Bonn: Federal Agency for Nature Conservation.

Rivera, N., F. Encina, R. Palma, and P. Mejías. 2009. La calidad de las aguas en el curso superior y medio del río Traiguén. IX Región-Chile. *Información Tecnológica* 20 (4): 75–84.

Roni, P., T. Beechie, R. Bilby, F. Leonetti, M. Pollock, and G. Pess. 2002. A review of stream restoration techniques and a hierarchical strategy for prioritizing restoration in Pacific Northwest watersheds. *North American Journal of Fisheries Management* 22: 1–20.

Schaefer, K., and J.W. Einax. 2010. Analytical and chemometric characterization of the Cruces River in South Chile. *Environmental Science and Pollution Research* 17 (1): 115–123.

San Martín, J., C. Ramírez, and C. San Martín. 1992. La flora de las dunas chilenas y sus adaptaciones morfológicas. *Bosque* 13: 29–39.

San Martín, C., C. Ramírez, J. San Martín, and R. Villaseñor. 2001. Flora y vegetación del estero Reñaca (V Región, Chile). *Gayana Botanica* 58: 31–46.

Segura, R., V. Arancibia, M.C. Zúñiga, and P. Pasten. 2006. Distribution of copper, zinc, lead and cadmium concentrations in stream sediments from the Mapocho River in Santiago, Chile. *Journal of Geochemical Exploration* 91: 71–80.

Soetaert, K., and P. van Rijswijk. 1993. Spatial and temporal patterns of the zooplankton in the Westerschelde estuary. *Marine Ecology Progress Series* 97: 47–59.

Teasdale, P.R., L. Allen, S.C. Apte, G.E. Batley, and G. Birch. 1998. In situ collection of diagenetic and induced oxyhydroxide precipitates from riverine and estuarine sediment. *Environmental Technology* 19: 1191–1201.

Tett, P., L. Gilpin, H. Svendsen, C.P. Erlandsson, U. Larsson, S. Kratzer, E. Fouilland, C. Janzen, J. Lee, C. Grenz, A. Newton, J. Gomes Ferreira, T. Fernandes, and S. Scory. 2003. Eutrophication and some European waters of restricted exchange. *Continental Shelf Research* 23: 1635–1167.

Uriarte, I., and F. Villate. 2004. Effects of pollution on zooplankton abundance and distribution in two estuaries of the Basque coast (Bay Biscay). *Marine Pollution Bulletin* 49: 220–228.

Urrutia, R., M. Yévenes, and R. Barra. 2002. Determinación de los niveles basales de metales traza en sedimentos de tres lagos andinos de Chile: Lagos Chungará, Laja y Castor. *Boletín de la Sociedad Chilena de Química* 47 (4): 457–467. doi:10.4067/S0366-16442002000400017.

Vila, I., L.S. Fuentes, and M. Saavedra. 1999. Ictiofauna en los sistemas límnicos de la Isla Grande, Tierra del Fuego, Chile. *Revista Chilena De Historia Natural* 72: 273–284.

Wedepohl, K. 1991. The composition of the upper earth's crust and the natural cycles of select metals. In *Metals and their compounds in the environment. Part I*, ed. E. Merian, 3–17. Weinheim: VCH.

Wong, M., and D. Townsend. 1999. Phytoplankton and hydrography of the Kennebec estuary, Maine, USA. *Marine Ecology-Progress Series* 178: 133–144.

Wood, E.J.F. 1967. *Microbiology of oceans and estuaries*, Elsevier Oceanography Series, 319. Amsterdam: Elsevier.

Zhang, Y., F. Guo, W. Meng, and X.Q. Wang. 2008. Water quality assessment and source identification of Daliao River basin using multivariate statistical methods. *Environmental Monitoring and Assessment* 152: 105–121.

Zhou, F., Y. Liu, and H.C. Guo. 2007. Application of multivariate statistical methods to the water quality assessment of the watercourse in the northwestern New Territories, Hong Kong. *Environmental Monitoring and Assessment* 132 (1–3): 1–13.

Chapter 9
Effects of the 2010 Tsunami on the Mataquito River Wetland

Pablo González, José Ortiz, Rodolfo Jerez, Marcelo Pavez, and Dagoberto Arcos

Abstract This chapter assesses the effects and subsequent changes of the Mataquito River estuary following the tsunami that struck the central Chilean coast on February 27th, 2010. The results indicate that after the tsunami the estuary established an extensive connection with the ocean, which in turn triggered significant changes in the physicochemical and biological patterns of the wetland. However, after 4 months, it was possible to observe clear signs of recovery in the estuary, gradually approaching the pre-tsunami state. Thus, these results indicate that the Mataquito estuary is capable of responding quickly to natural disturbances that modify its hydrological balance based on the interaction of continental and oceanic waters.

Keywords Coastal wetland • Central Chile • Estuary • Tsunami

Introduction

This chapter presents a hydrological, sediment and biotic characterization of the Mataquito River estuary (Fig. 9.1a) for the period after the February 2010 tsunami (March–October 2010), in order to establish the response of this system in the face of a natural disturbance/disaster. Based on a conceptual model generated from the results of our research presented in Chap. 8 (Fig. 9.1b), we intend to evaluate the degree of adjustment that this ecosystem exhibits to changes in its controllers.

P. González • J. Ortiz • R. Jerez • M. Pavez • D. Arcos (✉)
Regional Center for Environmental Studies (CREA), Universidad Católica de la Santísima Concepción, Talcahuano, Chile
e-mail: darcos@creaucsc.cl

J.M. Fariña, A. Camaño (eds.), *The Ecology and Natural History of Chilean Saltmarshes*, DOI 10.1007/978-3-319-63877-5_9

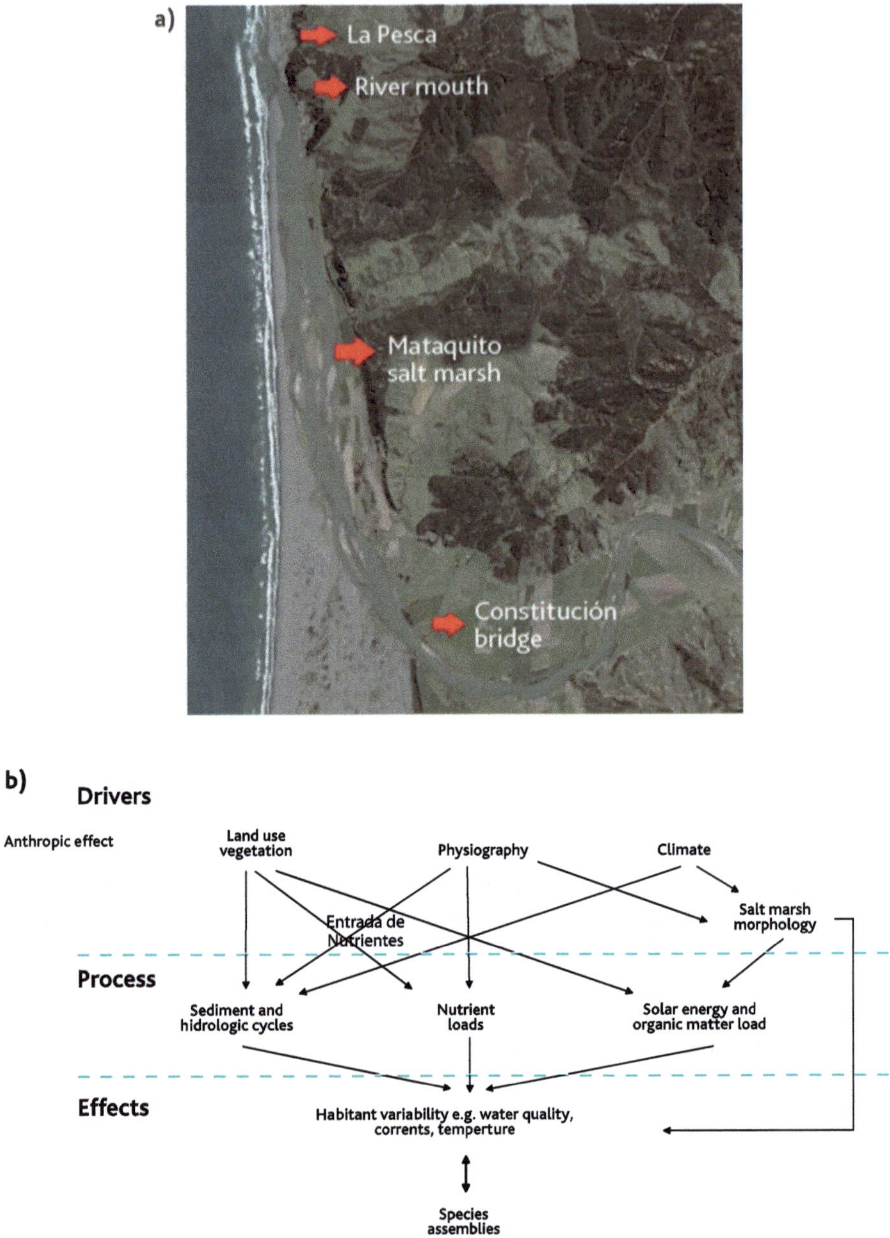

Fig. 9.1 (**a**) Coastal wetland in Mataquito River. (**b**) Conceptual model for the dynamic of the Mataquito estuary, where the different controllers are identified and their interactions are shown, together with the processes and effects that the cascade of interactions determines (modified from Roni et al. 2002)

General Background

Estuary of the River Mataquito

In the sector the river Mataquito mouth, specifically in La Pesca sector, an extensive, low depth estuary is formed, whose connection with the sea is limited by a bar of continental sands that develops towards the west (Fig. 9.1a). The sandbar determines the existence of a semi-enclosed interior lagoon, which has a restricted connection with the ocean. Sand that make up this bar have been transported to the river mouth by the runoff and have been deposited along the coast by the force of waves.

The estuary is strongly affected by the flood tides, especially during the summer and autumn, when the Mataquito River flow decreases. Hydrographic records indicate that the tidal prism can be extend up to 7 km upstream from the mouth (Fig. 9.2). However, the mixing processes that occur through the estuary determine the existence of a spatial gradient in the conductivity of its waters. Therefore, in its first 3 km, the system behaves like a "salt wedge" estuary, where the water entering from the sea penetrates the bottom stratum. Approximately 5 km upstream, the mixing processes associated with tides and wind determine a "partially homogeneous" estuary. While 7 km upstream, wind action tends to further dilute the influence of marine waters, giving way to a "vertically homogeneous" estuary.

The Mataquito River estuary has been characterized as a highly productive environment of low biological diversity, with a marked salinity gradient that occurs through the estuary, which determines significant spatial variations in the patterns of distribution, diversity and richness of flora and fauna species (see previous chapter). For example, the greatest abundance and specific richness of the planktonic and benthic communities has reached between 1.5 and 3 km upstream from the mouth. Indeed, in this sector there is a dilution of salinity, resulting from the mixture between inland and marine waters, allowing for the development of ecotonal communities. On the other hand, the greater penetration of ocean waters in the sector of the mouth determines the predominance of species of flora and fauna tolerant to the salinity, whereas towards the superior sector of the estuary, where the marine influence is not

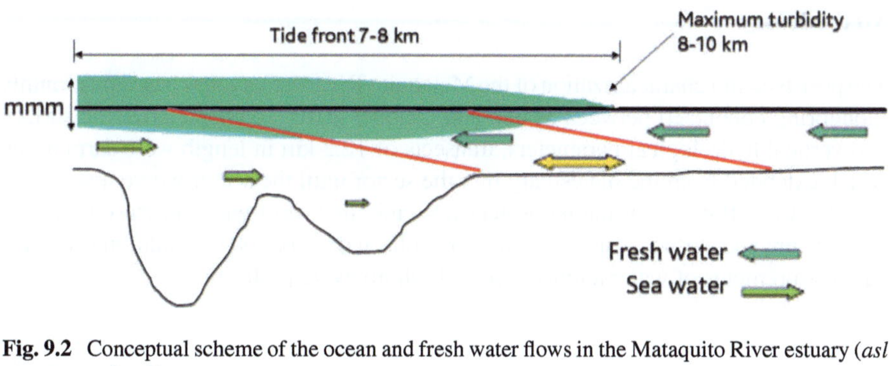

Fig. 9.2 Conceptual scheme of the ocean and fresh water flows in the Mataquito River estuary (*asl* average sea level)

as strong, the representation and abundance of organisms characteristic of inland waters and intolerant to salinity increases. Thus, the distribution and abundance of the flora and fauna species that inhabit the mouth are determined by seasonal fluctuations and the degree of penetration of the ocean waters and the volumes of the maximum and minimum flows of the river (controlled in turn by the regional climate, mainly precipitation and thawing, Fig. 9.1b).

Geomorphological Changes Associated with Tsunami 2010

According to Krishnakutty (2006), tsunamis constitute physical disturbances fundamental to the structuring of marine communities, as they alter biotic communities, quality/heterogeneity of the physical environment and distribution of nutrients. In Chile, tsunamis are recurrent along the tectonically active coast, with about 100 events of various magnitudes occurring during the period between 1550 and 2000 (Lagos-lópez 2000). Therefore, these events should be understood as agents of change inherent to natural variability, which determine the long-term structuring/dynamics of the coastal communities of our country.

Observations made in the estuary indicate that after February 27, 2010, significant changes occurred in the physical environment, resulting from profound modifications in the geomorphological characteristics of the coast and the Mataquito River basin. The large waves that struck the coast, destroyed 5 km of the extensive dune that separated the estuary from the ocean, which caused the dispersion of sandy sediments through the estuary (Fig. 9.3a, b). At the same time, the destruction of the dune, in turn, modified the course of the river, generating two mouths for which the new connections were created with ocean waters (Fig. 9.3b). The first of these connections corresponded to a modification of the old mouth located in the La Pesca sector, which was displaced towards the south for about 700 m, due to the uprising that suffered the beach and subsequent accumulation of sediments. For its part, the second connection corresponds to a prominent and mighty channel that was established about 3.5 km south of the La Pesca sector.

Methodology

The post-tsunami characterization of the Mataquito River estuary was based on monthly monitoring conducted between March and October 2010. To characterize the surface and vertical hydrological parameters, transects of 12.2 km in length were carried out, which extended from the old estuary into the sector until the sector where it was verified that the influences of marine waters were minimal (upstream). In these transects, temperature and surface water pH data were collected every 500 m. Additionally, vertical measurements of water temperature and salinity were performed.

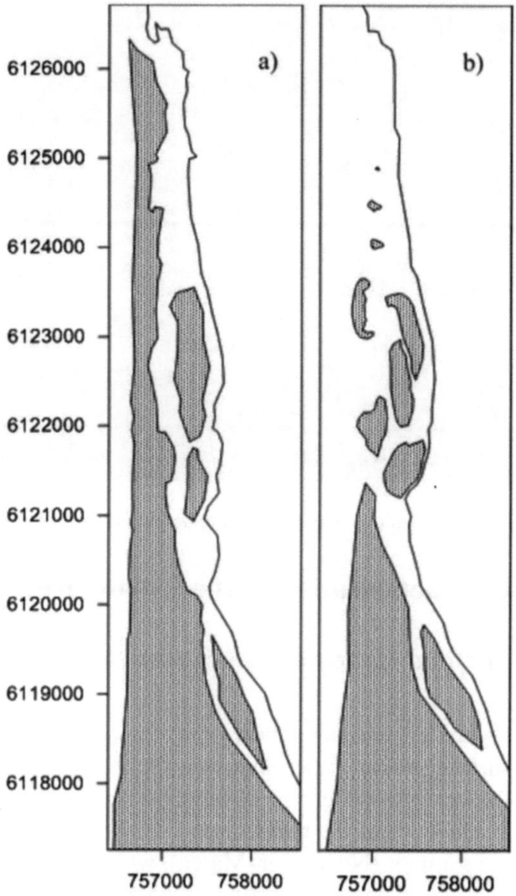

Fig. 9.3 Geomorphology of the terminal portion of Mataquito River in: (**a**) September 2009 (pre-tsunami) and (**b**) March 2010 (immediately after the Tsunami)

In order to characterize the biotic component during the post-tsunami period, monthly samples were taken for plankton (phytoplankton and zooplankton) and benthic macro-organisms in the four biological stations, most representative of the terminal portion (Fig. 9.4). In the biological stations located in the old and new mouths (E1 and E4, respectively, Fig. 9.4), sediment samples were collected, on which granulometric analysis, total organic matter (MOT) content, pH and Redox Potential were performed.

To determine the magnitude of the changes experienced by the tsunami estuary, we compared our results with those of previous studies conducted in the estuary by the Regional Environmental Studies Center (CREA-UCSC) between 2007 and 2008 (CREA-UCSC 2007a, b, 2008a, b, c, 2009).

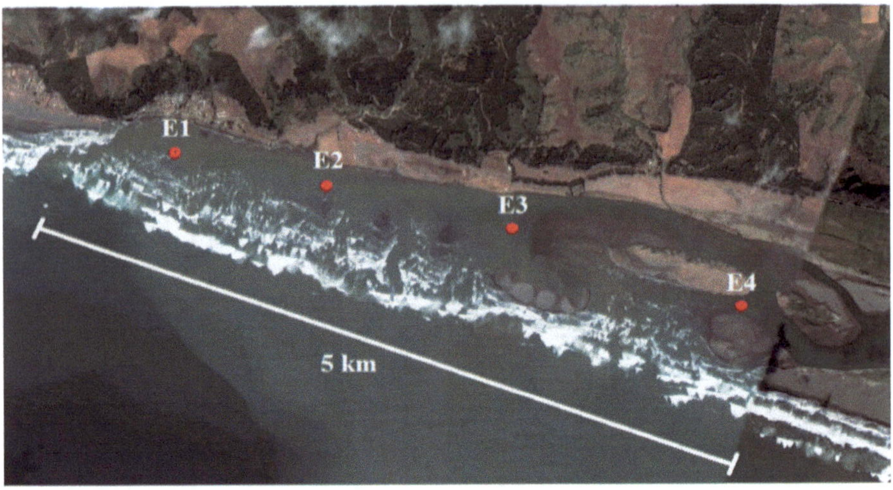

Fig. 9.4 Stations where plankton, benthic microorganisms and sediments were collected

Characterization of the Estructure

Surface Hydrological Parameters

Figure 9.5 presents a horizontal characterization of the temperatures and pH of the surface waters of the estuary during the post-tsunami period and the historical values for these physicochemical variables between 2007 and 2008. Throughout the series of available time, the temperatures showed a marked seasonal variation. In fact, during the summer (November–March), when there were low flows in the Mataquito River and the influence of marine waters is greater, surface water temperatures increased, reaching an average value, of the whole series, of 19.1 °C, with a maximum value of 23.3 °C recorded in March 2008 (Fig. 9.5a). Meanwhile, in the winter period, surface temperatures tend to decrease (Fig. 9.5a, average of the series = 12.5 °C, minimum value = 9.7 °C in April 2010) due to the lower influence of ocean waters. In the immediate post-tsunami campaign (March 2010), a surface temperature (18.1 ± 2.7 °C) was found to be slightly lower than the historic summer average. This is particularly evident in the sectors closest to the mouth (Stations E1 and E2, Fig. 9.5a), whose temperatures during March 2010 were around 13 °C. Compared to the standard deviations obtained for the average temperature of the other five post-tsunami months, the one obtained for the month of March 2010 is significantly higher. This is due to the significant spatial differences that occurred during this campaign between the surface temperatures of the waters near the mouth and the sector near the Mataquito Bridge.

In contrast, the pH of the surface waters shows no significant spatial or temporal variations (Fig. 9.5b), presenting an average of 7.7 (± 0.23). Exceptionally, in March 2010, the lowest pH value (5.2) is recorded in the portion of the estuary closest to the river mouth (Station E1, Fig. 9.5b).

In order to determine the evolution of the estuary after the tsunami, the averages of historical temperature and pH data were compared with the averages obtained in

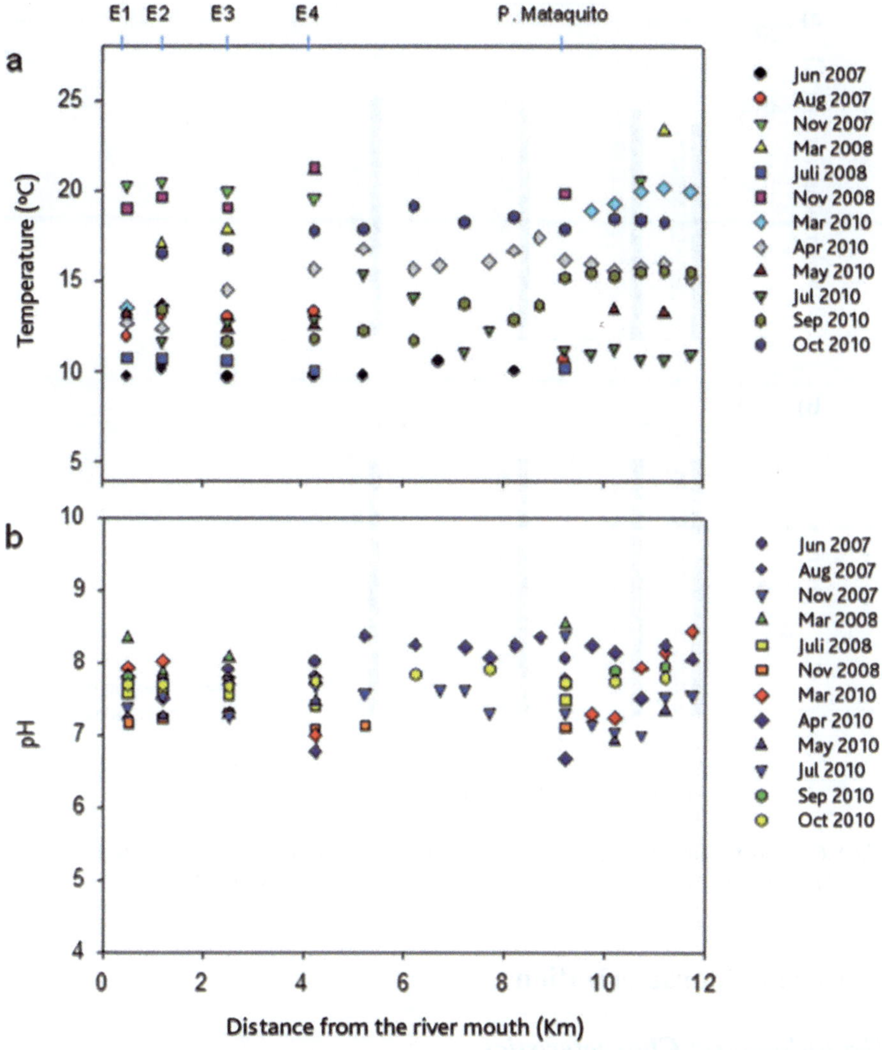

Fig. 9.5 Horizontal profiles for (**a**) temperature and (**b**) pH of superficial waters in the estuary, from "La Pesca" river mouth to 12.2 km upstream. Measurements are included for the 2007–2008 period (historical series) and post-tsunami (March–October 2010)

the post tsunami campaigns (Fig. 9.6). For this purpose, only stations with more than two monitoring stations per year were used to avoid biased results. It is observed that the pre-tsunami temperatures are higher than those of the post-tsunami seasons at all stations sampled (Fig. 9.6a). This could respond to the greater input of sea water with lower temperature, resulting from the destruction of the natural barrier in that area and the greater interchange between ocean and river. As for pH, no significant differences were observed between the two sampling periods, maintaining values between 7.3 and 7.7 for all the stations measured (Fig. 9.6b).

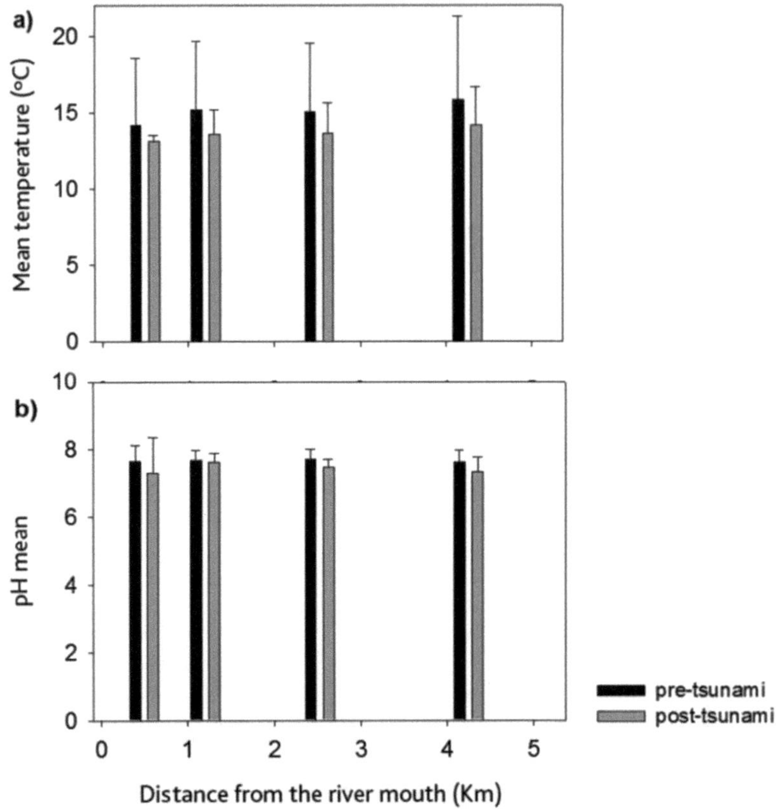

Fig. 9.6 Average values for (**a**) temperature and (**b**) pH for the first 5 km of the Mataquito estuary for the *pre-* and *post-tsunami* periods. The *bars* indicate one standard deviation

Sediment Characterization

Granulometric Characteristics

During the pre and post tsunami periods, the predominant grain type corresponded to fine sand. The middle size sand constitute a very small fraction of the sediments. Consequently, the average grain size for stations E1 and E4 during the period 2007–2008 ranges around 242 and 249 μm, respectively (Fig. 9.7). In the old mouth sector, grain size decreased after the tsunami, reaching an average of 156 μm for the 6 months of study (Fig. 9.7). In E1, no significant differences were found between the pre-tsunami period and the period immediately following the March–May 2010 tsunami (Mann Whitney $p = 0.275$). However, when comparing the historical data with the July–October period, the statistic values were at the limit of significance ($p = 0.06$). At the E4 station, average grain sizes of 269 μm were recorded for the post-tsunami period, which did not differ significantly from the historical average

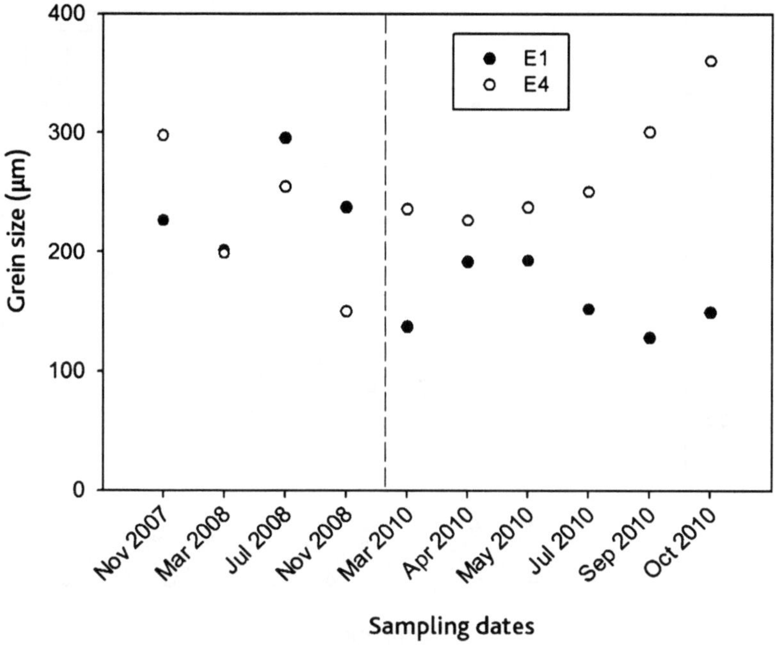

Fig. 9.7 Average grain size (µm) in sediments recovered from stations *E1* and *E4* during campaigns carried out through the 2007–2010 period. The *segmented line* separates pre- and post-tsunami campaigns

(Mann Whitney $p = 0.4$). However, as of July 2010, there is a steady increase in grain size of the sediments, reaching a maximum value of 360 µm in October 2010 (Fig. 9.7). Historical data (pre-tsunami) are not significantly different from the data from July to October 2010 (Mann Whitney $p = 0.157$), but the size of the grain found immediately after the tsunami (March–May 2010) with respect to the subsequent months, that is, July–October 2010 (Mann Whitney $p = 0.05$). This is due to a significant change in the composition of the sediments at station E4, which is related to the greater predominance of average sands during the period between September and October 2010. On the other hand, the size of the grain for E1 decreased progressively in the post-tsunami period, showing values lower than those found in the historical data (April and October 2010).

Total Organic Matter (TOM)

In Fig. 9.8, the total contents of historical and post-tsunami organic matter are given for the sediments of both stations. Between July 2007 and March 2008, the highest pre-tsunami values were observed, reaching maximums of 2.5% for E1 and 1.9%

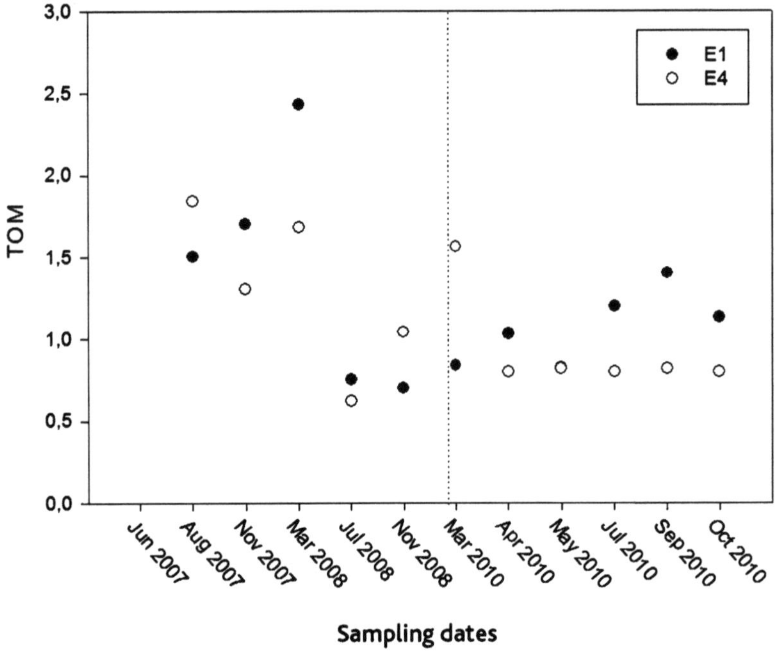

Fig. 9.8 Total organic matter (TOM) content (%), in sediments recovered from stations *E1* and *E4* during campaigns carried out through the 2007–2010 period. The *segmented line* separates pre- and post-tsunami campaigns

for E4 (Fig. 9.7). Out of the five samples taken in this period, in four of them the MOT percentage is higher in E1, which would be expected taking into account the greater exposure of this zone to the oceanic influence. As a consequence of the tsunami, the average TOM percentages decreased in both seasons (station E1 = 0.8%, station E4 = 1.0–1.4%, Fig. 9.8). Only in the campaign carried out immediately after the event (March 2010), the TOM content in the sediments of E4 is similar to those recorded in the pre-tsunami period, since these values then decreased. Statistical analysis indicates that post-tsunami changes in TOM in relation to pre-tsunami historical data are not significant (Mann Whitney $p = 0.7$ for E1, $p = 0.10$ for E4).

PH in Sediments

Figure 9.9 gives the pH readings on the sediments of stations E1 and E4, for each of the campaigns conducted before and after the tsunami. In general, both pH periods are neutral in the sediments (Fig. 9.9). For E1, the values do not show significant differences and are located within the range between 6.2 and 7.8 for both periods, pre and post-tsunami. However, for E4, there is an increase from March 2010 to

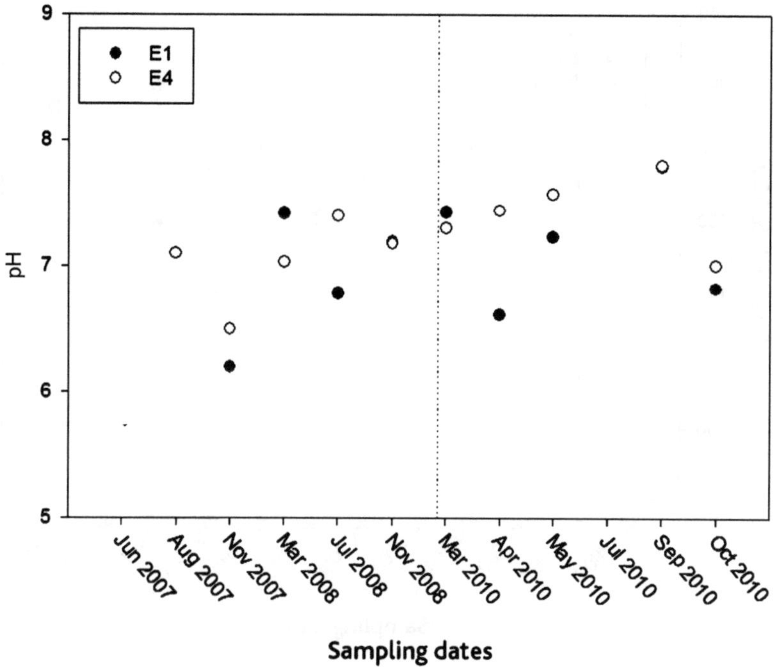

Fig. 9.9 pH values for stations E1 and E4 during campaigns carried out through the 2007–2010 period. The *segmented line* separates pre- and post-tsunami campaigns

September 2010, reaching a maximum of 7.8, higher than all historical values recorded in this sampling station. The Mann Whitney test proves it, indicating that there are significant differences ($p = 0.02$) in the pH of the pellets between the pre- and post-tsunami period.

Redox Potential

The redox potential did not significantly reflect the effect of the tsunami at any of the two sampling stations, although immediately after the tsunami the value found in E1 increased and in E4 it decreased, but these numbers are within the values of the historical series. The significance of the values found for both seasons is related to the homogenization of the post-tsunami environment. The average values of redox potential in the post-tsunami period are still positive (194.6 ± 91.3 mV) ranging from 150 to 250 mV (except E4 in September 2010), which speaks of well-oxygenated areas and good state for the development of aerobic sedimentary communities (Fig. 9.10).

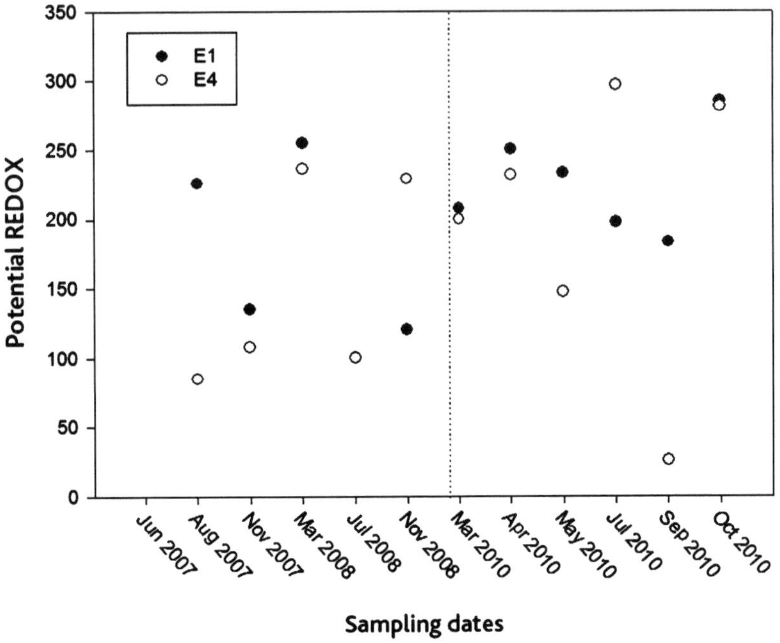

Fig. 9.10 Redox potential for stations *E1* and *E4* during campaigns carried out through the 2007–2010 period. The *segmented line* separates pre- and post-tsunami campaigns

Vertical Characterization of the Water Column Through a Spacial

Characterization During the Pre-tsunami Period

In Fig. 9.11 we present a vertical characterization of the water column through a spatial gradient for the summer (March 2008) and winter (July 2008) periods. Both characterizations are historical references that representatively capture the seasonal variations in temperature and salinity due to the differential influence of the tides and fluvial waters in the terminal sector of the estuary. During the summer (March 2008), warmer temperatures (>22 °C, Fig. 9.11a) were recorded. As a result of the influence of inland waters and the greater surface warming that they experience towards the summer period, temperatures in the upstream sector of station E4 are comparatively warmer than in the mouth sector (Fig. 9.11a), clearly observing vertical stratification with colder waters at the bottom (~14 °C). As for salinity, an intrusion of oceanic waters up to station 5 (~3 km), with values of up to 15 psu at that point (Fig. 9.11b), is clearly observed. From this point, the salinity decreases drastically up to <1 psu upstream, clearly defining the estuarine zone.

During the winter (July 2008), the water temperatures are cold, ranging between 9.8 and 10.7 °C. Towards the sector closest to the mouth, between 0 and 4.5 km, the surface waters have temperatures close to 10 °C, while the waters associated with the bottom stratum are around 11 °C (Fig. 9.11c). A great homogenization of the

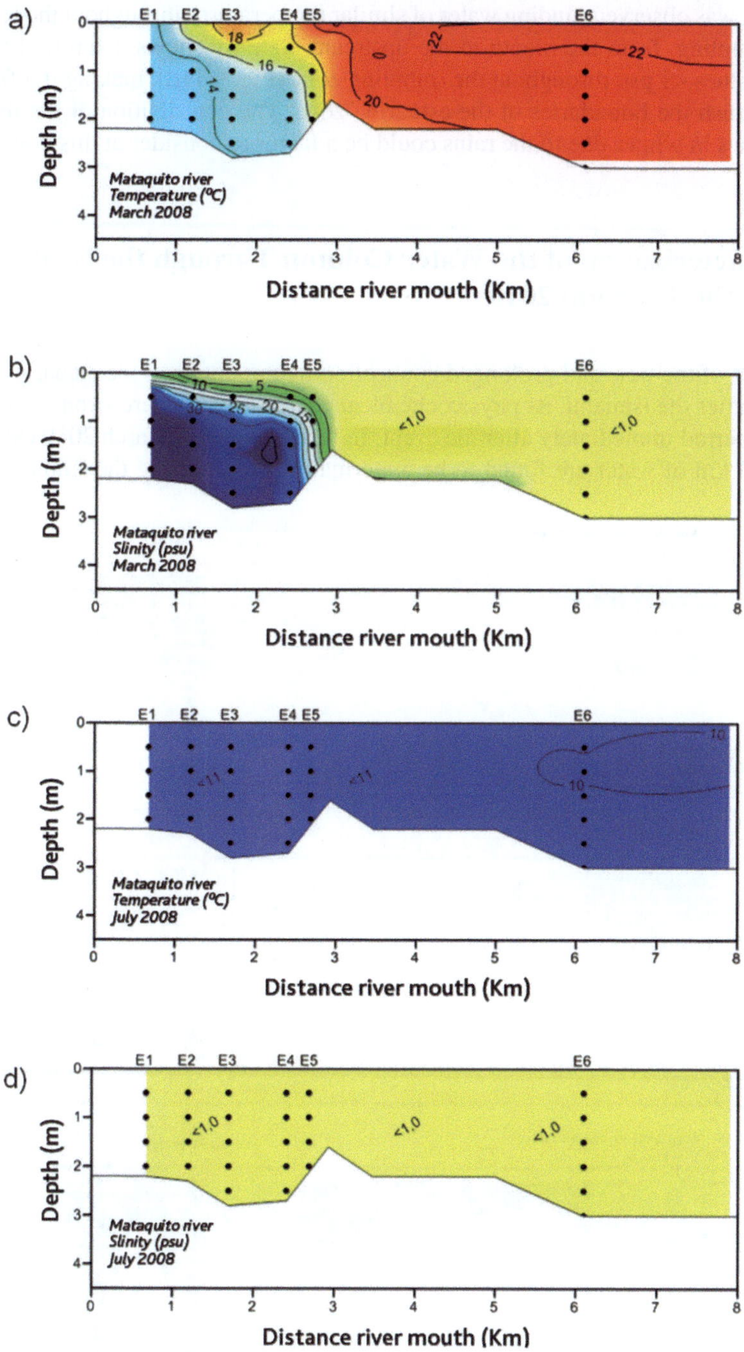

Fig. 9.11 Vertical profiles of water salinity and temperature for two contrasting periods in the year March 2008 (summer) and July 2008 (winter). Measurements were taken in the water column along the first eight kilometers of the Mataquito River estuary. (**a**) Summer temperature; (**b**) Summer salinity; (**c**) Winter temperature; (**d**) Winter salinity

system was observed, finding water of similar temperatures throughout the length of the sampling, from the mouth to the upstream. Salinity shows a similar behavior, with values <1 psu throughout the sampling zone (Fig. 9.11d), making it difficult to distinguish the boundaries of the estuarine zone. The contribution of the flow that increases in winter due to the rains could be a factor to consider at this point.

Characterization of the Water Column Through the Estuary After the Tsunami 2010

Product of the new and prolonged connection that maintained the estuary with the ocean after the tsunami, its physicochemical characteristics were strongly modified in the period immediately after the event. In fact, during the March 2010 campaign, up to 4 km of water are found to be abnormally colder during the summer period

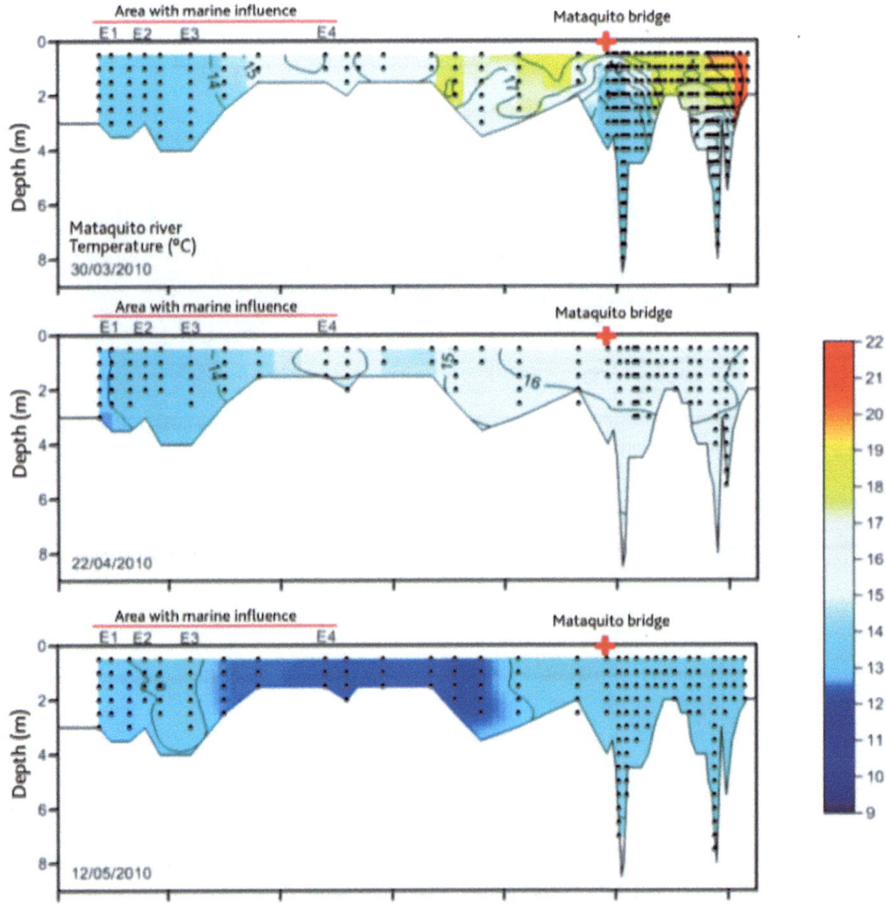

Fig. 9.12 Vertical and horizontal temperature profiles for the water column in the Mataquito River estuary during the post-tsunami period, March–October 2010

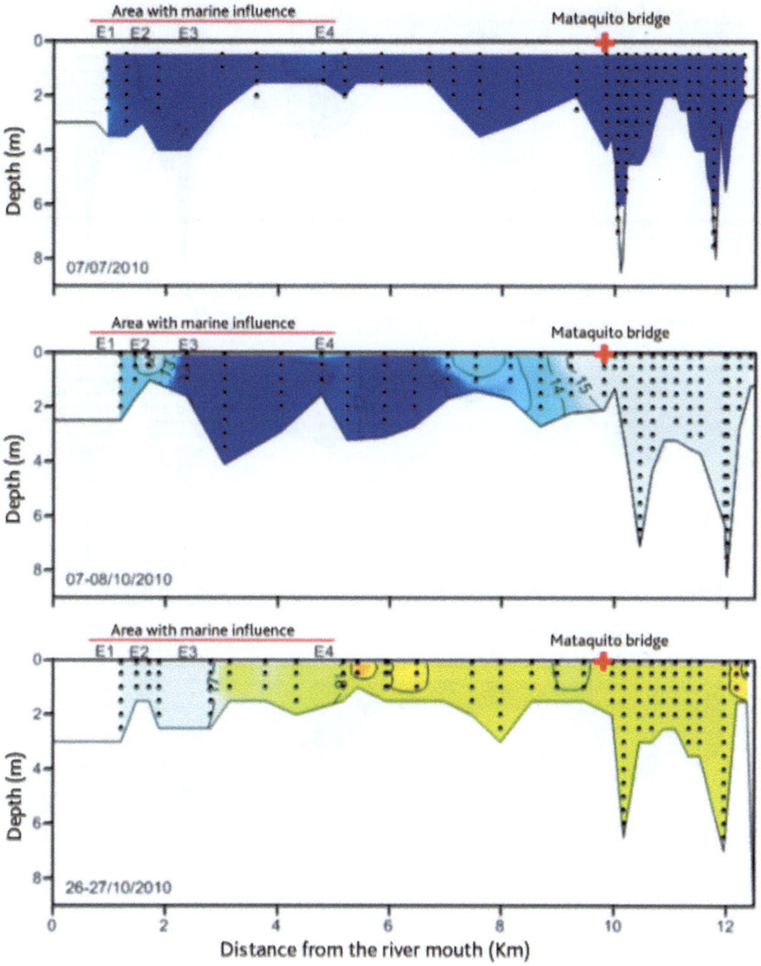

Fig. 9.12 (continued)

(<14 °C, Fig. 9.12). Nevertheless, the pattern of horizontal variation in the temperatures that characterizes the summer is maintained, registering more warmer temperatures, towards the upstream sector of the station E4. As for salinity, there are unusually salty waters in the estuary up to 12 km (>0.4 psu, Fig. 9.13). For example, in the most immediate sector at the mouth, there is a vertically homogeneous water column with salinity >31 psu. While upstream, the water column presents ascending vertical gradients in salinity (Fig. 9.13). Similar trends in temperatures and salinity are recorded during the month of April 2010.

In concordance with the fall in temperatures during the autumn, in the 2010 campaign, the temperature is comparatively cooler (13.1 °C, Fig. 9.12) than those recorded in March and April. During this campaign, horizontal variations in temperatures along the estuary are observed. In fact, a vertically homogeneous water column, with temperatures close to 12 °C, occurs between stations E1 and E2. The same happens upstream from station E3 up to 8 km. On the other hand, between

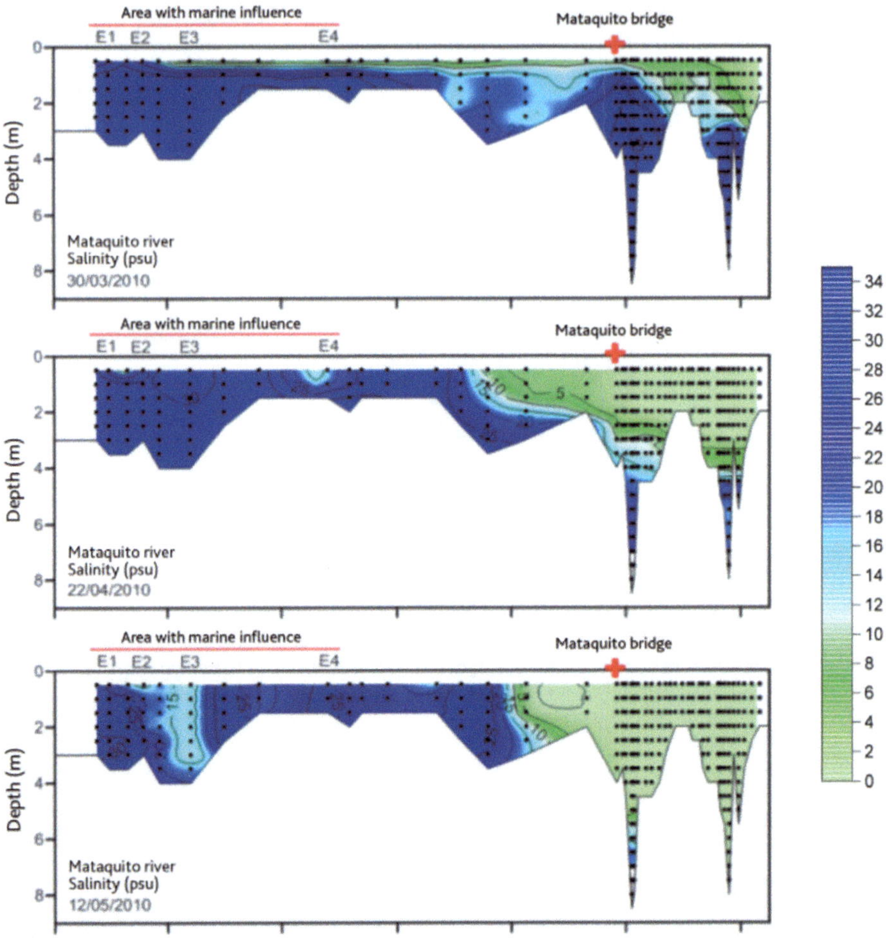

Fig. 9.13 Vertical and horizontal salinity profiles for the water column in the Mataquito River estuary during the post-tsunami period, March–October 2010

stations E2 and E4 and in the sector 8 km upstream, a poorly stratified water column develops, with temperatures above 18 °C (Fig. 9.12). Compared with previous seasons, during May 2010, abnormally saline waters (>31 psu) extended mainly up to 1.3 km upstream from the mouth (Fig. 9.13). Meanwhile, for over 8 km, there are waters with very low salinity detected, characteristic of fresh water and only slightly saline (Fig. 9.13).

During the winter (July 2010), the homogenization of the water column becomes evident, with very similar temperatures and salinities observed throughout the sampling range (up to 12 km) and in the vertical (Figs. 9.12 and 9.13). However, at the beginning of August 2010, there is a small warming from the 8 km upwards, and a clear decrease in salinity values in the same area.

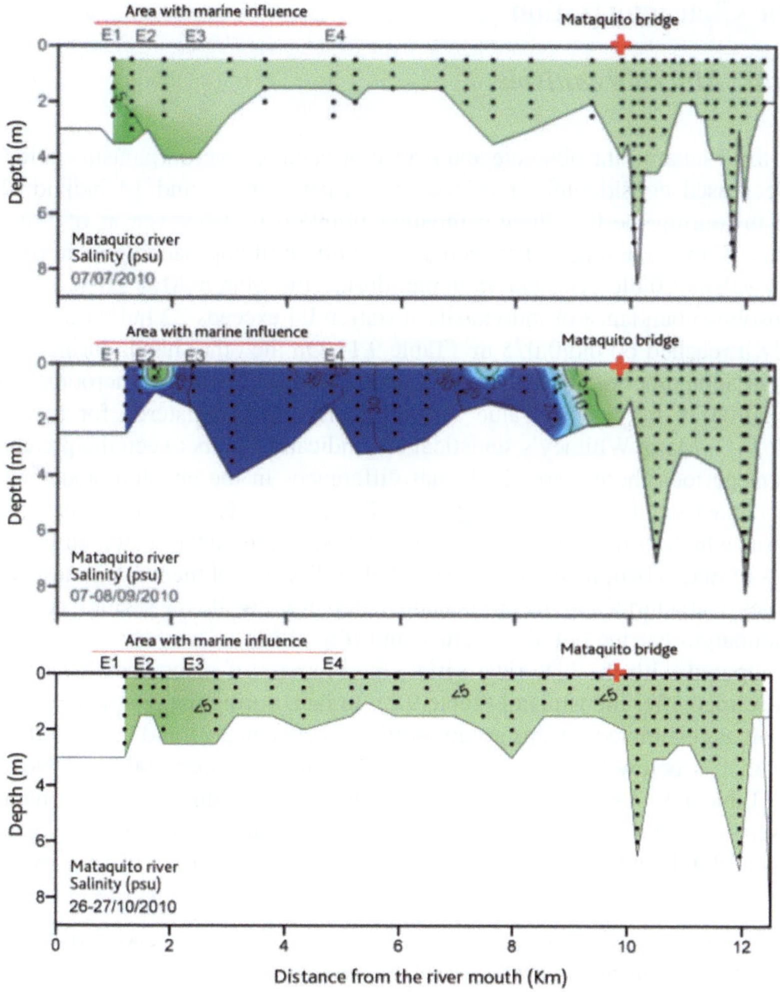

Fig. 9.13 (continued)

In the September and October campaigns, temperatures were warmer than during previous periods (>13 °C, Fig. 9.12). In general, during these campaigns, the estuary presented spatial variations in water temperatures very similar to those recorded in its historical references (Fig. 9.11). This is particularly evident in the October 2010 season. In the sector upstream of station E4, comparatively warmer temperatures (>18 °C) are detected than in the sector close to the old mouth (<17 °C; Fig. 9.12). However, during both months, the patterns of stratification in salinity that characterize the spring period were not observed. For example, abnormally saline waters (>30 psu) are recorded in the vicinity of the E4 station in September, which usually has a water column with continental characteristics, while in October 2010 the horizontal stratification resembles that recorded during winter (Fig. 9.13).

Biotic Characterization

Benthic Macroorganisms

After the Tsunami, the absolute abundance of benthic macroorganisms in the estuary decreased considerably, reaching values between 53 and 14 individuals per 0.025 m^2 during the first three campaigns (Table 9.1). However, as of July 2010, absolute richness has started to increase steadily until approaching its normal, historical values (Table 9.1). For example, during the March–May 2010 campaigns, the absolute abundance of individuals at station E4 exceeds 22 Ind/0.025 m^2, while in July it reached 60 ind/0.025 m^2 (Table 9.1). On the other hand, during the campaign in September of 2010, there was an abundance of benthic macroorganisms of 915 Ind/0.025 m^2, whose value surpassed any other registered for the estuary (Table 9.1). Mann Whitney's statistical test indicates that between the pre and post tsunami periods there were significant differences in the absolute abundances of benthic macroorganisms ($N = 40$, $p = 0.007$). The same is inferred from a similarity analysis, which shows two groups significantly different in terms of their total abundances of macro benthic organisms (Fig. 9.14). The first of them integrates the communities characterized in the pre-tsunami campaigns, while the second incorporates the campaigns carried out after the tsunami (Fig. 9.14).

Compared with the historical series, the biomass of macro benthic organisms also declined after the tsunami (Table 9.2). In fact, during the post-tsunami period, the biomass estimated for this group oscillated between 2.21 and 0.11 g/0.025 m^2, a range that is below what was recorded between November 2007 and November 2008 (Table 9.2). As for the wealth of taxa, an increase in this community parameter is observed in relation to the pre-tsunami period (Tables 9.1 and 9.3). For example, between March and September 2010, the number of taxa ($n = 9$ to 15, Table 9.1) is

Table 9.1 Abundance (ind/0.025 m^2) and macro benthic organism richness in the Mataquito River estuary for each sampling season during campaigns from 2007 to 2010

Period	Campaign/station	e1	e2	e3	e4	Average	S.D.	Richness
Pre-tsunami	November 2007	32	119	138	104	98	46	7
	March 2008	268	291	261	36	214	119	10
	July 2008	3	53	29	5	23	23	4
	November 2008	184	184	529	420	329	174	4
Post-tsunami	March 2010	19	11	2	22	13	9	13
	Abril 2010	3	7	3	1	4	3	9
	May 2010	2	6	13	1	6	5	12
	June 2010[a]	10			76			
	July 2010	11	11	120	60	51	52	15
	September 2010	31	99	915	19	266	434	11
	October 2010	265	143	615	16	260	258	6

sd standard deviation
[a]Complementary study developed in the Mataquito River estuary

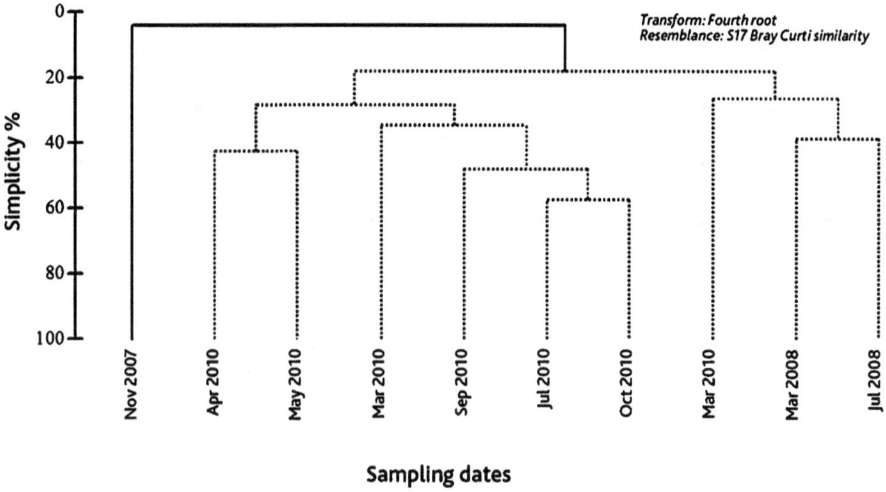

Fig. 9.14 Conglomerate analysis for total abundances of macro benthic organisms in the estuary for the pre- and post-tsunami periods

Table 9.2 Macrobenthic organisms' biomass (g/0.025 m^2) in the mataquito river estuary, for each sampling season, during campaigns from 2007 to 2010

Period	Campaign/station	e1	e2	e3	e4	Avg.	D.s.
Pre-tsunami	November 2007	0.03	0.20	0.29	0.13	0.16	0.11
	March 2008	0.35	0.80	1.50	0.04	0.67	0.63
	July 2008	0.00	0.05	0.06	0.01	0.03	0.03
	November 2008	0.07	0.08	4.36	0.19	1.17	2.12
Post-tsunami	March 2010	0.28	0.00	0.01	0.01	0.08	0.13
	Abril 2010	0.21	0.00	0.34	0.10	0.16	0.15
	May 2010	0.00	0.01	0.10	0.00	0.03	0.05
	June 2010[a]	0.75			0.04	0.39	0.50
	July 2010	0.01	0.01	0.09	0.05	0.04	0.04
	September 2010	0.03	0.10	2.06	0.02	0.55	1.00
	October 2010	0.16	0.08	0.58	0.01	0.21	0.26

sd standard deviation
[a]Complementary study developed in the Mataquito River estuary

significantly higher than that recorded in the historical series ($n = 4$–10, Table 9.1). However, during October 2010, this parameter decreased compared to the previous 6 months ($n = 6$; Table 9.1) and approximated the values recorded in the pre-tsunami campaigns (Table 9.1).

The increase in richness during the post-tsunami period is related to the scarce rate loss of taxa in the estuary after the tsunami, and the subsequent incorporation of new species, the latter being the most important factor. During the period March–September 2010, there were 18 new taxa in the estuary (Table 9.3): 12 taxa of Crustaceans, 3 taxa of Polychaetes; 2 taxa of Mollusks and 1 taxon of Nemertea.

Table 9.3 Taxonomic composition of macro benthic communities in the estuary, during the whole 2007–2010 sampling period. In the "registro" column, periods for appearance of each of the taxa found are indicated

Record	Phylum	Clase	Taxa	Nov 07	Mar 08	Jul 08	Nov 08	Mar 10	Abr 10	Mar 10	Jun 10	Jul 10	Sep 10	Oct 10
Pre and Post	Annelida	Poliqueta	*Nereis sp.*		+	+	+	+	+	+	+	+	+	+
Pre	Annelida	Poliqueta	*Platynereis sp.*	+										
Pre	Annelida	Poliqueta	*Microspio sp.*	+										
Pre and Post	Annelida	Poliqueta	*Lumbrineris chilensis*				+						+	
Pre and Post	Annelida	Poliqueta	*Capitella sp.*	+	+		+	+		+				
Pre	Annelida	Poliqueta	*Cossura chilenesis*		+									
Pre and Post	Annelida	Poliqueta	*Glycera americana*		+	+		+			+	+		
Pre and Post	Annelida	Poliqueta	*Spiophanes sp.*		+			+		+	+			
Pre and Post	Annelida	Poliqueta	*Prionospio sp.*			+		+	+		+	+	+	+
Pre and Post	Annelida	Poliqueta	*Paraprionospio pinnata*				+			+				
Pre and Post	Annelida	Poliqueta	*Goniada uncinigera*			+			+				+	
Pre and Post	Annelida	Poliqueta	*Neftys ferruginea*			+				+		+		
Pre	Annelida	Poliqueta	*Travisia sp.*									+		
Pre	Annelida	Poliqueta	*Scolelepis sp.*									+		

Pre	Annelida	Poliqueta	*Haploscoloplos kerguelensis*										+	
Pre and Post	Annelida	Oligoqueta	*Oligoqueto*	+	+	+	+	+				+	+	+
Pre and Post	Arthropoda	Crustacea	*Excirolana sp.*	+		+						+	+	+
Pre	Arthropoda	Crustacea	*Amphipoda ind.*		+	+	+							
Pre	Arthropoda	Crustacea	*Paracorophium sp.*	+										
Pre	Arthropoda	Crustacea	*Lysianassidae*	+										
Pre and Post	Arthropoda	Crustacea	*Gammarus sp.*		+						+	+		+
Pre	Arthropoda	Crustacea	*Megalopa*		+	+								
Pre	Arthropoda	Crustacea	*Hemigrapsus crenulotus*				+							
Pre	Arthropoda	Crustacea	*Cumacea*			+								
Pre and Post	Arthropoda	Crustacea	*Isopoda ind.*			+							+	
Post	Arthropoda	Crustacea	*Phoxocephalidae*					+	+	+		+	+	+
Post	Arthropoda	Crustacea	*Orchomenella sp.*					+						
Post	Arthropoda	Crustacea	*Hyalidae*					+						
Post	Arthropoda	Crustacea	*Eusiridae*					+						
Post	Arthropoda	Crustacea	*Idoteidae*					+						
Post	Arthropoda	Crustacea	*Ampeliscidae*								+			
Post	Arthropoda	Crustacea	*Blepharipoda spinintana*					+						
Post	Arthropoda	Crustacea	*Emierita sp.*					+	+	+	+			
Post	Arthropoda	Crustacea	*Ovalipes trimaculatus*						+	+	+			

(continued)

Table 9.3 (continued)

Record	Phylum	Clase	Taxa	Nov 07	Mar 08	Jul 08	Nov 08	Mar 10	Abr 10	Mar 10	Jun 10	Jul 10	Sep 10	Oct 10
Post	Arthropoda	Crustacea	*Bathymedon sp.*						+	+				
Post	Arthropoda	Crustacea	*Chaelilia sp.*									+		
Post	Arthropoda	Crustacea	*Tanaidacea ind.*										+	
Pre and Post	Arthropoda	Crustacea	*Macrochiridothea sp.*		+				+			+	+	
Pre y Post	Arthropoda	Insecta	*Chironomidae*	+	+	+				+				
Post	Mollusca	Mollusca	*Mesodesma donacia*						+					
Post	Mollusca	Mollusca	*Bivalvo juvenil*							+				
Pre and Post	Mollusca	Mollusca	*Mulinia sp.*		+							+		
Pre	Mollusca	Mollusca	*Mytilidae*		+									
Pre	Mollusca	Mollusca	*Cyclocardia compressa*	+										
Pre	Mollusca	Mollusca	*Nacella sp.*	+										
Pre	Platyhelminthes	Turbellaria	*Turbellaria ind.*		+									
Post	Nemertea	Nemertea	*Nemertea ind.*							+		+		

Pre pre-tsunami, *post* post-tsunami

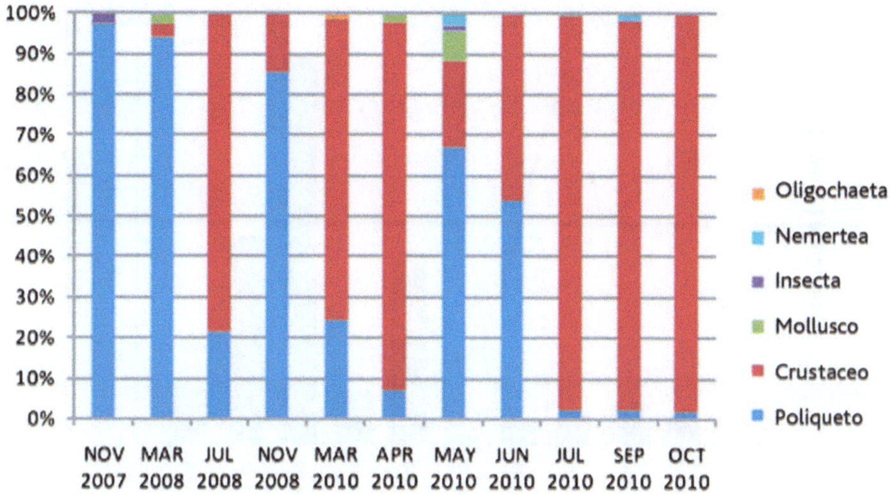

Fig. 9.15 Relative abundance for the different taxonomic groups present in macro benthic communities in the estuary during the pre- and post-tsunami periods

In Fig. 9.14, the dendrogram resulting from the numerical classification is presented on the basis of an array of similarity with the abundance data of the species present in each season. The classification analysis clearly showed the formation of two groups, which were one group for the campaigns carried out before and the other for the campaigns carried out after the tsunami, which relates to the composition of species in the community and the abundance registered for each of the campaigns. The campaign carried out during November 2007 is not significantly related to historical data from previous campaigns or to post-earthquake campaigns. In general, both groups present a clear grouping and similarity statistically significant around 30%. In a first cluster, there were campaigns carried out during November, July and March 2008, and in the other cluster all the campaigns carried out after the tsunami, between March and October 2010.

In parallel, during the post-tsunami period, the macro benthic communities of the estuary undergo a significant change in their structure. Indeed, during the pre-tsunami period, the macro benthic community was largely dominated by Polychaetes (90%), and secondary by Crustaceans, whereas during the post tsunami period, this pattern of dominance reverted, predominating Crustaceans above the Polychaetes (Fig. 9.15).

Zooplankton

During the post tsunami period, the absolute abundance and taxa richness in the zooplankton communities of the Mataquito estuary, there were significant changes compared to the previous period. Thus, during the campaign carried out in March

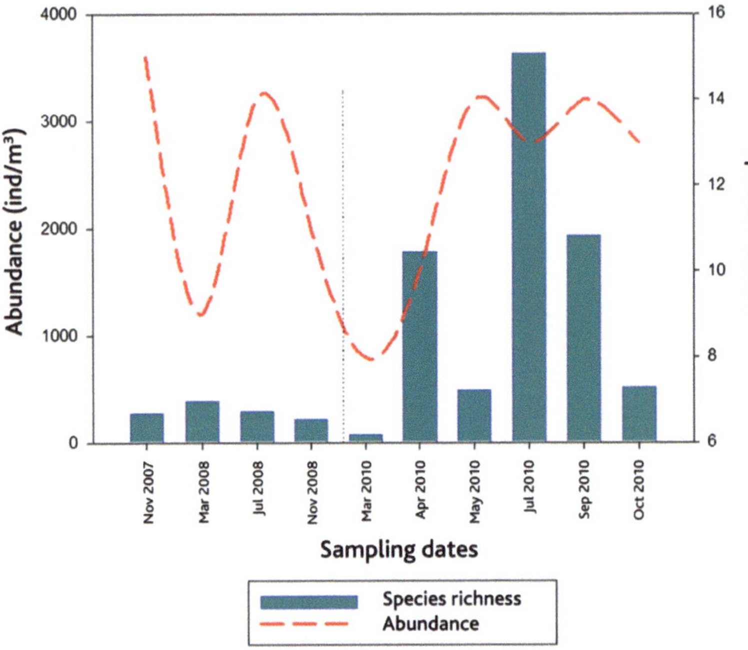

Fig. 9.16 Species richness and total abundance (ind/m^3) for zooplankton communities in the estuary during the pre- and post-tsunami periods

2010 there is a historical decrease in these two components (Fig. 9.16). However, as of April 2010, both components recorded a progressive increase, surpassing even the values recorded during the pre-tsunami period (maximum = 388 ind/m^3). For example, between April and October 2010, wealth of taxa ranged from 7 to 14 taxa (Fig. 9.15), while absolute abundance during the same period reached 488 ind/m^3 as the minimum value in May 2010, reaching a maximum of 3500 ind/m^3 during the month of July 2010 (Fig. 9.16).

After the tsunami, changes in the community structure are also observed as a result of variations in the absolute and relative abundances of the different taxonomic groups present in the zooplankton (Figs. 9.17 and 9.18). After the tsunami, those groups that were marginally represented in the historical series, such as Copepods, Cladocerans, Oligochaetes and Hydrozoans, are mainly dominant. During March–May 2010, the Copepods presented absolute and relative abundances above 1700 ind/m^3 and 90%, respectively (Figs. 9.17 and 9.18). During this same period, the Cnidarians constituted the second relevant group, reaching abundances around 56 Ind/m^3 and 10%, while the Cladocerans and Oligochaetes in July 2010 presented abundances that exceeded 70% and 20% respectively (Fig. 9.18).

Statistical analysis (Mann Whitney) indicates that between the pre and post tsunami periods there were significant differences in the absolute abundances of zooplanktonic organisms ($N = 40$ $p = 0.04$). Similar results are obtained from a cluster

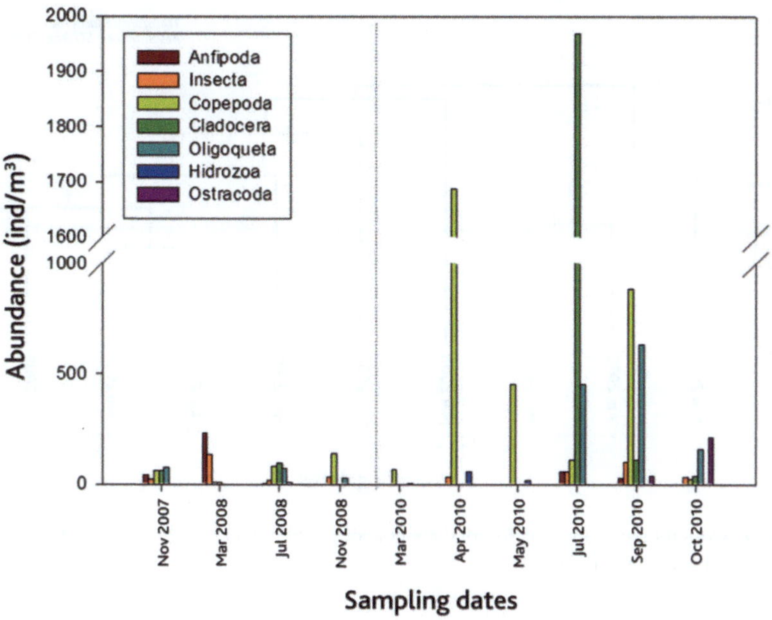

Fig. 9.17 Absolute abundances (ind/m^3) for the different taxonomic groups present in zooplankton communities in the estuary during the 2007–2010 campaigns. The *segmented line* separates pre- and post-tsunami campaigns

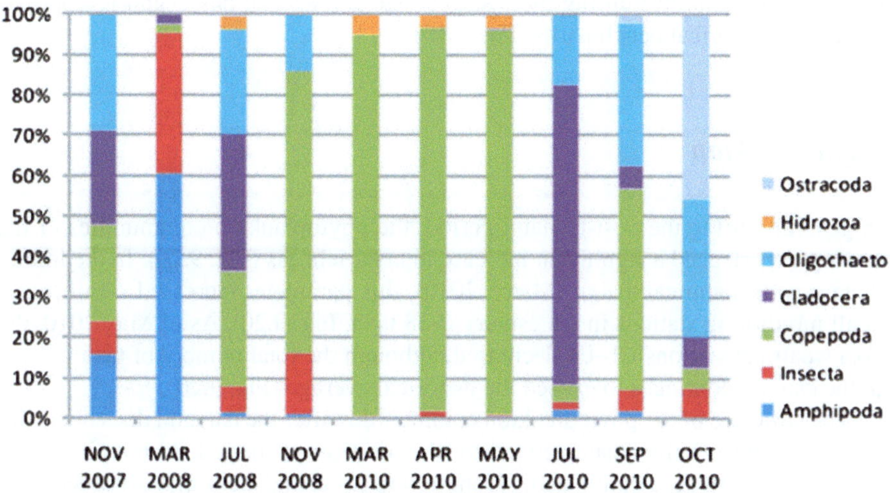

Fig. 9.18 Relative abundance for the different taxonomic groups present in zooplankton communities during the pre- and post-tsunami periods

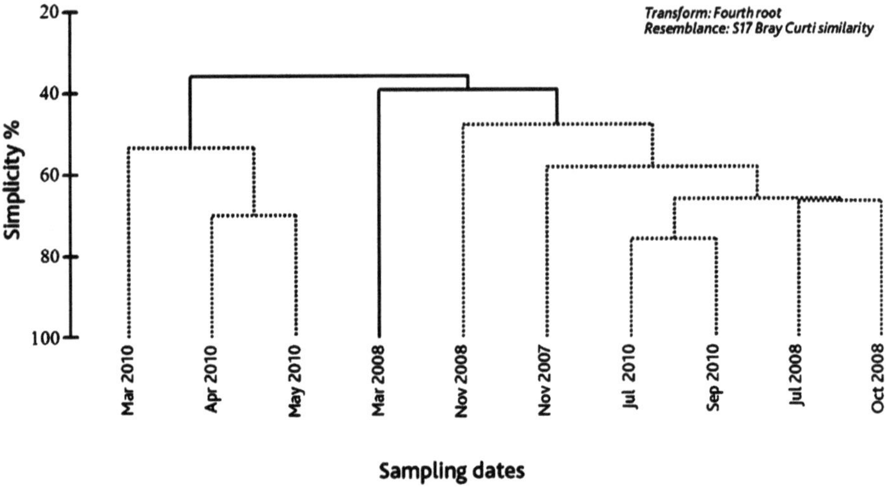

Fig. 9.19 Conglomerate analysis for total abundances of zooplankton organisms in the estuary for the pre- and post-tsunami periods

analysis, which shows (Fig. 9.18) two groups significantly different in terms of total abundances of zooplanktonic organisms (Fig. 9.18). The first conglomerate groups those with similarities by over 50% for those communities characterized during the campaigns of March, April and May 2010 (Fig. 9.19); while the second cluster integrates the remaining campaigns, with similarities around 50%. The high similarity between the 2010 and July 2008 campaigns (70%, Fig. 9.19) suggests that the structuring of communities during the last post-tsunami season has begun to approximate their historical benchmarks (Fig. 9.19).

Phytoplankton

In general, during the post-tsunami period, the phytoplankton communities of the estuary experienced a significant increase in taxa richness (Fig. 9.20). In the immediate post-tsunami campaign (March 2010), this parameter increased significantly at all monitoring stations in the estuary (>28 taxa, Fig. 9.20). As of May 2010, the taxa wealth at stations E1–E4 decreased, although the total number of taxa present in the estuary continues to exceed the historical average (Fig. 9.20).

The structure of the phytoplankton communities after the tsunami does not differ substantially from that characterized for the pre-tsunami period. At both intervals, the Bacillariophyceae dominated, reaching relative abundances that exceeded 95% during the post-tsunami period (Fig. 9.21). However, there are differences in the representation and abundance of other components of the communities. Cyanophyceae are not registered in any of the campaigns carried out between March and October 2010. On the other hand, the Dinophyceae and Chlorophyceae considerably decreased in their percentages (>5%, Fig. 9.21).

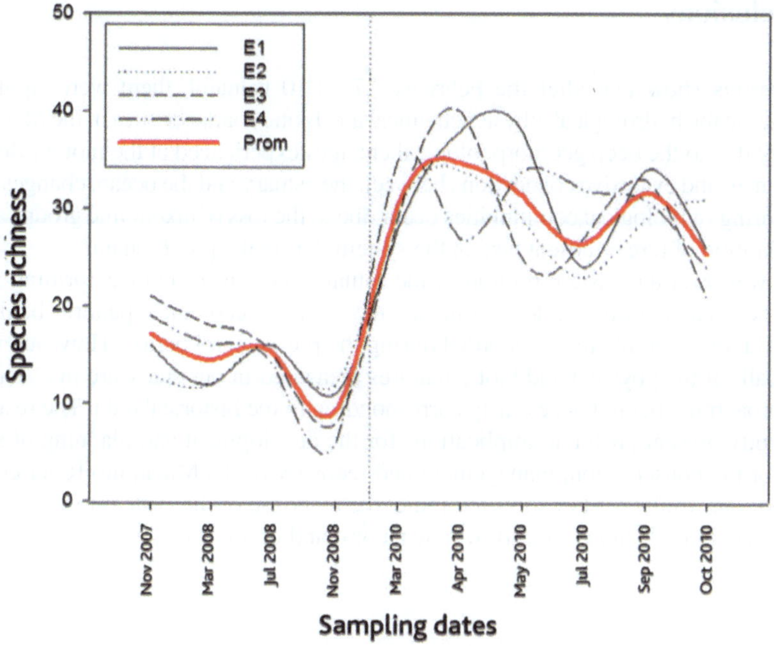

Fig. 9.20 Taxa richness for each sampling season and average for the estuary (*red line*) during the 2007–2010 campaigns. The *segmented line* separates pre- and post-tsunami campaigns

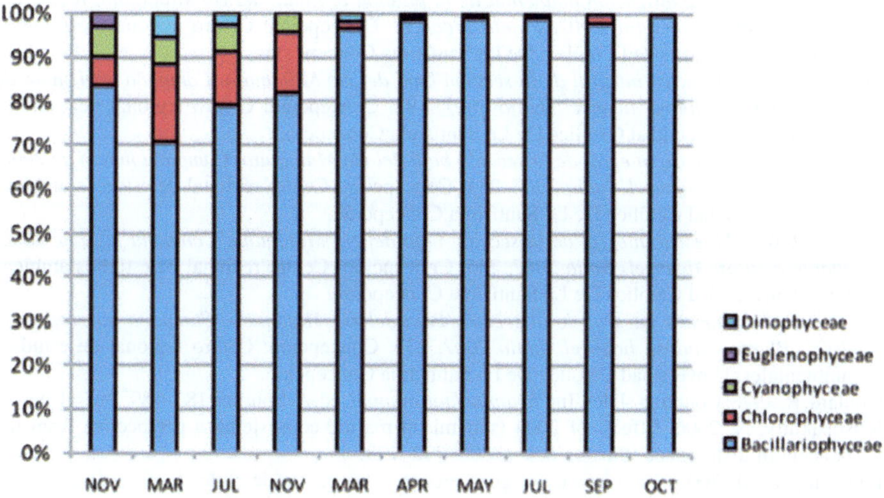

Fig. 9.21 Relative abundance for different taxonomic groups found in the phytoplankton communities in the estuary, during the pre- and post-tsunami periods

Conclusions

Our results show that after the February 27, 2010 tsunami, there were significant changes in the hydrological, physicochemical and biotic characteristics of the Mataquito estuary due to the deep geomorphological changes experienced at the mouth. Because of the new and extensive connections between the estuary and the ocean, changes in the structuring of biological communities occur due to the loss of taxonomic groups and/or proliferation of rare or absent taxa in the system during the pre-Tsunami.

However, at the eastern portion of the estuary, there is a sand bar forming again, with the oceanic influence decreasing and the estuary's ecological patterns beginning to gradually resemble those recorded during the pre-tsunami period. However, many, if not all, of the physical and biotic features measured in our study are in a state that is approaching, but not necessarily corresponding to the historical data. The results of this study present profound implications for the development and planning of strategies for the conservation, management and recovery of the Mataquito River estuary. Indeed, our results are key tools to predict the response of the system to variations in its main driving factors induced by new natural and/or anthropogenic disturbances.

References

CREA-UCSC. 2007a. *Caracterización ambiental de la porción terminal del río Mataquito, región del Maule. Campaña julio 2007 planta licancel.* Concepción: Centro regional de estudios ambientales, Universidad Católica De La Santísima Concepción.

———. 2007b. *Caracterización ambiental de la porción terminal del río Mataquito, región del Maule. Campaña octubre 2007. Planta Licancel.* Concepción: Centro regional de estudios ambientales, Universidad Católica De La Santísima Concepción.

———. 2008a. *Muestreo integral de la sección baja del río Mataquito. Campaña noviembre de 2007. Planta celulosa licancel. Junio 2007*, 192. Concepción: Centro regional de estudios ambientale, Universidad Católica De La Santísima Concepción.

———. 2008b. *Muestra integral de la sección baja del río Mataquito. Campaña marzo de 2008. Planta celulosa licancel. Junio 2007*, 231. Concepción: Centro regional de estudios ambientales, Universidad Católica De La Santísima Concepción.

———. 2008c. *Muestra integral de la sección baja del río Mataquito. Campaña julio de 2008. Planta celulosa licancel. Junio 2007*, 255. Concepción: Centro regional de estudios ambientales, Universidad Católica De La Santísima Concepción.

———. 2009. *Muestra integral de la sección baja del río Mataquito. Campaña noviembre de 2007. Planta celulosa licancel. Junio 2007*, 233. Concepción: Centro regional de estudios ambientales, Universidad Católica De La Santísima Concepción.

Guillard, R. 1978. Counting slides. In *Phytoplankton manual*, ed. A. Sournia, 182–189. París: Unesco.

Krishnakutty, N. 2006. Effects of 2004 tsunami on marine ecosystems–a perspective from the concept of disturbance. *Current Science* 90: 772–773.

Lagos-lópez, M. 2000. Tsunamis de origen cercano a las costas de chile. *Revista de geografía norte grande* 27: 93–102.

Roni, P., T. Beeche, R. Bilby, F. Leonetti, M. Pollock, and G. Pess. 2002. A review of stream restoration techniques and a hierarchical strategy for prioritizing restoration in Pacific Northwest watersheds. *North American Journal of Fisheries Management.* 22: 1–20.

Utermöhl, H. 1958. Zur vervollkommung der quantitativen phytoplankton methodik. *Mitteilungen Internationale Vereinigung fur Theoretische und Angewandte Limnologie* 9: 1–38.

Chapter 10
The Tubul-Raqui Coastal Wetland: A Chilean Ecosystem of High Conservation Value Severely Disturbed by the 2010 Earthquake

Claudio Valdovinos, Natalia Sandoval, Daniela Vasquez, and Viviana Olmos

Abstract Given the great extent of the Tubul-Raqui marshland, in addition to its high biodiversity and biological productivity, combined with a low degree of human disturbance, this wetland corresponds to one of the most important ones in Chile. This chapter describes its main features and effects caused by the February 2010 earthquake. The wetland continues to provide nesting and shelter sites for numerous resident and migratory bird species, many of which have conservation problems. The major anthropogenic threats to the conservation of the wetland and its different habitats have historically been bird hunting and deforestation of river shores that affect the sedimentation of the rivers that feed it. As for the structure of the physical habitat, water quality and aquatic biota of the wetland, since the earthquake of 2010 a series of relevant environmental changes have occurred. Some of them were expressed immediately during the earthquake, while others still continue to manifest themselves. The main changes resulting from the earthquake have resulted from its uprising of about 1.6 m above sea level, which has led to the total drying of the network of internal channels that irrigate the wetland, and partial one of the main channels of the rivers Tubul and Raqui. This has also limited the exchange with the sea, generating a decrease in the salinity of the waters to the interior of the wetland. The two most important benthic species for the artisanal fishery, the "Pelillo" seaweed (Gracilaria sp.), and the "navajuela" bivalve (Ensis macha), were severely affected and there is still no evidence of any degree of recolonization. Although the ecosystem was severely affected by the earthquake, it retains properties that allow it to remain within the country's most important coastal wetlands.

Keywords Biodiversity • Conservation • Marshland • Earthquake • Tubul • Raqui • Chile

C. Valdovinos (✉) • N. Sandoval • D. Vasquez • V. Olmos
Laboratory of Biodiversity and Conservation of Aquatic Resources. Center for Environmental Sciences EULA, University of Concepción, Casilla 160 -C, Concepción, Chile
e-mail: cvaldovi@udec.cl

J.M. Fariña, A. Camaño (eds.), *The Ecology and Natural History of Chilean Saltmarshes*, DOI 10.1007/978-3-319-63877-5_10

Introduction

Coastal wetlands are recognized worldwide as the ecosystems of particular interest for the conservation of biodiversity (Valiela et al. 2009). In addition, they provide multiple ecosystem services for our society, among which are the production of commercially important species, especially algae and mollusks (Valdovinos 2004). Along the coast of Chile, it is possible to find numerous types of estuarine wetlands located from the arid regions of the north, to the southern fjords. Each one of them presents particular ecological characteristics according to its latitude, coastal geomorphology, tidal regimes, fresh water contributions, recent geological history and degree of human intervention (Valdovinos 2004). A series of coastal and marine wetlands (see Stuardo and Valdovinos 1989) exist on the shores of the Biobio Region, located in south-central Chile, which, according to the classification system of the Ramsar Convention (2006), and corresponding to the typologies of "estuaries"[1] and "intertidal marshes and estuaries"[2]. The latter typology includes ecosystems that are considered to be among the scarcest and most important of Chilean coasts, especially for harboring numerous unique species, many of which present conservation issues (Stuardo et al. 1992). These ecosystems depend on a delicate balance between terrestrial, freshwater and coastal marine systems, and because are located on the coastal border, they are affected by the tidal cycles that allow the entry of marine waters into the continental area (Valdovinos 2004).

In the Biobio Region, these ecosystems are mainly represented by the Tubul-Raqui, Rocuant-Andalien, Lenga and Carampangue wetlands, which are home to a diverse avifauna, including an important set of seasonal migratory birds and extensive meadows or "Espartales" of the halophyte plant Spartina densiflora Brongn 1829. This species is of fundamental importance in these ecosystems, acting as an "bioengineering species" that models and stabilizes the river banks from the effects of the tides, and corresponds to the main energy source of the wetland, contributing large amounts of organic detritus that enters the aquatic and terrestrial trophic chains. The most important ecosystem of these coasts, due to its high biodiversity and services offered to society, is the Tubul and Raqui wetland (Fig. 10.1).

Associated with the coastal basin of the hydrographic system formed by the Tubul and Raqui rivers, located at the southern end of the Arauco Gulf (37 ° 13′S – 73 ° 26′W), 17 kilometers south of the city of Arauco (Valdovinos 2001), this wetland is considered the most important in the region, due to its large extent, which reaches approximately 2238 ha and is a hot spot of biodiversity.

In this wetland, the Tubul and Raqui rivers, of coastal origin and presenting a pluvial regime, flow into the sea. The wetland is typically an intertidal marsh-type estuary, in which there is a marked salinity gradient as a consequence of the transition between inland and coastal marine waters (Long and Mason 1983). Given these

[1] Type F, which includes permanent estuarine waters and estuarine delta systems.

[2] Type H, which includes marshes and flooded areas with salt water, halophytic grasslands, salt beds, flooded areas with salt water, freshwater and brackish areas flooded by the tide.

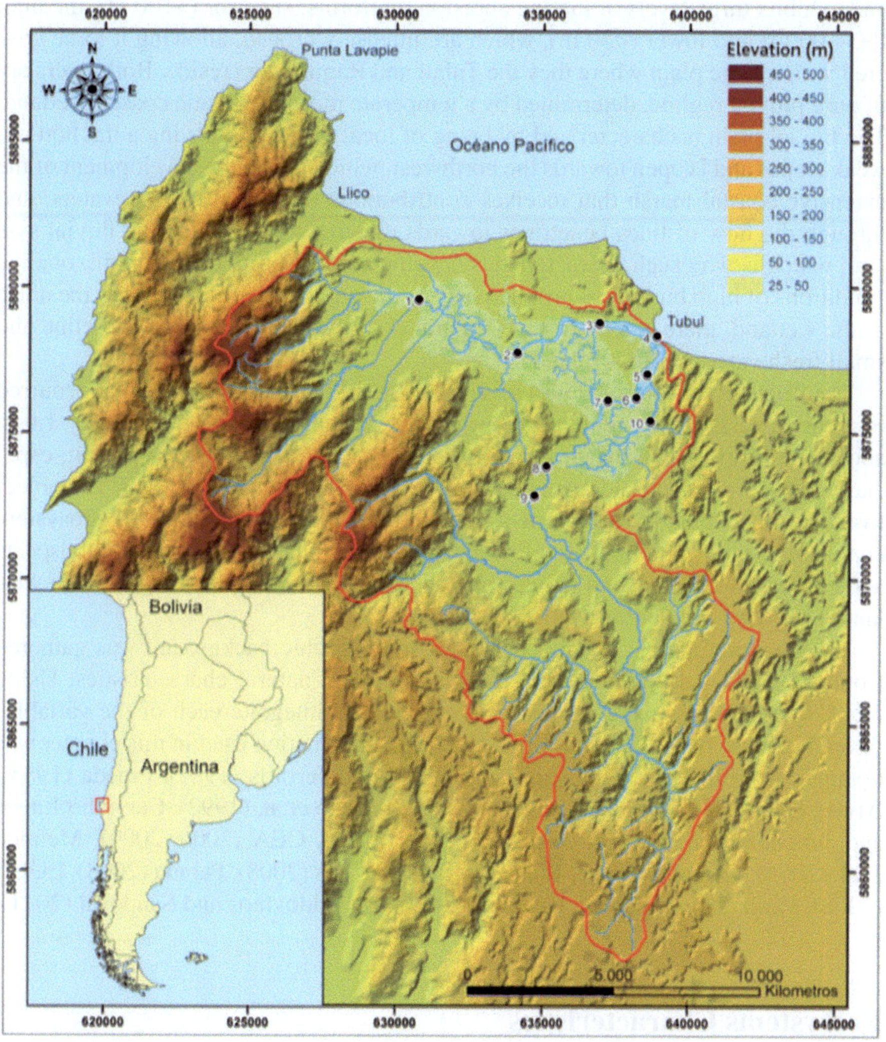

Fig. 10.1 Geographic location, relief and water network of the Tubul-Raqui wetland basin (Modified from Vasquez 2009), and location of sampling stations considered by Valdovinos and Sandoval (2011)

salinity conditions, the wetland has been an important part of the livelihood of the local human population concentrated mainly in the Tubul creek, through the cultivation and extraction of the algae Gracilaria spp., which until the February 2010 earthquake was intensively cultivated in the estuary.

According to Vasquez (2009), the system that hydrically feeds the wetland, is inserted in a sequence of platforms of marine erosion located in the vicinity of the western slope of the Coastal Mountain range. This unit has a high rate of forestation and exhibits three levels of marine abrasion platforms: superior (>100 m), medium (50–100 m) and lower (<50 m), which are intensely eroded, allowing it to be covered with a large plain where they the Tubul and Raqui rivers reside. Both rivers are of pure pluvial regime, determined by a temperate morphoclimatic coastal wetland. The fluvial plain is characterized by being of local origin and having a flat bottom, weak slopes and is open towards the northwest, which favors the development of the extensive coastal marsh that receives contributions of coastal marine waters, and through the flow of tides penetrates towards the continent through the fluvial system, which is why such characteristics develop at its river mouths with mixohaline conditions of high biological productivity. At the outer edge of the coast to the north of the wetland, there is a narrow plain parallel to the coast that has a shoreline and small freshwater wetlands inland.

The Tubul-Raqui wetland was severely affected by the earthquake that occurred on February 27, 2010, when it has suffered a vertical rise of approximately 1.6 m above sea level, which meant significant changes in its aquatic component, especially in the area of intertidal marshes that were partially out of the water. The objective of this chapter is to present a descriptive synthesis of the main ecosystem characteristics of the Tubul-Raqui wetland, with emphasis on its structural aspects and conservation value, and to present a synthesis of the major changes in the wetland, following the February 2010 earthquake.

For the development of this chapter, a bibliographic background was gathered from the study area, with emphasis on its physical-natural characteristics. Using those, cartographies were created to represent and integrate each of the variables through the use of the GIS platform. The basic information used in this chapter was obtained from the following studies: Biro (1979), Ferraris (1981), Pineda (1985), Alveal (1988), Werlinger and Alveal (1988), Stuardo et al. (1992), Carvalho-Lagos and Jimenez (2009), Carrasco-Lagos (2003, 2004), CEA (2006, 2009), Melnick et al. (2006), Conama (2007), Nielsen and Valdovinos (2008), Parada (2008), EULA (2008), Vergara et al. (2008), Vasquez (2009) and Valdovinos and Sandoval (2011).

Ecosystems Characteristics

Climate Framework and Geomorphology of the Basin

According to Vasquez (2009), the Tubul-Raqui wetland is located in a climatic transition zone between a warm temperate Mediterranean climate and a humid or rainy temperate climate, which develops immediately south of the Biobio River. Specifically, in the area of the Tubul-Raqui wetland, the climate is humid coastal temperate. According to Constabel (1993), in this zone there are anywhere from 2 to 5 months of cold and humid weather, and during those times, approximately 75% of the precipitations are concentrated, reaching 1300 mm annually between autumn and spring. The summer periods are quite dry, reaching between 5–6% of the annual precipitations. The relative humidity shows an oscillation around 80%, which is due in large part to the marine influence.

According to Pineda (1985), the Cretaceous-Tertiary sedimentary basin of Arauco, located in the present Arauco peninsula and the corresponding continental shelf, is characterized by an alternating the marine and continental sedimentary sequences. In this basin the following formations are recognized: Quiriquina (Senonian), Pilpilco (upper Eocene), Ranquil (Miocene) and Tubul (Pliocene). Vasquez (2009) has pointed out that the Arauco peninsula corresponds to an elevated continental slope, in which the Meso-Cenozoic sequences have been exposed. Pineda (1985) has proposed that during the Neogene (Pliocene) - Pleistocene, as a consequence of the pyogenic movements of the Pliocene, part of the area of the Gulf of Arauco basin has been raised and affected by a strong erosion, and at the same time as in the other areas, the sea transgressed on Miocene and Eocene sections. The marine sequences deposited during the Pliocene structured the Tubul Formation (see Nielsen and Valdovinos 2008). As Vasquez (2009) points out, from a geomorphological point of view, the Tubul-Raqui wetland has developed on an extensive fluvial-marine sedimentation plain of the coastal edge of the Arauco Province, an area that, in the last 50 years, has been severely affected by tectonic subsidence and sinking movements. The increase in flooded areas and/or the enhancement of flooding processes is due in large measure to the gradual obstruction of drainage caused by intense tectonic activity. This plain is developed with a flat morphology, with weak slopes, whose low plateau is only interrupted by the presence of a small marine terrace of lower level (<50 m), which helps to divide the hydrographic systems that are only interconnected in the opening, after surrounding the small terrace. Much of the wetland rests on a recent quaternary, in deposits of black sands, and river and beach sediments. Bordering this wide plain, there are three levels of terraces of marine abrasion (Fig. 10.2), which have been mainly modeled by the interaction of tectonic activity and climate. The main characteristics of these terraces are the following:

A. Lower level marine erosion terrace (<50 m): It is located around the wetland in the northeastern sector, presenting a semi-annular distribution, with a flat morphology at the top. This terrace is structured in sedimentary rocks of the Ranquil formation of Miocene age. Due to its morphology and the level of erosion that it exhibits, in its inner part a small plain of flat topography has developed, and it houses small marshy areas, fed by the drainage of a smaller estuary, that receives the contributions from numerous ravines activated only during the winter season. On this terrace there is also a scarce, semi-stabilized dune system.
B. Mid-level marine abrasion terrace (50–100 m): It is distributed in the east and southeast sector of the wetland. It is flatter than the previous one and is better preserved in terms of its morphostructure, but it is more intervened anthropically, due to the construction of access roads such as highways and the presence of dispersed human settlements. This terrace is dissected by valleys with flat bottoms that have taken advantage of fault lines to inscribe their morphology, exhibiting small, but well inscribed valleys that flow towards the Gulf of Arauco.

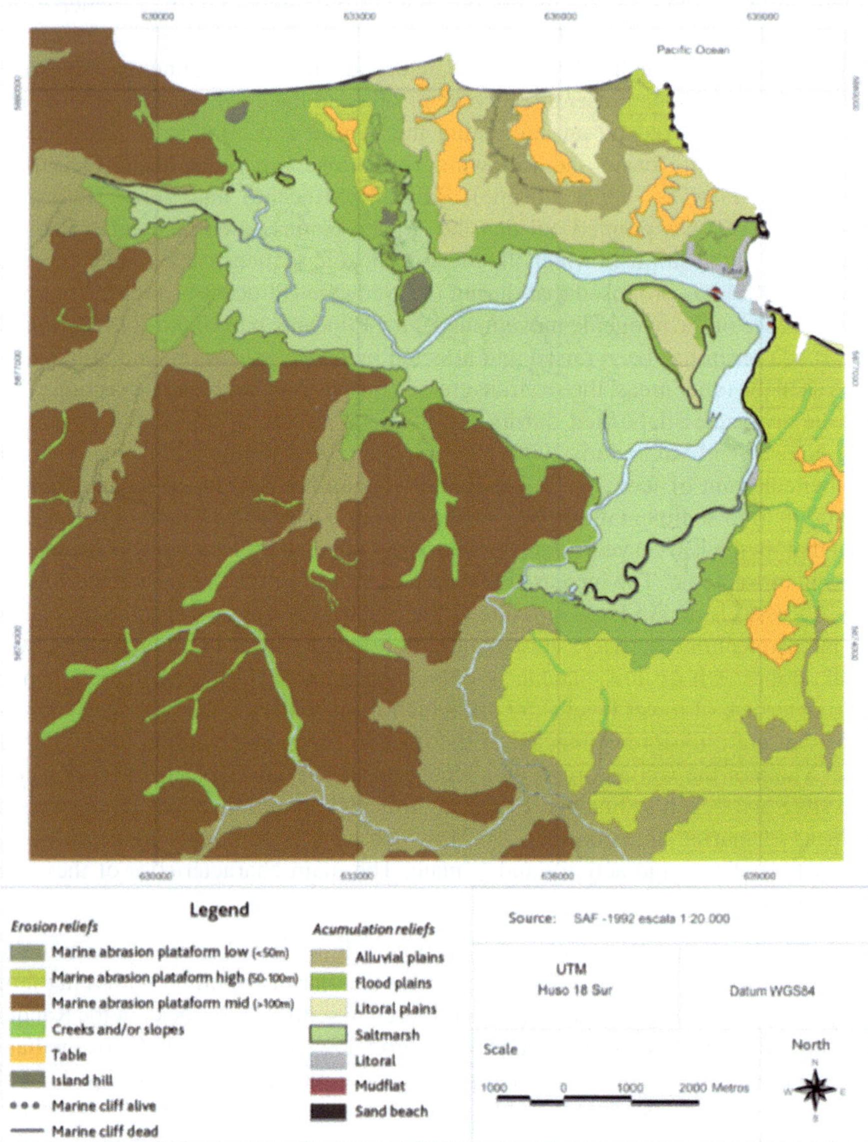

Fig. 10.2 Geomorphology of the Tubul-Raqui wetland basin (Modified from Vasquez 2009)

These valleys feed small strips of coastal marshes, which have no connection to the great wetland of Tubul-Raqui. This terrace denotes vestiges of the old direct contact with the ocean, with the presence of the dead high sea cliffs, which are currently used as nesting shelters for seabirds. The mid-level terrace is built on rocks of the Tubul formation, Pliocene age.

C. Top-level marine abrasion terrace (100–300 m): It is situated on the entire west part of the wetland, and is structured by sedimentary rocks of the Ranquen, Milongue and Trihueco formations, of Eocene age.

Water Network and Hydrology

The coastal basin formed by the hydrographic system of the Tubul and Raqui rivers is inserted in the corridors of the Arauco Gulf. According to Vasquez (2009), the basin as a whole has a total area of 274 km^2, with a perimeter of 81 km. The sub-basin of the Tubul River occupies 37.5% of the total area of the basin, and 64.3% of the River Raqui. The basin of the Tubul-Raqui wetland is located on the remains of marine terraces, modeled during repeated episodes of transgressions and marine regressions, opening to the sea with the combination of estuaries and marshes of the region (Fig. 10.2). The general characteristics of the formation of this coastal sector and its two sub-basins are considered by Pineda (1985) to be of Recent Pliocene origin, which by processes of filling, deposition of sediments and basements, were structured with the current characteristics.

According to Alveal (1988), who studied the distribution of Gracilaria sp. In the estuary, the Tubul River had an approximate course of 17–19 km long, with influence of the high tide over the first 6 km, from the mouth upwards. The Raqui River is about 15 km long and shallower than the Tubul, with maximum depths of up to 2 m, and water runoff throughout the year. The Tubul River presents sediments formed by the fine sand, in great part, from the bed of the river, and very fine sand in the margins in front of the estuary, with anoxic sludge in areas of scarce circulation. On the other hand, in the area near the mouth of the river Raqui, thinner sediments (very fine sands and silty clay) are developed in front of secondary streams, and a large part of the riverbed is covered with fine sand, especially on the left riverbank, and with medium sand in the central parts and to the right. The great estuary formed by the union of both rivers at its mouth, called the Tubul-Raqui estuary, corresponds to the micromareal regime, with seasonal contributions of fresh water. In general, it presents a greater contribution of fresh water through the Raqui River, even in summer, while in the Tubul River there is a greater marine influence (Stuardo et al. 1992). The morphometric analysis of the Tubul-Raqui basin conducted by Vásquez (2009) shows a compact index of 1.37, indicating that it is a basin whose shape resembles more to the circular form than to the elongated shape, which also explains the relatively rapid response to extreme rainfall events. It should be noted that the two main channels are heavily influenced by surface runoff and annual rainfall.

According to the drainage hierarchy proposed by Strahler (1981), the Tubul River is of the order 3, whereas the Raqui River is of order 4, reason why this river is better fed than the first, receiving the confluence of a series of smaller estuaries, including Las Peñas and Los Puentes. On the other hand, the Tubul-Raqui coastal basin presents the typical dendritic drainage pattern for the exortic basins of continental Chile. This drainage pattern is characterized by an irregular branching of tributaries, and is typically present in those places where the rocky substrate is relatively homogeneous and does not offer greater resistance to river erosion, its shape being determined mainly by the slope direction of the ground. This type of drainage pattern can be observed in some sectors of the headwaters of the basin, and especially in the lower course where both hydrographic systems meet, due to the fact that there is a greater predominance of moderate slopes here. On the other hand, and due to the antecedents of the basin previously indicated, it can be recognized that in much of the middle and lower course there is a predominance of the sub-parallel drainage pattern, which is determined by the remarkable structural control of the zone described by Pineda (1985). As mentioned previously, this area was the subject of alternating epigenetic descents and ascents, possibly during the Eocene era, characterized by a strong tectonics of extension faults, which are currently active, and which are responsible for giving rise to structures of "horts" and "gra-ben" type. However, it is also necessary to take into account that a large number of drains of the lower course, that present a rectilinear physiognomy, have been altered by human action and channeled to be used in agricultural and livestock activity in the area.

From a hydrological point of view, the estuarine system formed in a large part of the marsh, receives its main contributions of fresh water through the river Raqui, even in the summer season, and in the Tubul River is generally found a greater marine influence, as described by Stuardo et al. (1992). As it will be discussed below, this situation has been greatly disrupted following the February 2010 earth-quake, which produced major changes in the interaction between inland and marine waters, as a result of the waterbed elevation. According to Stuardo *et al.* (1992), the maximum depths of the system exceeded 2 m, and today, after the earthquake, they are approximately 0.4 m.

Soils and Vegetation

According to the CIREN-CORFO (1999; fide Vasquez 2009), the largest studied area is occupied by soils belonging to class VII, with very severe limitations, which makes them unsuitable for crops, but not for forestry and grazing. Much of the area occupied by this type of soil is severely degraded or eroded. According to Mardones (1971), this is mainly due to intensive agricultural and grazing practices of the late ninetieth and early twentieth centuries, and to the planting of exotic forestry species, replacing more than 80% of the original native vegetation. Due to this, approxi-mately 54% of the soils are in an advanced state of erosion, despite not having strong slopes sectors, except in the river ravines. The main soil series present in the area of the Tubul-Raqui wetland basin corresponds to the following: Miscellaneous Swamp (MP), Caripilun Series (CR), Las Puentes Series (LPU) and Larkote Series (LQT).

In the Tubul-Raqui wetland basin, part of the original forest cover has been heav-ily modified and exploited more than a century ago (Constabel 1993), as well as

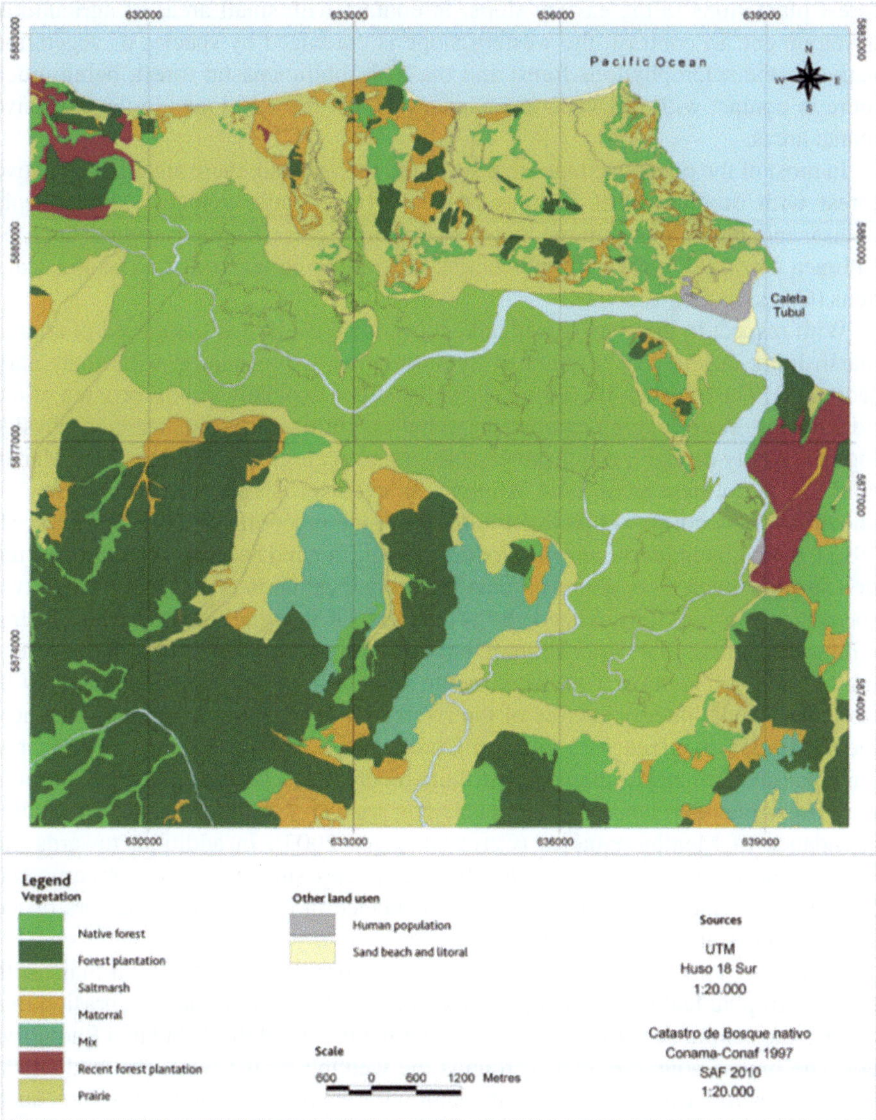

Fig. 10.3 Vegetation units in the Tubul-Raqui wetland basin (Modified from Vasquez 2009)

being heavily intervened and fragmented, currently occupying the ravines and some of the highest marine platforms (Fig. 10.3). Much of the native woodland vegetation has been replaced by plantations of Pinus radiata and Eucalyptus globulus, and may now be relegated to small patches of native vegetation mixed with forest plantations and/or in the interior of the ravines. The small platform that separates the Tubul and Raqui rivers in the lower course is an example of this, presenting small stretches of forest plantations in the eastern slope, in contact with small areas of agricultural development. In contrast, the western slope is colonized by species of vegetation native to the sclerophyllous forest and the hydrophilic coastal forest, being much more in contact with the wetland environment that has developed in the extensive marsh areas.

In most of the platform slopes that surround the wetland, there are relics of native forest with dominance of Peumus boldus, Myrceugenia exsucca, Myrceugenia obtusa and Podanthus mitiqui that also includes less abundant species such as Lithraea caustica, Aetoxicum punctatum, Aristotelia chilensis and Griselina scandens (Carrasco-Lagos 2003).

With respect to the coastal wetland environment itself, it is composed of a large marsh lying on the coastal plain of fluvial-marine sedimentation with quaternary sediments. The marshes are salt marshes that suffer two daily floods, product of the tidal cycles. Waterlogging creates anaerobic conditions in sediments, and physiological salinity drought, which generates extreme ecological conditions (San Martin et al. 1992). In these areas, only salinity – adapted species survive, such as the one known as Sarcocornio – Spartinentum densiflorae (according to San Martín et al. 1992), which includes the species Spartina densiflora and Sarcocornia fruticosa, the first species much more abundant than the second (Stuardo et al. 1992). These two species are very closely related to the marshes of the south central Chile, being arranged in strata. The upper stratum tends to be dominated by S. densiflora, reaching maximum heights of 1 m, whereas the stratum basal <1 m, is dominated by S. fruticosa. To the interior sectors of the rivers, and with minimal marine influence, the vegetation is mixed with grass of Cotula coronopifolia, Polyonum persicaria, Taraxacum officinale, Rumex sp., Trifolium repens, Eleocharis pachycarpa, Cyperus eragroslis, Plantago lanceolata, Flypochaeris sp. Carex sp. species, and in smaller abundance of Mentha aquatica (Carrasco-Lagos 2003). In addition, towards the head of the rivers, it is possible to observe species such as Juncus procerus and Scirpus californicus, the latter of great abundance in the innermost sectors of the Tubul River (Carrasco-Lagos 2003).

According to Vasquez (2009), the wetland is developed in a coastal plain with flat topographic features, with a presence of a small terrace of marine abrasion that divides the central sector of the wetland near the mouth of the Tubul and Raqui rivers. The high topography that surrounds the wetland by the west, east and south, corresponds to terraces of marine abrasion with maximum heights of 300 m, such as the internal terraces towards the continent. However, these slopes lead to the wetland sector with extremely moderate slopes, which are probably due to the intense erosion work carried out by the ocean during the phases of marine regressions and transgressions. Even so, it is still possible to observe that the geomorphological unit immediately adjacent to the wetland corresponds to flood plains, strongly influenced by the hydrology of the wetland itself, and the hydrology of the entire basin of the Tubul and Raqui rivers. These plains are characterized by sweet pasture species and are mainly used for the development of the agricultural and livestock activities that are carried out in the area, presenting a considerable change regarding the humidity level between the months of autumn-winter and spring-summer. During the winter, these plains are extremely humid, even becoming completely saturated by surplus water from the recurrent rainfall recorded throughout the south-central zone, unlike the dry months, when the plains receive water sporadically through concentrated, small ravines surrounding them, and in some cases, are even affected by water stress.

The vegetation of the Tubul-Raqui marshes is constituted by emerged, non-arboreal communities of vascular plants with roots, that occupy territories where the salinity, derived from the coastal marine influence, generally exceeds 1–4 UPS, and where the periodic emptiness generated by the tides, happens at least once a year, normally during the equinoctial tides. The perennial species and almost all halophytes dominate in the marshes, although there are other species of wide distribution, but those exhibit different character (Adam 1990). The scarce number of species able to tolerate the high salinity of these ecosystems determines the amount of small, but diverse vegetation (Chapman 1974; Adam 1990; Jiménez 1996). In the Tubul-Raqui wetlands, S. densiflora dominates (Fig. 10.4), and corresponds to a salt-tolerant grass, although it is also possible to find succulent halophyte Sarcoconnia fruticosa in some sectors, and in many cases it is possible to observe a combination of both species, which gives rise to complex mosaics of vegetation.

The species Spartina densiflora, because of its wide coverage, density and biomass, is a fundamental species in the structuring of the Tubul-Raqui wetland (Figueroa and Valdovinos 1997). This is one of the 16 species of the genus Spartina

Fig. 10.4 General view of the *grassland* dominated by *Spartina densiflora* that is observed with a darker tone (Photograph: C. Valdovinos, April 2010)

Schreber described in the world, all perennial rhizomatous and tolerant to salinity (Mobberley 1956; Jiménez 1996). This genus is widely distributed in both hemispheres, inhabiting mainly the coastal systems of America, Africa and Europe (Chapman 1977). According to Jimenez (1996), the species of this genus possess excretory glands of salt, and although they develop preferentially in saline soils, it is not only limited to them. According to Chapman (1977), the marshes dominated by Spartina represent very relevant communities within the estuaries, being very successful in the colonization of recent sedimentary deposits. Species of this genus are vigorous herbaceous plants, which can form large prairies in areas of estuarine sand and sludge (Chapman 1974). Its sediment retention capacity has motivated its planting in many areas of the world, having been introduced by man, as settlers, in different sectors of the Asian coast and Oceania (Chung 1990). According to Begon et al. (1987), the genus Spartina is one of the few groups of C4 plants that, despite the high temperatures required for their optimum growth, extend their geographic range to temperate climate latitudes, occupying saline ecosystems, in which the osmotic conditions especially favor these species with a high efficiency in the use of water.

Spartina densiflora Brong is of South American origin and is very abundant in certain parts of the Chilean coast, as in the Tubul-Raqui wetland, where it constitutes almost monospecific communities (Mobberley 1956; Figueroa and Valdovinos 1997). The taxonomic status of this species has been the subject of some controversy. In this paper, we have followed Mobberley (1956) and Cabrera (1970), who consider Chauvinia chilensis Steud., S. montevidensis Arech., S. patagonica Speg. and S. juncea Willd. Var. Montevidensis St. Yves as synonyms of this species, According to Costa and Davy (1992), S. densiflora is geographically distributed along a wide latitudinal gradient on the American coasts, encompassing both hemispheres and the Pacific and Atlantic Oceans (ca 32 ° N – 51 ° S). Within this territory it presents a shared distribution, although clearly skewed towards the Atlantic coast of the Southern hemisphere, where it extends uninterruptedly by the marshy coastal environments for thousands of kilometers. In the southern hemisphere, it is found in much of the temperate and subtropical zone of South America, on the coasts of southern Brazil, Uruguay, Argentina, in addition to central and southern Chile. In the Northern Hemisphere its distribution is limited to small nuclei on the shores of the Gulf of Mexico. Both regions are separated by a wide strip, with limits very close to the tropics. Among the 86 vascular species considered in the review of coastal marsh communities in Latin America by Costa and Davy (1992), S. densiflora is the one that encompasses a greater latitudinal gradient. Other marsh species that present a similar distribution pattern in this geographical area are Juncus acutus (L.) and Sarcocornia fruticosa.

The "espartales" of the Tubul-Raqui wetland are a key element for the conservation of the wetland, being a bioengineering species that strongly models the structure of aquatic habitats. Fortunately this is a species tolerant to strong environmental changes, and it was not affected by the of 27 February earthquake. A different situation would have occurred in the event that this species had been affected, causing a series of physical, chemical and biological changes in the wetland.

Terrestrial and Riverine Fauna

With regard to amphibians, the species are restricted to the freshwater wetland sectors, with abundant palustrine vegetation dominated mainly by "Junco" (Juncus procerus) and "totora" (Scirpus californicus). This group has been recorded in the wetland by Vergara et al. (2008), as being the most frequent species to observe in these habitats, like the "Four Eye Froggy" (Pleurodema thaul), which are common to find in its larval states in pond areas. Another species registered is Batrachyla taeniata, which according to SAG (2009) qualifies in the category of "Vulnerable". Reptiles present in the wetland are suitingly distributed in the sunny areas surrounding the "espartales", although it is also possible to observe them in the vega sectors, during the summer. According to Vergara et al. (2008), there are four species of lizards belonging to the genus Liolaemus (L. cyanogaster, L. tenuis, L. lemniscatus and L. chiliensis), and two species of snakes, corresponding to Tachymenis chilensis and Phylodrias chamissonis. Out of the reptile species, the only species classified in a conservation category is L. tenuis, which according to SAG (2009) qualifies as "Vulnerable".

The Tubul-Raqui wetland is a marshy intertidal estuary, with a marked salinity gradient, which allows the existence of strong environmental gradients, and an important diversity of avifauna habitats. That is why this wetland is one of the main nesting and shelter sites for numerous species of birds, resident and migratory, of the Biobio Region. The special geographic and ecological conditions of this wetland permit the presence of a fluctuating number of migratory and resident bird species, among which there are 31 families and 83 bird species (Carrasco-Lagos 2003, 2004; Vergara et al. 2008; Carrasco-Lagos and Jiménez 2009). The most diverse species in the wetland are Anatidae (8 spp.), Laridae (7 spp.), Furnariidae (7 spp.), Tyrannidae (7 spp.), Ardeidae (6 spp.) And Scolopacidae (4 spp.). Among the migratory birds are the Rynchops riger, Numenius phaeopus, and the Limosa haemastica. Regarding the conservation status of wetland species, the Hunting Law of the Agricultural and Gaming Services.

SAG (2009) established that the species Cygnus melancoryphus and Coscoroba coscoroba as "endangered", while the Theristicus melanopis qualifies as "Vulnerable". On the other hand, according to the International Union for the Conservation of Nature (IUCN)[3], the species Sterna elegans, Phalacrocorax gaimardi, Pelecanus thagus, and Anas specularis qualify in the category of "near threat", that is, they could soon be classified as "Vulnerable".

According to Vergara et al. (2008), mammals are represented by four species of rodents, corresponding to: Abrothrix olivaceus, Abrothrix longipilis, Oligoryzomys longicaudatus, and Myocastor coypus. According to SAG (2009), only Myocastor coypus qualifies within the category of "Vulnerable".

[3] http://www.iucnredlist.org/.

Aquatic Flora and Fauna

The aquatic flora and fauna presents strong gradients in the wetland, which are closely associated with the variations of salinity. In general, it is possible to recognize two large areas, one sweet water with saline <4 UPS and another estuarine or mixohaline, with strong fluctuating salinities in a range of 4 and 30 UPS, depending on the location within the estuary, and its tidal condition .

Estuarine Area

The microalgae of the estuarine zone have been studied by CEA (2006), indicating within the peripheral community the presence of 38 species of diatoms, most of which correspond to marine environments. In quantitative terms, the species of greatest abundance registered in the study are Opephora pacifica and Opephora martyi, which have reached respective densities of 373,713 and 175,913 cel/mm^2, inside the estuary. In a second study carried out by CEA (2009), 44 phytobenthic species were recorded, the majority of them being diatom species typical of environments. In this study, dominant species in the community were Fragilaria pinnata and Achnanthes thermalis. As for macroalgae, the dominant species until before the February 2010 earthquake was the Rodophyta Gracilaria chilensis, which as mentioned above, was of great economic importance for the inhabitants of the area (Dellacroce and Valdovinos 1994). Another species present in the area is Chlorophyta Ulva (= Enteromorpha) intestinalis, which develops in the quieter areas of the protected areas of the estuary. This species was also affected by the earthquake, though to a lesser extent than the aforementioned species.

In relation to estuarine zooplankton, this was also studied by CEA (2009), finding as a general pattern, a greater diversity of taxa in the estuary area of the Tubul River than in that of the river Raqui. The taxa recorded in the area corresponded to copepods Calanoidae, Cyclopoidae, and copepodite and nuplius states. Among the taxa found, they were notable for their abundance Tumediaptomus (= Diaptomus) diabolicus and Microcyclops sp.

With respect to the benthic invertebrates of the estuarine area, these are less diverse than those in the freshwater zone located in the upper part of the basin. However, their abundances, biomass, whether per unit area or throughout the ecosystem, are clearly higher. Within the sessile or sedentary species present in the estuary, are basically recognized as those that inhabit hard river beds, and others that inhabit soft river beds.

Before the earthquake, the dominant hard river beds species corresponded mainly to riparian root habitats. These correspond to the red wines of Spartina densiflora located in the limits of the spartal and which were, before the earthquake, affected by the tide. Subsequent to the earthquake, these habitats were completely desiccated due to the severe lifting of the wetland, which meant the total loss of its aquatic fauna and its replacement by terrestrial invertebrates (Valdovinos and Sandoval 2011). Before the earthquake, the dominant species in these habitats were the estuarine Bryozoa estuarine Conopeum sp., The Amphipoda Paracorophium hartmannorum and polychaetes, especially Prionospio (Minuspio) patagonica. These habitats were also occupied as refuge and feeding areas by juvenile stages of the crustacean Decapoda Hemigrapsus crenulatus, which were generally abandoned after reaching 20 mm cephalothorax lengths to occupy the soft bottom habitats of the wetland. Another relevant component of hard-bottomed benthos corresponded to Cirripedia Elminius kingii, which was also severely affected by wetland drying. This filtering species inhabited preferably woody and rocky substrates present in the intertidal zone of the wetland.

Prior to the earthquake, the intertidal and subtidal soft beds of the wetland were occupied by large densities of polychaetes of the Spionidae and Nereidae families, which reached high biomass, and were especially dominated by P. (M.) patagonica and Perinereis gualpensis. In these environments were also located great concentrations of the bivalve Tagelus dombeii that were exploited in the area for commercial purposes. These depths were also occupied by the adult crustaceans H. crenulatus, reaching high densities, and in some sectors it was possible to find the small estuarine bivalve Kingiella chilenica. As it will be discussed later, populations of polychaetes and H. crenulatus were significantly reduced after the earthquake, due to the desiccation of much of the water floors. On the other hand, the banks of T. dombey and K. chilenica disappeared completely from the wetland. Little is known about the fish in the estuarine area of the wetland. General observations made in the area by our research group before the earthquake of 2010 confirmed the presence of estuarine and marine species, such as "Elegonops maclovinus", "Mugil cephalus, Galaxias maculatus and Odontesthes mauleanum.

Freshwater Zone

Since the Raqui River sub-basin is the most important in extension, draining part of the high sectors of the mountain range of Nahuelbuta, its fluvial and caudal geomorphology allows the existence of an abundant and diverse aquatic flora and fauna. This is not observed in the case of the Tubul River, which has lower tributaries that originate at low altitude. Upstream of the estuary, in the river Raqui, it is possible to recognize two clearly defined zones. A ritron zone is in the high part of the basin, characterized by waters of high speed, rocky beds and a typical sequence of rapids and backwaters. The other area is the "potamon" one, located on the plain, characterized by slower flows, sand and gravel beds, and without the sequence of rapids and backwaters. The "ritron" area is very poor in periphytic microalgae, since the river is mainly a gallery and a shadow, due to the important influence of riparian vegetation, both native and exotic species (Pinus radiata and Eucalyptus globulus). There are no systematic studies of the aquatic biota in this area of the basin, and what is described here corresponds to general observations made by our research group during April 2011. Within the aquatic and riparian flora of this sector, the presence of moss of the genus Sphagnum, covering part of some semi-submerged rock substrates is highlighted. This area is very rich in benthic macro-invertebrates, which are represented by immature states of 32 insect families, mainly belonging to the orders Trichoptera (Glossosomatidae, Helicopsychidae, Hydrobiosidae, Hydropsychidae, Hydroptilidae, Limnephilidae and Seicostomatidae), Plecoptera (Austroperlidae, Diamphipnoidae, Eustheniidae, Gripopterygiidae, Notonemouridae and Perlidae) and Ephemeroptera (Leptophlebiidae, Baetidae, Amelotopsidae, Caenidae and Nesameletidae). The crustaceans are represented by the shrimp Samastacus spinifrons and the crab Aegla sp. Further studies are needed to determine which species this crab belongs to, considering the high diversity of species (and genetics) in this area of the Coastal Mountain range (Habit and Victoriano 2005). It is important to mention that the representatives of Aeglidae are among the most threatened freshwater invertebrates in the country (Pérez-Lozada et al. 2002). As for fish, general observations indicate the dominant presence of the "rainbow trout" (Onchorhynchus mykiss), which is an exotic invasive species, and of the native species "painted catfish" (Tricomycterus areolatus). The fauna of the ritual area of the river Raqui has a high dependence on its riparian forests. Foliar detritus from these forests has been identified as one of the main energy components of the river ecosystem (e.g. various species of Nothofagus, among others), being an important source of food for invertebrates, which are the food base of fish species observed in the area.

In the potamal zone, which is open given the scarce riparian arboreal vegetation in the area, there is sufficient light to develop the community of periphytic microalgae. Qualitative observations made in the area show the dominant presence of diatoms of the genera Achanthes, Aulacoseira, Cocconeis, Cyclotella, Cymbella, Diatoma, Fragilaria, Gomphonema, Nitzschia, Pinnularia and Synedra. As for the benthic invertebrates, the diversity is considerably lower than the one existing in the potamal zone of the river, registering a total of 15 families of Mollusks (Chilinidae, Amnicolidae, Physidae, Sphaeriidae), Platyhelminthes (Dugesiidae), Oligochaeta (Lumbriculidae, Naididae), Crustacea (Hyalellidae, Parastacidae), Coleoptera (Dytiscidae, Elmidae, Gyrinidae, Hydrophilidae), Diptera (Chironomidae) and Hemiptera (Corixidae).

As for the fish in this area, in addition to recordeding the same two species present in the ritral zone, the presence of the native species Percichthys trout, Geotria australis and Galaxias maculatus.

Human Settlements and Economic Activities

According to what was documented by Vasquez (2009), the wetland area is inhabited only by rural population, which amounts to approximately 2500 inhabitants, concentrated mainly in the Tubul creek. According to this census, the poverty level of this cove is 56%. The second locality of population importance is the creek Las Peñas, located on the banks of the River Raqui, counting with a total of 261 inhabitants (according to data from the CASEN, commune of Arauco). The rest of the human presence adjacent to the wetland corresponds to small dispersed settlements, including the small localities of Raqui and Raqui Alto.

The population associated with the wetland is mainly engaged in agriculture and the raising of poultry, cattle and pigs, mainly developed in meadows and/or floodplains. All of them on a small scale, mainly for family consumption, with a small surplus for marketing, where one of the products that have stood out as of lately is the production of handmade cheeses. Agriculture has received some stimulus from the state, through the promotion of private investment in irrigation and drainage. This is why it is possible to observe a large number of pipes that drain part of the wetland. The livestock activity developed today is incipient, since in general, the area does not possess soil conditions favorable to the development of natural or artificial prairies, vital condition for all livestock economy.

A relevant productive activity carried out by the population of the Tubul creek, but associated with the proper coastal marine ecosystem, is the artisanal fishing activity. In Tubul, the main resource extracted is the "huepo" or "razor" (Ensis macha) and the "navajuela" (Tagelus dombeii), in addition to other mollusks and bivalves. These products are sold to individuals who export them mainly to USA, Japan and Spain. Another important activity is forestry, due to the presence of extensive plantations of exotic species, such as Pinus radiata and Eucaliptus globulus, which have replaced the original native vegetation. The erosion terraces adjacent to the Tubul-Raqui wetland have been gradually colonized since the beginning of the 1990s, due to the predominance of soils IV and VII, whose capacity of use allows the development of this type of activity.

An important economic activity developed in the wetland, prior to the earthquake of February 2010, was the intensive cultivation of "pelillo" (Gracilaria spp.). Due to the severe wetland rise caused by the earthquake, a large part of the channels where the crops were made was desiccated, which currently prevents the development of this productive activity. Until February 2010, the main economic activity inside the wetland was the cultivation of this seaweed, practiced solely by the inhabitants of the area, bringing together a total of 500 people. The extraction of "pelillo" has been regulated by the Undersecretary of Navy of the Ministry of Defense through a maritime concession of 320 hectares. The production of "pelillo" harvested before the earthquake was approximately 2000 tons of wet algae, which were exported mainly to Japan.

Livestock is also developed within the basin, through the breeding of cattle that are used for the production of meat and milk for the production of cheeses. Since the dominant vegetation in the wetland is characterized by thick and saline pastures, livestock is limited mainly to areas located between mountainous and "espartal" sectors. Agricultural activity is incipient within the basin and is mainly intended for self-consumption.

Use of Soil in the Wetland Area

According to the photo interpretation work carried out by Vasquez (2009), the area of the Tubul-Raqui wetland and its surroundings, presents a total of 10 classes of land uses, corresponding mainly to "forest plantations" (Pinus radiata and Eucalyptus globulus), covering a 33% of the total area of the area, followed by "agricultural use" intended mainly for rotating crops of wheat and oats, corresponding to 22% of the area. The rest of the developed economic activities has an insignificant area compared to the above mentioned ones, not exceeding 1%; among them the "aquaculture use" for the cultivation of seaweed Gracilaria spp. (Even before the earthquake of 2010) and "recreational use", represented by a small resort located in the coastal area. The third dominant land use currently corresponds to "grassland shrubs", which are located mainly in the remnants of mid-level terraces (50–100 m) and in those sectors of the upper level platforms (>100 m) that have not been reached by the expansion of forestry, and represent 19% of the study area.

The area corresponding to "wetland" has represented up to 16% of the area before the earthquake (there are nowadays extensive drying areas not yet quantified), while the "native forest interspersed with scrub" does not reach 10%, corresponding mainly to remnants of "olivillo", "oak" and "boldo" forests. The "human settlements and beaches" also do not represent 1% of the area, due to the scarce development of the sandy coast in the area and because most of the human settlements are dispersed.

Water and Sediment Quality

The main study of the sediments and their dynamics in the wetland is done by Constabel (1993). This study describes, preferably sandy bottoms in the wetland, and in the vicinity of the estuary mouth, in areas with high hydrodynamics due to tidal currents. On the other hand, in areas of less hydrodynamics in the interior of the wetland, there are predominantly muddy bottoms. In close relation with the dynamics and the contributions of autochthonous and allochthonous organic matter of the system, Stuardo et al. (1992) showed high levels of the organic matter content of the sediments, which vary from 2–4% in the vicinity of its mouth and 10–20% in the interior of the wetland.

Regarding water quality, the various studies conducted in the area prior to the 2010 earthquake (Stuardo et al. 1992; Constabel 1993; CEA 2006; Valdovinos and Sandoval 2011) generally showed a system with strong salinity changes resulting from tidal fluctuations and eutrophic conditions. Also, conditions of slightly lower salinity were observed in the Raqui River rather than in the Tubul River, due to the higher fresh water intake from its drainage basin. An important aspect highlighted by Stuardo et al. (1992) was the marked stratification of the waters at some times of the year, despite the low depth of the estuary. This stratification was due to marked density gradients generated by differences in continental and marine density. It should be noted that, despite this stratification and the presence of sediments rich in organic matter at the bottoms, there were no conditions of hypoxia or anoxia in the brackish waters.

After the February 2010 earthquake, Valdovinos and Sandoval (2011) describe the changes observed over time in both the sediments and the estuary water column, which will be described in more detail below. Table 10.1 and Fig. 10.5 present the water quality data of the Tubul and Raqui rivers (obtained by Valdovinos and Sandoval 2011), in the 10 sampling stations indicated in Fig. 10.1.

These data were obtained before the earthquake (December 2008 and August 2009), and after the event (April, August and December 2010). This table compares the values of temperature, salinity, pH, dissolved oxygen, chlorophyll-a, ammonia, total nitrogen, total phosphorus and suspended solids. This Table shows that the most relevant change occurred after the earthquake was a decrease in salinity, especially in the dry season.

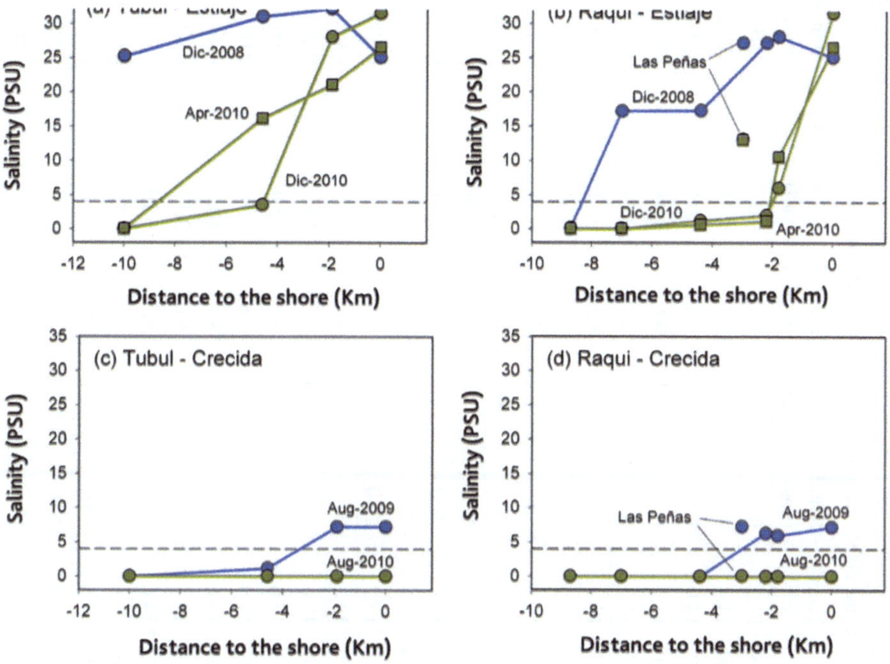

Fig. 10.5 Water quality in Tubul river, and in the 10 sampling stations indicated in Fig. 10.1. These data were obtained prior to the earthquake (December 2008 and August 2009), and after the event (April, August and December, 2010)

Table 10.1 Characterization of water quality in Tubul (stations 1–4) and Raqui (stations 5–9) rivers and Las Peñas (station 10) estuary, prior to the earthquake, December 2008 and August 2009, and after the event; April, August and December, 2010

Station	Date	Temp (°c)	Salinity (ups)	pH	oD (mg/l)	Chlorophyll (ug/l)	Ammonia (mg/l)	P-total (mg/l)	N-total (mg/l)	sst (mg/l)	ssi (mg/l)	sso (mg/l)
1	Dec-08	14.5	0	7.1	9.8	0.05	0.03	0.02	0.35	2.10	1.79	0.32
	Aug-09	13.4	0	7.7	9.7	0.04	<0.02	0.01	0.40	5.50	4.51	0.99
	Apr-10	14.0	0	6.2	9.8	0.08	0.20	0.09	0.80	3.20	2.80	0.40
	Aug-10	11.2	0	6.1	9.6	0.05	0.05	0.03	0.43	6.70	5.49	1.21
	Dec-10	14.2	0	7.5	9.7	0.09	0.02	0.03	0.09	6.40	4.50	1.90
2	Dec-08	14.7	31.0	7.0	9.5	0.12	0.03	0.04	0.39	8.20	4.67	3.53
	Aug-09	12.9	1.2	7.6	9.6	0.09	0.02	0.05	0.40	10.10	6.16	3.94
	Apr-10	16.1	3.5	6.7	9.7	6.31	0.26	0.51	0.80	39.30	20.70	18.60
	Aug-10	12.9	0	6.7	5.0	2.68	<0.02	0.09	0.41	17.60	12.67	4.93
	Dec-10	18.3	16.1	8.0	9.6	7.58	0.14	0.23	0.99	39.00	23.80	15.20
3	Dec-08	15.4	32.2	7.0	9.2	1.45	0.09	0.03	0.41	9.30	6.70	2.60
	Aug-09	14.1	7.2	7.6	9.3	1.31	0.02	0.05	0.40	10.20	6.94	3.26
	Apr-10	17.2	28.0	7.6	9.5	13.10	0.28	0.28	1.48	40.60	29.20	11.40
	Aug-10	13.1	0	7.1	4.8	3.45	0.02	0.08	0.42	19.40	13.39	6.01
	Dec-10	18.6	21.0	8.1	9.3	15.32	0.07	0.24	0.99	46.00	26.60	19.40
4	Dec-08	15.6	25.1	7.0	8.9	3.00	0.09	0.07	0.55	9.80	7.84	1.96
	Aug-09	11.9	7.2	8.2	8.8	2.10	0.02	0.09	0.49	20.50	15.99	4.51
	Apr-10	15.6	31.5	7.4	8.9	10.20	0.44	0.31	0.64	80.20	63.60	16.60
	Aug-10	13.8	0	7.1	6.6	3.12	0.02	0.12	0.59	30.40	24.02	6.38
	Dec-10	14.6	26.5	7.7	8.7	6.10	0.21	0.28	0.62	74.20	56.60	17.60
5	Dec-08	15.6	28.1	7.0	9.8	1.20	0.08	0.06	0.43	8.30	5.81	2.49
	Aug-09	13.6	6.0	8.2	10.0	1.10	0.03	0.05	0.40	17.20	12.21	4.99
	Apr-10	16.3	6.0	7.3	10.1	12.30	0.37	0.30	0.84	46.40	37.00	9.40
	Aug-10	15.1	0	7.3	8.8	2.95	0.04	0.06	0.30	19.90	18.40	1.50

Station	Date	Temp (°c)	Salinity (ups)	pH	oD (mg/l)	Chlorophyll (ug/l)	Ammonia (mg/l)	P-total (mg/l)	N-total (mg/l)	sst (mg/l)	ssi (mg/l)	sso (mg/l)
	Dec-10	18.5	10.5	8.2	9.9	5.30	0.11	0.28	1.27	23.80	12.20	11.60
6	Dec-08	15.1	27.2	7.0	9.7	1.32	0.04	0.04	0.32	7.90	5.61	2.29
	Aug-09	13.7	6.3	7.3	9.9	1.12	0.03	0.05	0.32	16.60	12.12	4.48
	Apr-10	17.1	2.0	7.2	10.3	6.25	0.47	0.30	0.46	30.40	23.40	7.00
	Aug-10	14.2	0	7.2	7.2	2.72	0.07	0.06	0.31	13.60	10.88	2.72
	Dec-10	17.8	1.1	8.1	9.8	4.80	0.13	0.12	0.64	16.40	12.60	3.80
7	Dec-08	15.1	17.3	7.0	9.9	1.01	0.02	0.06	0.67	5.20	3.80	1.40
	Aug-09	13.0	0	7.3	9.9	0.91	0.03	0.05	0.41	12.20	9.52	2.68
	Apr-10	17.0	1.2	7.0	10.3	3.45	0.49	0.28	0.78	36.80	30.00	6.80
	Aug-10	13.9	0	7.1	8.4	1.25	<0.02	0.09	0.48	14.20	11.36	2.84
	Dec-10	17.8	0.6	8.3	9.9	4.70	0.14	0.11	0.60	18.80	14.20	4.60
8	Dec-08	14.8	17.2	7.0	9.8	0.09	<0.02	0.05	0.71	4.80	3.79	1.01
	Aug-09	13.8	0	7.2	9.9	0.07	<0.02	0.03	0.32	9.00	6.93	2.07
	Apr-10	16.9	0	7.0	10.0	0.15	0.30	0.10	0.80	25.30	21.10	4.20
	Aug-10	13.4	0	7.0	9.7	0.09	<0.02	0.04	0.27	9.40	7.61	1.79
	Dec-10	17.3	0	7.6	9.9	0.20	0.13	0.06	0.37	11.80	9.20	2.60
9	Dec-08	13.6	0.1	7.1	9.5	0.09	<0.02	0.05	0.75	3.60	2.56	1.04
	Aug-09	12.9	0	7.1	9.7	0.08	<0.02	0.04	0.30	9.10	6.83	2.28
	Apr-10	16.8	0	6.9	9.8	0.10	0.24	0.08	0.82	19.60	16.20	3.40
	Aug-10	13.4	0	7.0	9.7	0.08	<0.02	0.05	0.32	10.20	9.38	0.82
	Dec-10	17.1	0	7.5	9.1	0.10	0.04	0.08	0.49	12.20	9.80	2.40
10	Dec-08	14.9	27.2	7.0	8.7	5.12	0.08	0.09	0.92	10.10	6.16	3.94
	Aug-09	13.5	7.3	7.3	8.9	3.12	0.04	0.10	0.61	21.00	12.39	8.61
	Apr-10	17.1	13.1	7.2	10.2	15.20	0.52	1.08	2.11	38.80	20.40	18.40

(continued)

Table 10.1 (continued)

Station	Date	Temp (°c)	Salinity (ups)	pH	oD (mg/l)	Chlorophyll (ug/l)	Ammonia (mg/l)	P-total (mg/l)	N-total (mg/l)	sst (mg/l)	ssi (mg/l)	sso (mg/l)
	Aug-10	14.3	0.1	7.2	5.5	4.14	0.18	0.14	0.68	24.20	5.49	19.01
	Dec-10	18.3	13.0	8.4	7.6	12.40	0.06	0.25	1.36	26.80	21.60	5.20

n-total total nitrogen, *do* dissolved oxygen, *p-total* total phosphorous, *iss* inorganic suspended solids, *oss* organic suspended solids, *tss* total suspended solids, *temp* temperature

Disturbances Caused by the 2010 Earthquake

In order to present the characteristics of the aquatic and riparian component, it is necessary to make a comparative analysis between the situation existing before the February 2010 earthquake and the current one, since the changes occurred in the wetland have been of great magnitude. This review is based fundamentally on the comparison of the 10 sampling stations that have studied previously the earthquake, by Stuardo et al. (1992), in 1990–1992, and by Valdovinos and Sandoval (2011), in December 2008 and August 2009 (Fig. 10.1). These same authors repeated the study during the same seasons after the earthquake, in April, August and December of 2010. According to the results obtained by Valdovinos and Sandoval study (2011), regarding the structure of the physical habitat, water quality and aquatic biota of the wetland, a series of environmental changes resulting from the earthquake (Table 10.1) have occurred since the February 27, 2010 earthquake.

Some of them occurred simultaneously during the earthquake, while others are still manifesting. Those generated during these events have been inferred from indirect evidence obtained at 40 days post-earthquake. On the other hand, the changes observed at 40 days, 6 and 10 months, correspond to observations and direct measurements made in the field. The following is a brief discussion of the results obtained by the authors, following the temporal sequence of events and field sampling.

Changes that Occurred During the Earthquake

The earthquake magnitude Mw8.8 had an approximate duration of 180 seconds, period in which occurred the lifting of the wetland of ca. 1.6 m above mean sea level. Farias et al. (2010) have described in detail the changes that occurred along the coast of central Chile. This survey would not have been homogeneous throughout the area of the wetland, since some areas would have experienced an uprising at almost two meters. During the earthquake, the cracking of many sectors of the "espartal" ocurred, cracks that in some areas of the river Raqui had a depth of 1.5 m and a width of 0.3 m. During the telluric movement, some of these cracks have released muddy sediments, which covered part of the plants of S. densiflora. With respect to water quality, it is likely that during the earthquake, the violent oscillations of the terrain and the water body of the wetland increased the turbidity by re-suspension of the bottom sediments. In relation to the aquatic biota, it is possible that this one has not suffered significant effects, except the burial of some macro invertebrates, by the transport of sediments caused by the elevation swell to the interior of the estuary.

Changes that Occurred in the First 10 Hours of the Earthquake

Given the magnitude of the rise of the wetland above sea level, it can be deduced that immediately after the earthquake strong currents of water were generated towards the sea, eroding part of the sand banks of the mouth. Evidence of this has remained on the banks of the lower reaches of the Tubul and Raqui rivers, where much of the sand and gravel fractions were transported downstream, leaving many boulders and large rocks deposited on the banks. The detailed characterization of the sediments and their associated processes had been described for the estuary by Constabel (1993).

Once the wetland was emptied, it was affected by the devastating entrance of tsunami, which reached a maximum of 12 m. This tsunami not only meant the entry of coastal marine water, but also of large amounts of sand deposited on the "espartal" and in the network of inner channels of the wetland, a situation that is frequent in this type of phenomena, as it was verified Cisternas et al. (2000) for the Maullin River estuary, which was affected by the 1960 Tsunami. Given the predominantly NS orientation of the Tsunami and the configuration of the coastline (see Ferraris 1981), the most affected area of the wetland was the sector of the river Tubul, where the tsunami entered by ca. 3 km. Along its Tubul River course, the body of water ripped from the shore entire pieces of espartal, up to 50 m^2 of surface, transporting them upstream, and these are the ones that currently make up islets in the middle of the channel. After the tsunami, the wetland continued to empty, until it was virtually disconnected from the sea by a sand bar, which limited the normal exchange with the coastal waters of the Gulf of Arauco. With regard to water quality, it is logical to think of an increase in salinity during the Tsunami. In addition, an increase in water turbidity is likely due to the re-suspension of fine sediments from the bottom of the wetland, adding to an increase in the organic sexton from the "wash" of the "espartal".

In relation to the aquatic biota, the most visible destruction was the total destruction of the "Pelillo" (Gracilaria sp.) seaweed plantations, and a high mortality of the estuarine crustacean Hemigrapsus crenulatus, by dragging these benthic organisms to the terrestrial system and their subsequent desiccation. This crab is one of the most abundant epifaunistic species in Chilean estuarine ecosystems (Grandjean 1985), also living in similar systems in New Zealand (Retamal 1981). It is able to regulate the volume and osmolarity of its body fluids, a physiological condition that allows it to live in estuaries (Retamal 1967; Taylor and Seneviratna 2005). In crustaceans, such regulation involves an energy cost that can generate changes in the oxygen consumption of the animal and the rates of excretion. Exposure to low salinities may also alter ingestion, another of the physiological aspects associated with the behavioral response of organisms. Recently, Urbina et al. (2010) studied the physiological energy of H. crenulatus, suggesting that the increase in energy expenditure at low salinity is due to an increase in osmoregulation costs, which are compensated by an increase in the rate of ingestion, which contributes to the success of individuals of this species in dynamic environments, such as estuaries. Subsequent to the Tsunami, it was frequent to observe, in the forest environs bordering the wetland areas, high densities of these dead organisms.

In relation to the "Pelillo", the loss of these plantations was total, which has had important social and economic consequences for the area. The relevance to local communities of this resource has been discussed by Alveal (1988), Werlinger and Alveal (1988), and recently by EULA (2008). Figure 10.6 shows dried remnants of the prairie of this algae.

With regard to birds, Valdovinos and Sandoval (2011) did not consider the study of the aquatic avifauna of the wetland, whose composition had previously been studied in general by Carrasco-Lagos (2003, 2004), and more quantitatively by Parada (2008). Within 40 days of the earthquake, it was possible to verify the presence of some dead birds, presumably directly or indirectly caused by the tsunami. In addition, a high presence of carrion birds as the "Gallinazo" (Cathartes aura), have suggested the existence of dead organisms in the area.

Changes Registered 40 Days After the Earthquake

At the beginning of April 2010, there was virtually no water runoff from the wetland to the sea. The area of the mouth was sealed by a wide bar of sand, which could not be crossed by small craft. It is necessary to indicate that due to the

Fig. 10.6 Dry remainders of an algae prairie "Pelillo" (*Gracialaria sp.*), in Raqui river (Photograph: C. Valdovinos, April 2010)

earthquake, the great bridges on the rivers Tubul and Raqui were destroyed. As a way to improve the connectivity of the Tubul creek, prior to the construction of mechanical bridges, a fill in or an embankment was built on the Raqui River that allows vehicle traffic. Therefore, in the first days of April, until the 6th of the same month, the Raqui River was momentarily fragmented, further favoring the drying up of the lower section of the river.

Through aerial and ground inspections, using a remote controlled model equipped with a camera, it was possible to verify the partial drying of 100% of the inner canals that irrigate the wetland, preserving only isolated pools of standing water, or with limited runoff towards the main channels. With respect to the latter, the desiccation of up to 85% of the bottoms of the main channels of the rivers Tubul, Raqui and the Las Peñas estuary was observed. In the remaining waters, a significant decrease in depth of up to ca. 1.8 m, which made it impossible to cross these rivers using a Zodiac, since it was stranded in many sectors of the wetland. All the ground emanating from the wetland due to the coastal uprising showed different degrees of drying cracks and whitish spots corresponding to the deposition of salts. With respect to the Spartina root aquatic habitats, all of them disappeared from the wetland because of the lack of water. The dry bottoms of the wetland have been transformed into corridors for dogs, allowing them access to feed, and for waterfowl nesting areas.

Due to the fact that at the 40 days mark after the earthquake the total loss of the tidal regime inside the wetland, as well as the exchange with the coastal marine waters was maintained, there was a decrease in salinity, especially in the Raqui river sector, but on the other hand, there was an increase of nutrients (ammonium, total nitrogen and total phosphorus) and of chlorophyll-a inside the wetland. This condition of water quality was atypical for this seasonal period of the year, and if these values are compared with those of December 2008, or those described in the literature by Stuardo et al. (1992) and CEA (2006, 2009). There were no conditions of hypoxia or anoxia in the main channels of the estuary, allowing the normal development of aquatic life in those areas where water was conserved. In contrast, in the isolated pools of stagnant water that remained in the network of inner channels of the wetland, there was hypoxia associated with the decomposition of the high levels of organic matter produced by the eutrophication of the system. Due to the desiccation of S. densiflora root habitats, high mortality of the aquatic macro invertebrates that inhabited them, such as Bryozoa, Crustacea and Annelida, occurred. This is relevant, as these habitats are areas of shelter and growth of juveniles of the estuarine crab Hemigrapsus crenulatus, whose adults are an important part of the diet of birds. In relation to these crabs in an adult state, a 100% mortality was recorded inside the wetland, as they normally seek refuge in the dense formations of the "Pelillo" seaweed, which was also severely affected in the wetland.

With respect to soft bottoms, the total mortality of bivalve banks was recorded.

The most affected species, given their high abundance and biomass in the area, was the "navajuela" (Tagelus dombeii), whose thousands of empty shells are still visible in the sediments, indicating a rapid death in the emerged bottoms (Fig. 10.7). This is a very important commercial specie that is present in the intertidal plains of

Fig. 10.7 Mortality in the bivalve banks of the "navajuela" (*Tagelus dombeii*), of commercial importance (Photograph C. Valdovinos, April 2010)

southern Chile (Clasing et al. 1994; Jaramillo et al. 2007). According to the measurements made of this species by Navarro et al. (2008), it is mainly a suspended feeding organism, reason why its absence would have important implications in the characteristics of the phytoplankton community of the wetland.

Another relevant aspect observed 40 days after the earthquake was that in the main channels of the rivers Tubul, Raqui and Las Peñas estuaries, in which some water was conserved, and benthic communities were registered, although with smaller abundances and species richness than what was reported prior to the earthquake. With regards to the hard substrates present in the wetland, such as bridge pillars and piers, whether wood, concrete or metal, as well as artificial castings, there was the total mortality of the Elminius kingii estuarine crustacean reported in all of them. These aquatic organisms normally live in the upper limit of the intertidal zone. Given the severe rise of the coast, all the intertidal bands that conformed these organisms were exposed to the desiccation.

Another organism that disappeared from the area was the small, shallow-bottom estuarine bivalve Kingiella chilenica (Soot-Ryen 1957), which had been recorded prior to the earthquake in the stations of the Tubul and Raqui rivers, although always in very low densities. This is an endemic bivalve that inhabits soft bottoms of marshes and estuaries in the south of Chile (Gallardo et al. 2006).

It is a small clam that reproduces by breeding its embryos, being semelparous with an annual cycle, and whose recruits must survive an inhospitable winter before they grow and reach reproductive maturity in the following summer season (Gallardo et al. 2006). This bivalve has very low scattering capacity because it is not planktonic, and its re-colonization capacity is extremely low.

With respect to the terrestrial system of the wetland, it is important to note that 40 days after the earthquake, although all the "espartal" rose significantly above mean sea level, the plants that make up this formation did not show evidence of having been affected. Species of the genus Spartina are vigorous herbaceous plants, which can form large prairies in areas of estuarine mud and sand (Chapman 1974). Its sediment retention capacity has motivated its planting in many areas of the world, having been introduced by settlers, in different sectors of the Asian coast and Oceania (Chung 1990). According to Begon et al. (1987), the genus Spartina is one of the few groups of C4 plants that, despite the high temperatures required for their efficient growth, extend their geographic range to temperate climate latitudes, occupying saline ecosystems, in which the osmotic conditions especially favor these species with a high efficiency in water use.

Changes Registered Six Months After the Earthquake

In the samplings carried out by Valdovinos and Sandoval (2011), 6 months after the earthquake, it was verified that the habitat conditions registered at 40 days after the earthquake were maintained, in regards to the drying of the inner canals, the aquatic root habitats and the lack of exchange with coastal waters. However, there was a slight increase in the level of the water mirror, due to a higher amount of fresh water coming from the drainage basin derived from the winter rains. The observations of the marks of the maximum level of the waters registered on the borders have shown that during the great winter increase there was a wide waterlogging of the wetland. This would probably have been favored by the embankment of the mouth of the estuary, which hindered the flow of fresh water into the sea. The maximum level of flooding of the wetland was evidenced by accumulations of dried remains of S. densiflora, which formed clear bands inside the "espartal". In the middle areas of the wetland there was an erosive process, while in the lower part a thin sedimentation process occurred.

Another relevant change registered in the 6 months of the earthquake was that in several places of the "espartal" there was a sinking of between 8 to 15 cm, and a slight horizontal deformation, due to the loss of water and compaction in the soils in which it develops. This sinking was found by measuring the difference in height between the old maximum tidal level given by the upper limit of distribution of E. kingii on poles contiguous to the "espartal", and the upper limit of the horizon of S. densiflora roots. The difference in height in these two levels is indicative of the sinking of the "espartal". Under natural conditions, there are no significant differences in height between these two limits.

In contrast to the April sampling conducted in the dry season, and in the same way as that reported in previous years for the winter period, in the August sampling the whole wetland presented freshwater conditions in terms of salinity and other parameters of water quality. However, something that had not been observed in previous samples was an increase of metals (Fe and Mn), in the waters that flowed from the "espartales". The presence of these metals was evidenced by the red-brown tonality of the waters. Laboratory analysis of some of these samples showed that the metals had outer range concentration that could pose a risk to human health and the ecosystem (Mn 0.1–0.2 mg/L and Fe 4–5 mg/L).

In relation to the aquatic biota, 6 months after the earthquake, the de-faunation of the desiccated areas was maintained. A very significant aspect in the ecosystem, given its relevance in the diet of waterfowl, was a partial re-colonization of amphipods and polychaetes in the areas of the main non-desiccated channels, despite fresh water at the time of sampling. There was no degree of re-colonization by T. dombei, E. kingii and H. crenulatus, which had previously disappeared from the area. But one of the species that showed re-colonization was the estuarine polychaete Perinereis gualpensis, which in terms of abundance and biomass corresponds to one of the most important polychaete species of estuaries in central and southern Chile (Bertrán 1989; Diaz-Jaramillo et al. 2010). A similar situation occurred with the amphipod Paracorophium hartmannorum, which is an important component in the estuarine trophic chains, especially because they are consumed by estuarine fish such as the "Robalo" (Eleginops maclovinus (Cuvier, 1830) (2000), Jaramillo et al.

Six months after the earthquake, the "espartal" also showed no evidence of being affected. However, in many areas the recolonization of areas flooded by winter rains by Parastacus pugnax, which is typically sweetwater, was found. The dominant presence of the remains of this shrimp (C. Valdovinos, personal obs) was found in many aquatic bird species observed in the riparian areas of the wetland. On the other hand, in the root zones of the "espartal" that before the earthquake constituted habitats for aquatic organisms, it was observed in some sectors that it was being colonized by lycosidae spiders, constructors of fabrics easily visible. Finally, in the same month of August, freshwater fish were recorded near the mouth of the estuary, a situation that had not previously been reported in the area. This is relevant since several of the reported species present conservation problems (see Valdovinos 2006a, b) and that would have been favored in the winter, due to the increase in the availability of freshwater habitat in the wetland.

Changes Registered 10 Months After the Earthquake

Within 10 months of the earthquake, some indicators showed practically no signs of recovery, but others are moving towards a new condition. In the last sampling it was found that the habitat conditions were maintained, such as the desiccation of the inner canals and the root habitats. However, the inner canals that irrigate the wetland, in which some pools isolated from stagnant water were conserved, were practically dried up in their totality in December.

Other habitat variables showed clear evidence of a slight return to the pre-earthquake condition. For example, an increase in the water mirror level in ca. 0.3 m, and thus an increase in the area of flooded bottoms and habitat availability for benthic organisms. In addition, there was an increase in the exchange with sea waters, which was expressed in slight fluctuations in tidal and salinity levels, more evident in the Tubul River than in the Raqui River. Associated with the above, there was an increase of the depth in the mouth of estuary, which facilitated the navigation and the interchange between the estuarine and marine waters. The increase in depth would also be associated with the extraction of sand with heavy machinery in this area, to allow the movement of boats. With respect to water quality, in the.

Sectors that remained flooded had a clear increase in the salinity, depth and changes of level and hydrodynamics associated with the tide cycles. This has favored the development of the aquatic biota in parts of the areas that were originally desiccated by the coastal uprising. In relation to the aquatic and terrestrial biota, during the 10 months of the earthquake, the defaunation of the desiccated areas was maintained. However, the increased recolonization of amphipods and polychaetes was maintained in the areas of the main dried up channels, associated with a return of the brackish water condition. There was no recolonization of T. dombeii and E. kingii, but of the crab H. crenulatus, which is indicative of a trend towards recovery of the estuary. Given the existence of brackish waters in the estuary in the last sampling, unlike the one observed in the August sampling, the presence of freshwater fish in the middle and lower part of the wetland was not recorded.

Benthic invertebrates from estuarine ecosystems inhabit environments where the water column is subject to constant changes due to regular factors (e.g. tides and storms), and more unpredictable or stochastic events (e.g. precipitation and wind) (Urbina et al.2010). Thus, the different properties of the water column (e.g. temperature, salinity and seston) undergo strong changes in time and space. Salinity is one of the most obvious environmental factors in estuarine systems. In fact, salinity is so important in estuaries that it has come to be considered the "main ecological factor" (Kinne 1966), for its fundamental role in the dynamics and distribution of populations (Kinne 1971; Davenport et al. 1975; Bayer et al. 1976), and in the modulation of reproductive processes (Cunha et al. 1999) and behavioral ones (Spicer and Strömberg 2003), as well as many of the physical processes (Shuhong 2006; Urbina et al. 2010).

With respect to terrestrial vegetation, the "espartales" continued to show symptoms of an appropriate physiological state, even a massive flowering of S. densiflora was observed throughout the wetland. Another aspect recorded in the last sampling was the colonization of the bottoms that were desiccated by terrestrial vegetation that covered in some places up to 60% of the emerged terain.

Conservation of Wetland Biodiversity

The great extension of the Tubul-Raqui wetland, the most important one in the center and in the south of Chile; its high biodiversity and biological productivity, combined with a low degree of human disturbance, make this wetland one of the most important in Chile in terms of biodiversity conservation. Although severely affected by the February 2010 earthquake, it maintains properties that allows it to stay within the country's most important coastal wetlands. The wetland continues to provide nesting and shelter sites for numerous resident and migratory bird species, many of whom have conservation problems. This situation has been recognized at the regional level and has been identified as one of the six priority sites established in the Regional Strategy and Action Plan for the Conservation of Biodiversity of the Biobio Region.

The major anthropogenic threats to the conservation of the wetland and its environments have historically been bird hunting, and deforestation of the basins, which affects the sedimentation of the rivers that feed it. In relation to the above, the main measures adopted are, firstly, the hunting or capturing ban for amphibians, reptiles, birds and wild mammals, in a territory of 7822 ha. This ban is valid for 30 years from June 2006 and was established by the Supreme Decree No. 285 of the Agricultural and Livestock Service. The second measure of great importance was the declaration for conservation purposes of 350 ha of the Raqui island tax exemption estate, through the Supreme Decree No. 454 of the Ministry of National Assets, which aims to conserve habitats and their associated biota. A third measure that is currently being implemented, obviously the most important one because of its extension and ecosystem approach, corresponds to the wetland postulating as a Ramsar site, to form part of the network of wetlands of international importance. The total area requested to be declared as a Ramsar Site is 562 ha, of which 350 correspond to tax exemption land, and the other 212 are bodies of water in the rivers Tubul, Raqui and Las Peñas estuary, which correspond to the area of aquaculture concession owned by the Tubul Trade Association. This was granted in 1994 by the Directory of Maritime Territory and Merchant Marine, by the Supreme Decree No. 296. The lands surrounding the area being requested to be established as a Ramsar site are privately owned. It is clear that the designation of the wetland as a Ramsar site would be a major step forward for the conservation of this wetland, considering that their main objective is the conservation and wise use of wetlands through local and national actions, supported by international cooperation.

Conclusions

This chapter has presented a synthesis of the main ecosystem characteristics of the Tubul-Raqui wetland, as well as the most important changes in the wetland following the February 2010 earthquake.

Given the great extent of the Tubul-Raqui marsh, in addition to its high biodiversity and biological productivity, combined with a low degree of human disturbance, this wetland corresponds to one of the most important in Chile in terms of biodiversity conservation.

Although severely affected by the February 2010 earthquake, it maintains properties that allow it to be within the country's most important coastal wetlands. The wetland continues to provide nesting and shelter sites for numerous resident and migratory bird species, many of which have conservation problems.

The main anthropogenic threats to the conservation of the wetland and its surroundings have historically been the hunting of birds and deforestation of the basins, which affects the sedimentation of the rivers that feed it.

As for the structure of the physical habitat, water quality and aquatic biota of the wetland, from the February 27, 2010 earthquake, a series of relevant environmental changes resulting from the earthquake occurred. Some of them occurred immediately during the earthquake, while others still continue to manifest.

The major changes in the wetland caused by the earthquake have resulted from its rising at ca. 1.6 m above sea level, which has led to the total drying up of the network of internal channels that irrigate the wetland, and partial drying of the main channels of the rivers Tubul and Raqui. This has also limited the exchange with the sea, generating a decrease in the salinity of the waters to the interior of the wetland.

The desiccation of an important part of the wetland's bottoms produced the mortality of benthic invertebrates that inhabited them. Many of them survived in the shallow areas where water was conserved, and have shown a tendency to recover their populations over time. However, the two most important benthic species for the artisanal fishery, "Pelillo" (Gracilaria sp.) and "navajuela" bivalve (Ensis macha) were severely affected and there is no evidence of any degree of recolonization.

The extensive saline or "spartan" pastures dominated by Spartina densiflora have shown no signs of being affected by the coastal uprising which is important, considering that it is a species of great relevance in the wetland, its coverage and biomass.

Although the wetland was severely affected by the earthquake, it already shows signs of recovery. It is clear that these conditions will not return to what they were prior to the February 27, 2010 earthquake, however, it will continue to constitute the main coastal wetland with "Espartales" of Central Chile and probably the temperate Pacific coast of South America.

Acknowledgements The authors would like to thank the ARAUCO company's Management for Environment, Safety and Occupational Health for the facilities given during the preparation of this chapter, as well as Andres Camano, Carolina Vargas, Oscar Parra and Adolfo Acuna (†) for their suggestions on the manuscript. A special recognition to Patricia Carrasco, Cecilia Jiménez, Cristian Cornejo and the members of the Tubul-Raqui Mesa de Trabajo (Working Table), for their support for carrying out this research. This publication is part of the activities of Group 8, Impact and Environmental Recovery, of the University of Concepción.

References

Adam, P. 1990. *Saltmarsh Ecology*. Cambridge: Cambridge University Press.

Alveal, K. 1988. Gracilaria de Tubul. Historia y significado de un recurso. *Gayana Botánica* 45: 119–140.

Bayne, B.L., C.J. Bayne, T.C. Carefoot, and R.J. Thompson. 1976. The physiological ecology of *Mytilus californianus* Conrad. 1. Metabolism and energy balance. *Oecologia* 22: 211–228.

Begon, M., J.L. Harper, and C.R. Towsend. 1987. *Ecología, individuos, poblaciones y comunidades*. 2da ed. Barcelona: Omega.

Bertrán, C. 1989. Zonación I dinámica temporal de la macroinfauna intermareal en el estuario del río Lingue (Valdivia, Chile). *Revista Chilena de Historia Natural.* 62: 19–32.

Biró, L. 1979. Contribucióńal Conocimiento de la Formación Tubul, Plioceno Superior, Provincia de Arauco (37°14′ Lat. Sur), Chile. *II Congreso Geológico Chileno* 3: H33–H44.

Cabrera, A.L. 1970. *Flora de la Provincia de Buenos Aires. Parte II.* Ed. A.L. Cabrera, J.C. Hernández, and J.M. Caro. Buenos Aires: INTA.

Carrasco-lagos, P. 2003. Variaciones de la estructura de ensamble de aves a lo largo de un gradiente ambiental en el humedal Tubul-Raqui, VIII Región, Chile. Seminario de Investigación para optar al Grado de Licenciado en Biología, Facultad de Ciencias Naturales y Oceanográficas, 56, Universidad de Concepción.

———. 2004. Propuesta de manejo para la conservación del humedal Tubul-Raqui, VIII Región, Chile. Tesis para optar al título profesional de Biólogo, 76. Universidad de Concepción, Chile.

Carrasco-lagos, P., and M.C. Jiménez. 2009. Ficha Informativa de los Humedales de Ramsar (FIR): Humedal Tubul-Raqui. Propuesta Unidad de Biodiversidad, 39, Departamento Proteccióńde Recursos Naturales, CONAMA Regióńdel Biobío.

Chapman, V.J. 1974. *Salt marshes and salt desert of the world.* 2da ed. Lehre: V von J Cramer.

———. 1977. In *Introduction. En: Wet coastal ecosystems. Ecosystems of the world,* ed. V.J. Chapman, 29. Amsterdam: Elsevier.

Chung, C.H. 1990. Twenty-five years of introduced Spartina anglica in China. In *Spartina anglica – A research Rewiew*, ed. A.J. Gray and P.E.M. Benham, 72–76. Institute of Terrestrial Ecology, Natural Environment research Council. China

Cisternas, M., A. Aranda, and I. Contreras. 2000. Reconocimiento y caracterización de la facies sedimentaria depositada por el Tsunami de 1960 en el estuario del río Maullín, Chile. *Revista Geológica de Chile.* 27: 3–11.

Clasing, E., T. Brey, R. Stead, J.M. Navarro, and G. Asencio. 1994. Population dynamics of *Venus antiqua* (Bivalvia: Veneracea) in the bahía de Yaldad, Isla de Chiloé, Southern Chile. *Journal of Experimental Marine Biology and Ecology.* 177: 171–186.

Conama. 2007. Solicitud de auto destinación de 350 hectáreas del Estuario Tubul- Raqui de propiedad del Ministerio de Bienes Nacionales, 22. Departamento de Recursos Naturales, CONAMA Región del Bío-Bío.

Constabel, S. 1993. Ambientes sedimentarios de los estuarios Tubul y Raqui, VIII Región, Chile. Tesis para optar al grado de Doctor en Ciencias Ambientales, 96, Universidad de Concepción.

Costa, C.S.B., and A.J. Davy. 1992. In *Coastal saltmarsh communities of Latin America. En: Coastal Plant Communities of Latin America*, ed. U. Seeliger, 179–199. Academic Press, Inc. USA

Cunha, M.R., J.C. Sorve, and M.H. Moreira. 1999. Spatial and seasonal changes of brackish peracaridan assemblages and their relation to some environmental Variables in two tidal channels of the Ria de Aveiro (NW Portugal). *Marine Ecology Progress Series* 190: 69–87.

Davenport, J., L.D. Gruffydd, and A.R. Beaumont. 1975. An apparatus to supply seawater of fluctuating salinity and its use in the study of the salinity tolerance of larvae of the scallop *Pecten maximus* L. *Journal of the Marine Biological Association of the United Kingdom* 55: 391–410.

Della-croce, N., and C. Valdovinos. 1994. Caracterización ambiental del área marina costera bajo la influencia de la cuenca hidrográfica del río Biobío, Chile central (1989–1992). En: *El río Biobío y el área marina adyacente*, ed. N. Della-Croce, 85–163. Italia: Univ. Di Genova.

Díaz-jaramillo, M., J.L. Ferreira, L.L. Amado, J. Ventura-lima, A. Martins, M.R. Retamal, R. Urrutia, C. Bertrán, R. Barra, and J.M. Monserrat. 2010. Biomonitoring of antioxidant and oxidative stress responses in *Perinereis gualpensis* (Polychaeta: Nereididae) in Chilean estuarine regions under different anthropogenic pressure. *Ecotoxicology and Environmental Safety* 73: 515–523.

Eula. 2008. *Guía de manejo para terreno fiscal con alto valor en biodiversidad en la Región del Bío-Bío.* Sector humedal Tubul-Raqui. Informe técnico Centro de Ciencias Ambientales EULA, Universidad de Concepción. 143 pp.

Farías, M., G. Vargas, A. Tassara, S. Carrethier, S. Baize, D. Melnick, and K. Bataille. 2010. Land-Level Changes Produced by the Mw 8.8 2010 Chilean Earthquake. *Science* 329: 916. doi:10.1126/science.1192094.

Ferraris, F. 1981. Hoja Arauco– Lebu, Regióndel Biobío, escala 1: 250.000. Mapas Geoló- gicos Preliminares de Chile, N° 5, 10, Instituto de Investigaciones Geológicas.

Figueroa, R., and C. Valdovinos. 1997. Productividad de pastizales salinos del estuario Lenga (Chile) a escala de paisaje ecológico: análisis de imágenes Landsat TM y experimentos in situ. *Boletin de la Sociedad de Biología de Concepción, Chile* 68: 7–11.

Gallardo, C.S., M. Filún, and C. Manque. 2006. *Kingiella chilenica* (Bivalvia: Cyamiidae); populationdynamics, rates of survival, embryoproductionandannual recruitment of a semelparous brooding clam. *Journal of the Marine Biological Association of the United Kingdom* 86: 757–766.

Grandjean, M.M. 1985. Efecto de la temperatura sobre el balance energético y el crecimiento en *Hemigrapsus crenulatus* (Milne-Edwards, 1837) bajo condiciones de laboratorio. Tesis de magíster. Universidad Austral de Chile, Valdivia, Chile.

Habit, E., and P. Victoriano. 2005. Peces de agua dulce de la Cordillera de la Costa. En: *Historia, Biodiversidad y Ecología de la Cordillera de la Costa de Chile,* eds. C. Smith-Ramírez, J. Armesto and C. Valdovinos, 392–406. Santiago: Editorial Universitaria.

Jaramillo, E., H. Contreras, and C. Duarte. 2007. Community structure of the macroinfauna inhabiting tidal flats characterized by the presence of different species of burrowing bivalves in Southern Chile. *Hydrobiology* 580: 85–96.

Jiménez, J. 1996. Aspectos ecológicos en *Spartina densiflora* Brong. Memoria para optar al grado de doctor en Biología, Facultad de Biología, Universidad de Sevilla, España.

Kinne, O. 1966. Physiological aspects of animal life in estuaries with special reference to salinity. *Netherlands Journal of Sea Research* 3: 222–244.

Kinne o. 1971. Salinity, animals: invertebrates. Environmental factors. Part 2. In A comprehensive, integrated treatise on life in oceans and coastal waters. Volume I. *Marine Ecology* 1: 821–995.

Long, S.P., and C.F. Mason. 1983. *Saltmarsh Ecology*. Glasgow: Blackie.

Mardones, M. 1971. Contribución al estudio de la ganadería. Análisis del libro de cuentas de la feria agrícola-ganadera de Arauco, Concepción, Chile, 4-16. Universidad de Concepción.

Melnick, D., B. Bookhagen, H. Echtler, and M. Strecker. 2006. Coastal deformation and great subduction earthquakes, Isla Santa María, Chile (37°S). *Geological Society of America Bulletin.* 118: 1463–1480.

Mobberley. 1956. Taxonomy and distribution of the genus Spartina. *Iowa State College journal of science* 30: 471–574.

Navarro, J., E. Clasing, M. Lardies, and R. Stead. 2008. Feeding behavior of the infaunal bivalve *Tagelus dombeii* (Lamarck, 1818). Suspension vs. Deposit feeding. *Revista de Biología Marina y Oceanografía.* 43: 599–605.

Nielsen, S.N., and C. Valdovinos. 2008. Early Pleistocene Mollusks of the Tubul Formation, South-Central Chile. *Nautilus* 122: 201–216.

Parada, M. 2008. Proyecto Eólico Raqui – Quidico: Evaluación Ornitológica. Informe técnico de biornat Ltda., para Arauco Generación S. A, 54

Pérez-losada, M., C. Jara, G. Bound-backup, and K. Crandall. 2002. Conservation phylogenetics of Chilean freshwater crabs *Aegla* (Anomura Aeglidae): Assignig priorities for aquatic habitat protection. *Biological Conservation* 105: 345–353.

Pineda, V. 1985. *Evolución Paleogeográfica de la Cuenca Sedimentaria Cretácico– Terciaria de Arauco. Geología y Recursos Minerales de Chile*, 438 pp. Chile: Editorial de la Universidad de Concepción.

Ramsar. 2006. Manual de la Convención de Ramsar: Guía a la Convención sobre los Humedales (Ramsar, Irán, 1971), 4a. Edición. Secretaría de la Convención de Ramsar, 124, Gland (Suiza).
Retamal, M.A. 1967. Estudios bionómicos en una población de *Hemigrapsus crenulatus* (H. Milne Edwards), 1837, Lenga (36° Lat. S.). Informe de tesis para optar al título de Licenciado en Biología, 68. Instituto Central de Biología, Departamento de Zoología, Universidad de Concepción.
———. 1981. Catálogo ilustrado de los crustáceos de Chile. *Gayana Zoológica.* 44: 1–110.
Servicio agrícolayganadero (sag). 2009. Ley de caza y su reglamento. Legislación de fauna silvestre. Ministerio de Agricultura. Servicio Agrícola y Ganadero. División de Protección de los Recursos Naturales Renovables, 10 Ed., Santiago.
Sanmartín, C., D. Contreras, J. Sanmartín, and C. Ramírez. 1992. Vegetacióndelasma- rismas del centro-sur de Chile. *Revista Chilena de Historia Natural* 65: 327–342.
Shuhong, Z. 2006. The influence of salinity, rhythm and day length on feeding behavior in *Meretrix meretrix* Linnaeus. *Aquaculture* 252: 584–590.
Spicer, J.I., and J.O. Stromberg. 2003. Metabolic responses to low salinity of the shipworm *Teredo navalis* (L.). *Sarsia* 88: 302–305.
Strahler, A.N. 1981. *Geografía Física*, 767. Barcelona: Editorial Omega S. A.
Stuardo, J., and C. Valdovinos. 1989. Estuarios y lagunas costeras: Ecosistemas importantes de Chile Central. *Ambiente y Desarrollo.* 5: 107–115.
Stuardo, J., C. Valdovinos, R. Figueroa, and A. Occhipinti. 1992. Ambientescosteros del Golfo de Arauco y áreas adyacentes. *Publ. EULA-Chile. Ser. Mon. Cient.* 9: 1–157.
Taylor, H., and D. Seneviratna. 2005. Ontogeny of salinity tolerance and hyper- osmoregulation in embryos of intertidal crabs *Hemigrapsus edwarsii* and *Hemigrapsus crenulatus* (Decapoda, Grapsidae): survival of acute hyposaline exposure. *Comparative Biochemistry and Physiology. Part A* 140:495–505.
Urbina, M., K. Paschke, P. Gebauer, and O. Chaparro. 2010. Physiological energetics of the estuarine crab Hemigrapsus crenulatus (Crustacea: Decapoda: Varunidae): responses to different salinity levels. *Journal of the Marine Biological Association of the United Kingdom.* 90: 267–273.
Valdovinos, C. 2001. Biodiversidad en fondos blandos sublitorales: Macroinvertebrados del Golfo de Arauco (Chile). En *Sustentabilidad de la Biodiversidad: Bases científico-técnicas, teorizaciones y proyecciones*, ed. K. Alveal and T. Antezana, Univ. De Concepción. 33:635–662.
———. 2004. Ecosistemas estuarinos. En *Biología marina y oceanografía: Conceptos y procesos*, ed. C. Werlinger, Trama Impresores S. A., Chile. 2:395–414.
———. 2006a. Humedales dulceacuícolas y biodiversidad. En *Atlas social y ambiental El Humedal costero tubul-raqui: un ecosistema chileno de alto valor de conservación • CAPÍTULO 10 Del área metropolitana de Concepción. Región del Bio-Bío, Chile*, eds. J. Rojas, G. Azócar, M.D. Muñoz, C. Vega, A. Kindler and S. Kabisch, 104–124. Chile: Editorial Universidad de Concepción.
———. 2006b. Invertebrados dulceacuícolas. En *Conama*, eds. Biodiversidad de Chile, 204–225. Patrimonio y desafíos. Santiago de Chile: Ocho Libros Editores.
Valdovinos, C., and N. Sandoval. 2011. Cambios ambientales del humedal Tubul- Raqui derivado del alzamiento cosísmico y Tsunami, asociado al terremoto Mw 8,8. Informe técnico de la Unidad de Sistemas Acuáticos, del Centro de Ciencias Ambientales EULA, Universidad de Concepción, Chile.
Valiela, I., E. Kinney, J. Culberston, E. Peacock, and V.A. Smith. 2009. Global losses of mangroves and saltmarshes. En: Carlos M. Duarte, Ed. Global loss of coastal hábitats. Fundación BBVA, Bilbao, España. 4:107–138.
Vásquez, D. 2009. Delimitación y zonificación de los humedales costeros de Lenga y Tubul-Raqui: Orientaciones para el Ordenamiento Territorial de la Región del Bío – Bío, Tesis para optar al título de Geógrafo, Universidad de Concepción, 66–85.
Vergara, O., P. Carrasco-pagos, M.F. Saavedra, and J.C. Ortiz. 2008. *Fauna del humedal Tubul-Raqui, Provincia de Arauco. Sitio Prioritario para la Conservación de la Biodiversidad*, 72. Concepción: CONAMA Regióndel Biobío. Ed. Impresos del Sur.
Werlinger, C., and K. Alveal. 1988. Evaluación de algas en ambientes restringidos del Golfode Arauco(Chile): Punta Fuerte Viejoa Río Tubul. *Gayana Botánica.* 45: 461–474.

Epilogue

This book is the result of efforts of different scientific teams and institutions, which together give us a new demonstration of the enormous value that collaborative work represents in pursuit of a common interest.

Wetlands are a subject of enormous relevance in our social and environmental development. We know very well who we work with in Arauco, because we have been the center of a strong national debate as a result of the environmental crisis of the Carlos Andwanter wetland in the Cruces River, which has highlighted the lack of dialogue and research that exists in this area in our country.

As a company, we have understood that in order to address in all its complexity the strategic challenge of making our business more sustainable, we must participate in social dialogue in different areas and support the generation of knowledge.

We are interested in encouraging and participating in the discussion, research and learning, especially in the environmental field, because the more a society knows, and the more collaborative are the measures to protect its ecosystem services, the more capacities we will have to share our resources with and make grow our country.

For Arauco, sustainable development is not a rhetorical concept, but a constant challenge at the heart of our business strategy. That is why we have implemented the Environmental Management system based on the ISO 14001 standard that allows us to identify and control the environmental risks of our operations; and also because of this, we take on the challenge of going beyond preventive environmental management through research into the ecosystems in which we operate.

As a result of this conviction, we have expanded our High Conservation Value Areas (AAVC) in Chile from 3 thousand to more than 62 thousand hectares, taking into consideration not only environmental, but also social and cultural aspects.

J.M. Fariña, A. Camaño (eds.), *The Ecology and Natural History of Chilean Saltmarshes*, DOI 10.1007/978-3-319-63877-5

This book, which brings together a number of valuable scientific works, provides us with a new and important body of information, especially for more informed decision-making that requires the conservation of coastal wetlands in this case, and the services they provide for future generations.

Matias Domeyko

Executive Vice President

Celulosa Arauco and Constitucion S.A. (Arauco)

Index

J.M. Fariña, A. Camaño (eds.), *The Ecology and Natural History of Chilean Saltmarshes*, DOI 10.1007/978-3-319-63877-5

MIX
Papier aus verantwortungsvollen Quellen
Paper from responsible sources
FSC® C105338
FSC
www.fsc.org

If you have any concerns about our products,
you can contact us on
ProductSafety@springernature.com

In case Publisher is established outside the EU,
the EU authorized representative is:
Springer Nature Customer Service Center GmbH
Europaplatz 3, 69115 Heidelberg, Germany

Printed by Libri Plureos GmbH
in Hamburg, Germany